T0406742

Progress in Colloid and Polymer Science

Recently Published and Forthcoming Volumes

Progress in Colloid and Polymer Science

Editors: F. Kremer, Leipzig and W. Richtering, Aachen

Volume 134 · 2008

Surface and Interfacial Forces – From Fundamentals to Applications

Volume Editors:
Günter K. Auernhammer
Hans-Jürgen Butt
Doris Vollmer

The series Progress in Colloid and Polymer Science is also available electronically (ISSN 1437-8027)

- Access to tables of contents and abstracts is *free* for everybody.
- Scientists affiliated with departments/institutes subscribing to Progress in Colloid and Polymer Science as a whole also have full access to all papers in PDF form. Point your librarian to the Springerlink access registration form at http://www.springerlink.com

ISSN 0340-255X
ISBN-10 3-540-68018-7
ISBN-13 978-3-540-68018-5
e-ISBN 978-3-540-68023-9
DOI 10.1007/978-3-540-68023-9
Library of Congress Control Number 2008932367
Springer Berlin, Heidelberg, New York

Springer is a part of Springer Science + Business Media

http://www.springer.com

Printed in Germany

Cover design: eStudio Calamar S. L., F. Steinen-Broo, Pau/Girona, Spain

Typesetting and production: le-tex publishing services oHG, Leipzig

Printed on acid-free paper

Progr Colloid Polym Sci (2008) 134: V

PREFACE

43rd Biennial Meeting of the German Colloid Society

This volume contains selected papers presented at the 43rd Biennial Meeting of the German Colloid Society held at the Schloß Waldthausen near Mainz, October 8–10, 2007. The meeting's emphasis was given to "Surface and Interfacial Forces – From Fundamentals to Applications" but also provided a general overview on current aspects of colloid and polymer science in fundamental research and applications.

The contributions in this volume are representative of the richness of research topics in colloid and polymer science. They cover a broad field including the application of scanning probe techniques to colloid and interface science, surface induced ordering, novel developments in amphiphilic systems as well as the synthesis and applications of nano-colloids.

The meeting brought together people from different fields of colloid, polymer, and materials science and provided the platform for dialogue between scientists from universities, industry, and research institutions.

Günter K. Auernhammer
Hans-Jürgen Butt
Doris Vollmer

Progr Colloid Polym Sci (2008) 134: VI–VII

CONTENTS

Progr Colloid Polym Sci (2008) 134: 1–10
DOI 10.1007/2882_2008_075

Published online: 1 May 2008

Torbjörn Pettersson
Zsombor Feldötö
Per M. Claesson
Andra Dedinaite

The Effect of Salt Concentration and Cation Valency on Interactions Between Mucin-Coated Hydrophobic Surfaces

Torbjörn Pettersson · Zsombor Feldötö ·
Per M. Claesson (✉) · Andra Dedinaite
Department of Chemistry, Surface Chemistry, Royal Institute of Technology,
Drottning Kristinas väg 51,
SE-100 44 Stockholm, Sweden
e-mail: per.claesson@surfchem.kth.se

Per M. Claesson (✉) · Andra Dedinaite
Institute for Surface Chemistry (YKI),
Drottning Kristinas väg 51, P.O. Box 5607,
SE-114 86 Stockholm, Sweden
e-mail: per.claesson@surfchem.kth.se

Abstract The AFM colloidal probe technique has been utilized in order to investigate the forces acting between preadsorbed mucin layers on uncharged, hydrophobic mercaptohexadecane-coated gold surfaces. Layers with some highly extended tails are formed when the adsorption proceeds from 25 ppm mucin solution in 30 mM $NaNO_3$. The effects of salt concentration and cation valency on the interactions have been explored using NaCl, $CaCl_2$, and $LaCl_3$ in the concentration range 1–100 mM. It will be shown that the results in NaCl, where the tail length decreases as the salt concentration is increased, can be rationalized by considering the polyelectrolyte nature of mucin and the screening of intralayer electrostatic interactions between charged groups, mainly anionic sialic acid. When multivalent cations are present in solution a significant compaction of the mucin layer occurs even at low concentrations (1 mM), suggesting binding of these ions to the anionic sites of mucin. The results are discussed in relation to previous data from quartz crystal microbalance measurements on the same systems.

Keywords Adsorption · Atomic force microscope · Cation binding · Mucin · Multivalent ion · Surface force · Swelling

Introduction

Mucins are high molecular weight glycoproteins that form an essential structural element on many internal surfaces in animals and humans, where they fulfil a biological function as a protective and lubricating layer. The polypeptide backbone of mucin is decorated by oligosaccharide side chains, both linear and branched, containing typically 1–20 carbohydrate residues [1] that are clustered in heavily glycosylated regions separated by less densely glycosylated, "naked", regions [2]. The global bulk conformation of mucin is described by Carlstedt et al. as a train of stiff rods (the heavily glycosylated regions) interspaced with more flexible regions (the naked regions) [3]. Mucins have an anionic polyelectrolyte character due to the presence of sialic acid groups, reported to have a pK_a value [4] of 2.6, and a small fraction of sulfate groups ($pK_a < 1$). The carbohydrates constitute approximately 80 wt. % of typical mucins, and the typical carbohydrate residues found in mucins are *N*-acetyl-glucosamine, *N*-acetyl-galactosamine, galactose, fucose, and sialic acid [1].

The mucin molecule can thus, from a physicochemical point of view, be described as an anionic polyelectrolyte, where the heavily glycosylated regions have a bottle-brush structure with the anionic charges situated primarily on the side chains. These anionic bottle-brush regions are separated by less charged and more flexible polypetide regions containing cationic, anionic, and uncharged segments; Fig. 1 illustrates the typical mucin structure. The properties of polyelectrolytes are in general strongly influenced by salt since electrolytes screen the repulsion between the charged segments. Thus, flexible polyelec-

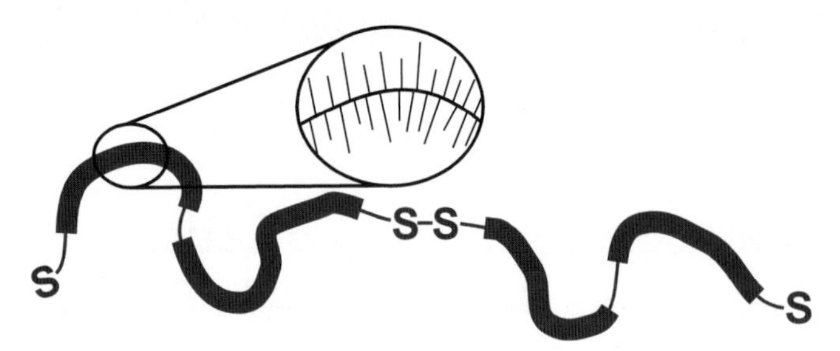

Fig. 1 Typical structure of a mucin molecule consisting of several subunits connected by disulfide bridges. Note the presence of dense regions of oligosaccharide side chains (of varying lengths) separated by naked regions exposing the protein backbone

trolytes adopt more compact conformations in bulk solution as the salt concentration is increased [5]. In contrast, the conformation of brush-like polyelectrolytes are less affected by salt addition due to steric repulsion between the side chains [6]. There are thus reasons to expect that both bulk and interfacial properties of mucins will be strongly affected by the concentration and type of salt present in solution. Indeed, it has been demonstrated that calcium ions increase the intrinsic viscosity of salivary mucin solutions [7], and the presence of trivalent cations causes mucin aggregation at solid/liquid [8, 9] and air/liquid interfaces [10]. Recently, we also used the quartz crystal microbalance technique with dissipation monitoring, QCM-D, to elucidate the effects of NaCl, $CaCl_2$, and $LaCl_3$ concentration on the properties of adsorbed mucin layers [11]. In the present report we describe how these ions influence the interactions between mucin layers. We also correlate the force data with our previous QCM data to obtain information related to both extended tails and the inner region of the adsorbed layer.

Materials and Methods

Materials

Bovine submaxillary gland mucin (mucin), BSM, catalogue number M3895, with a molecular weight of about 7×10^6 g mol^{-1} as determined by static light scattering [12], was purchased from Sigma and used as received. The BSM, according to the specification of the manufacturer, contains 12% of sialic acid residues. In passing, we note that this commercial mucin contains some traces of other proteins, e.g. bovine serum albumin [13]. The following chemicals were used as received: pure ethanol 99.5% from Kemetyl, sodium nitrate ($NaNO_3$) suprapur from Merck, sodium chloride (NaCl) pro analysi from Merck, calcium chloride ($CaCl_2$) pro analysi from Merck, lanthanum chloride hepta hydrate ($LaCl_3 \times 7H_2O$) reagent grade from Scharlau, 1-hexadecanethiol (mercaptohexadecane) 95% from Fluka. All solutions were made with water pretreated with a Milli-RO 18 unit, followed by purification with a Q-PAK unit. The outgoing water, resistivity $> 18\,\mathrm{M\Omega\,cm}$, was filtered through a 0.2 μm membrane. The organic content of the outgoing water was less than 2 ppb.

Methods

All force curves were determined with an atomic force microscope from Veeco Instrument, Nanoscope Multimode III Pico Force, using a fused silica liquid cell. Rectangular tipless cantilevers (MikroMasch) with an approximate dimension of 250 μm in length and 35 μm in width, and a normal spring constant in the range 0.02–0.2 N m^{-1} were used for all experiments. The exact values of the normal spring constants, k_N, were determined by the method based on thermal noise with hydrodynamic damping [14] using AFM Tune IT v2.5 (ForceIT, Sweden). Gold particles of approximately 10 μm in diameter, determined for each particle using optical microscopy, were made under a flow of filtered nitrogen gas by an electrical discharge induced by bringing together two gold wires connected to a current power supply generating approximately 10 A. A similar procedure has been described previously [15, 16]. The particles were attached to the end of the cantilever, with the aid of an Ependorf Micromanipulator 5171 and a Nikon Optiphot 100S reflection microscope, using a small amount of a two-component epoxy (Araldite Rapid, Casco Nobel, Sweden) that was allowed to cure overnight. The gold particles were used as the colloidal probes [17–19]. Cantilevers carrying a colloidal probe were immersed in 1 mM mercaptohexadecane solution (99.5% pure ethanol as solvent) for 24 h, after which they were transferred into 99.5% pure ethanol until use, but not kept for longer than one week.

The flat surfaces were modified gold QCM crystals (QSX-301, Q-Sense, Sweden). Prior to use, the surface of the crystals was placed in contact with chromosulfuric acid (BIC) for 3×5 min, washed with Millipore water, then with 99.5% pure ethanol, and finally dried with a gentle jet of filtered N_2 gas. Thiol modification of these surfaces followed the same procedure as for the gold colloidal probes.

Force curves were measured at a constant driving velocity of 10 μm s^{-1} (triggered at a given applied maximum force of approximately 8 mN m^{-1}). A schematic drawing of the AFM set-up is shown in Fig. 2. The force displacement curves were analyzed by using the deflection sensitivity, α (m v^{-1}), as recorded for the bare gold probe vs. the bare gold flat surface. This value was used when analyzing the force curves for surfaces coated with mucin. This method was chosen since the adsorbed layers may have a finite compressibility even at the highest loads applied. The measured forces, F, are normalized by the radius, R, of the colloidal probe. This normalization allows experiments with different probes to be directly compared, and the normalized force is related to the free energy of inter-

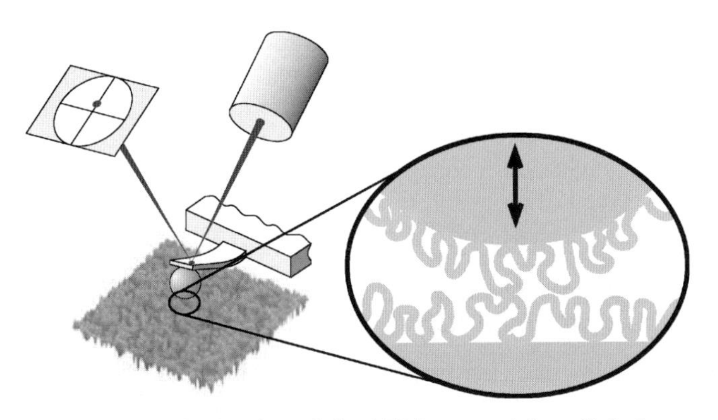

Fig. 2 Schematic drawing of the AFM set-up. A laser light is reflected from the back of the cantilever that holds the colloidal probe. The cantilever deflection is measured using a split photodiode

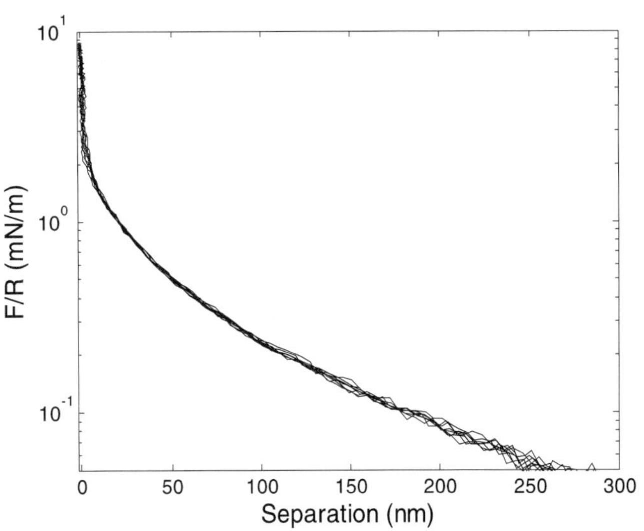

Fig. 4 Ten subsequent measurements of interactions between mucin-coated mercaptohexadecane-modified gold across 30 mM $NaNO_3$

action per unit area, G_{flat}, between flat surfaces as [20]:

$$\frac{F}{R} = 2\pi G_{flat} \,. \quad (1)$$

The experiments with adsorbed mucin layers were conducted in the following manner: The fused silica liquid cell was cleaned with Deconex 11 Universal solution (Borer Chemie, Switzerland) and rinsed with plenty of purified water, flushed with ethanol, and dried under a flow of nitrogen gas. A mercaptohexadecane-modified QCM crystal and a cantilever with the colloidal probe were washed with ethanol and dried in a flow of filtered nitrogen gas and mounted into the AFM. Next, the probe was brought close to the flat surface and 30 mM $NaNO_3$ solution was injected into the liquid cell and allowed to equilibrate for 10 min before measuring ten force displacement curves.

The probe was retracted from the surface (approximately 60 μm) and 4 mL of a 25 ppm mucin solution in 30 mM $NaNO_3$ was run through the liquid cell ($V \approx$ 0.1 mL). The mucin was allowed to adsorb for 60 min, after which 8 mL of 30 mM $NaNO_3$ was flushed through the cell. The system was then allowed to equilibrate for 10 min. Ten force curves were measured; the sample was then moved slightly to find a new spot and ten additional force curves were recorded. The probe was retracted slightly from the surface and a new electrolyte solution was injected (8 mL). For each type of electrolyte the following concentrations were used sequentially: 1, 10, 50, and 100 mM.

Both the flat gold surface and the gold colloidal probe were imaged by AFM, and typical height images obtained with tapping mode in air are shown in Fig. 3. We note that the gold colloidal probe is significantly rougher than the flat gold surface. This demonstrates the importance of documenting the roughness of both surfaces in AFM colloidal probe measurements. The significant surface roughness precludes any firm conclusions to be drawn about the short-range interactions, even though the general features of the surface interaction between the gold surface and the probe were found to be reproducible (see "Results"). However, the main emphasis in this report is on the long-range

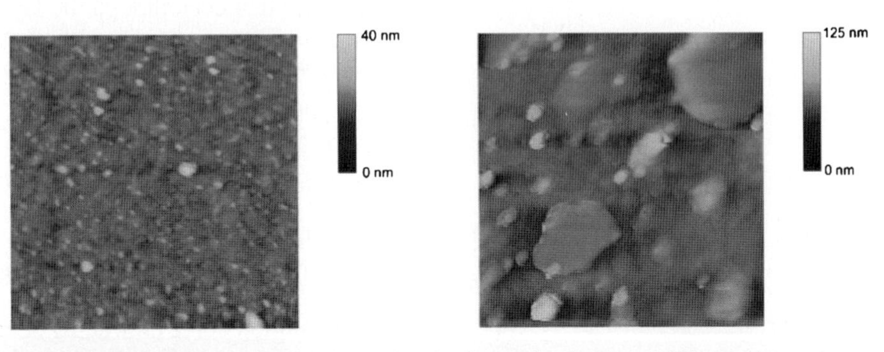

Fig. 3 *Left*: An AFM height image of a QCM gold crystal showing the fine-grained structure. The roughness parameters are $S_q = 1.98$ nm, $S_a = 1.34$ nm. The area shown is 1 μm^2, and the height scale is 40 nm. *Right*: A similar picture of the gold probe, showing a larger grain structure than the flat surface. The roughness parameters are $S_q = 8.68$ nm, $S_a = 6.28$ nm. The area shown is 1 μm^2, and the height scale is 125 nm

interactions, and these are expected to be unaffected by the surface roughness. Indeed, the reproducibility of the forces measured between mucin-coated surfaces was found to be good (see Fig. 4). The force curves reported in the following figures were selected as one representative curve from such a family of curves.

Results

Forces Between Mercaptohexadecane-Coated Gold Surfaces

The forces acting between two mercaptohexadecane-coated gold surfaces across aqueous 30 mM $NaNO_3$ are shown in Fig. 5. We note the absence of any long-range repulsive force, demonstrating that steric forces due to compression of surface irregularities are unimportant at large separations and that electrostatic double-layer forces can be ignored. The Debye length in 30 mM 1 : 1 electrolyte is 1.8 nm, so the fact that we do not observe any double-layer force does not prove that the surfaces are uncharged. However, previous measurements of the forces acting between mercaptohexadecane-coated gold surfaces across pure water have demonstrated the absence of any measurable double-layer force [21], allowing us to state that these surfaces are essentially uncharged. On approach, an attractive force is observed from separations of about 15 nm, and on separation a normalized adhesion force (F/R) of 30 $mN\,m^{-1}$ was measured. The short-range repulsion present below about 4 nm is attributed to the roughness of the surfaces employed in this investigation. The general features described above were reproducible, whereas the exact range of the attractive force, the short-range repulsion, and the adhesion varied somewhat. These observations are consistent with those reported for other hydrophobic and rough surfaces, where it has been argued that a cavitation mechanism is responsible for the attractive force [22].

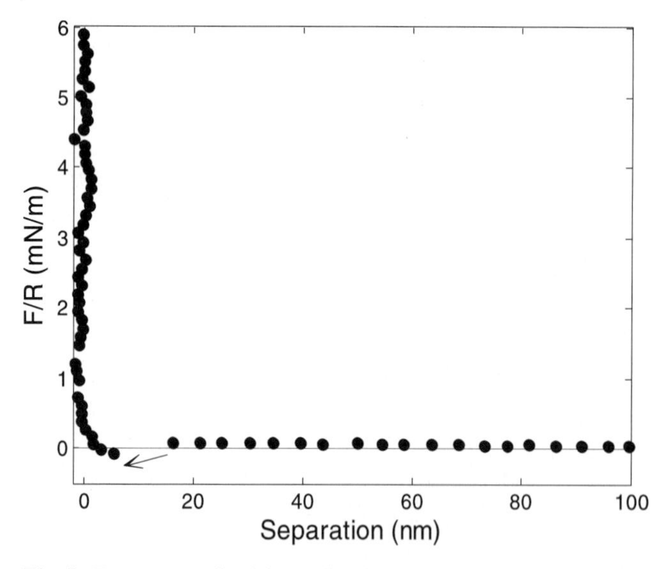

Fig. 5 Force normalized by radius between mercaptohexadecane-modified gold surfaces (a flat and a colloidal probe) as a function of separation. The *arrow* indicates a jump into contact due to the action of attractive surface forces

Adsorption of and Interaction Between Mucin-Coated Surfaces in 30 mM $NaNO_3$

BSM adsorbs extensively onto hydrophobic surfaces, e.g. the adsorbed mass on hydrophobic silanated silica has been determined to be 2 $mg\,m^{-2}$ when the adsorption was allowed to proceed for 60–70 min from a 25 ppm mucin solution containing 30 mM $NaNO_3$ [23]. The adsorption is largely and practically irreversible with respect to dilution, leaving about 1.8 $mg\,m^{-2}$ BSM on the surface after rinsing [23]. This observation is consistent with the slow desorption kinetics expected for macromolecules [24]. Considering the extensive adsorption of mucin on hydrophobic surfaces, it is not unexpected that the forces between mercaptohexadecane-coated gold surfaces carrying a preadsorbed mucin layer (adsorbed from 25 ppm mucin solution containing 30 mM $NaNO_3$ for 60 min) are completely different from those observed prior to mucin adsorption. The forces observed between the mucin layers across mucin-free 30 mM $NaNO_3$ are shown as crosses in Fig. 6. The forces measured on approach reach a measurable strength at a separation just below 300 nm, indicating that some tails extend at least 150 nm from the surface. The outermost part of the force increases exponentially with decreasing separation (decay-length $\approx$ 90 nm), and this exponential distance dependence is observed down to a separation of 100 nm. At shorter separations the repulsive force becomes steeper. On retraction, significantly less long-range repulsive forces were encountered and in many cases a weak attractive minimum ($\approx$ 1 $mN\,m^{-1}$) was observed. These data will be discussed in detail in a forthcoming publication. We note the reproducibility of the measurements (see Fig. 4), which demonstrates that even though the layers adopt a flat conformation under compression they return to their original extension within the time between two force curves ($\approx$ 1.8 s).

Effect of NaCl Concentration

After the force–distance relationship was established in 30 mM $NaNO_3$, the solution was changed to 1 mM NaCl. This resulted in massive swelling of the adsorbed layer, with a measurable force now being detected at a separation of 700 nm, as seen in Fig. 6. Thus, the longest tails now extend to at least 350 nm. It seems clear that the swelling is due to an increase in intralayer electrostatic repulsion between anionic groups, mainly sialic acid, caused by the

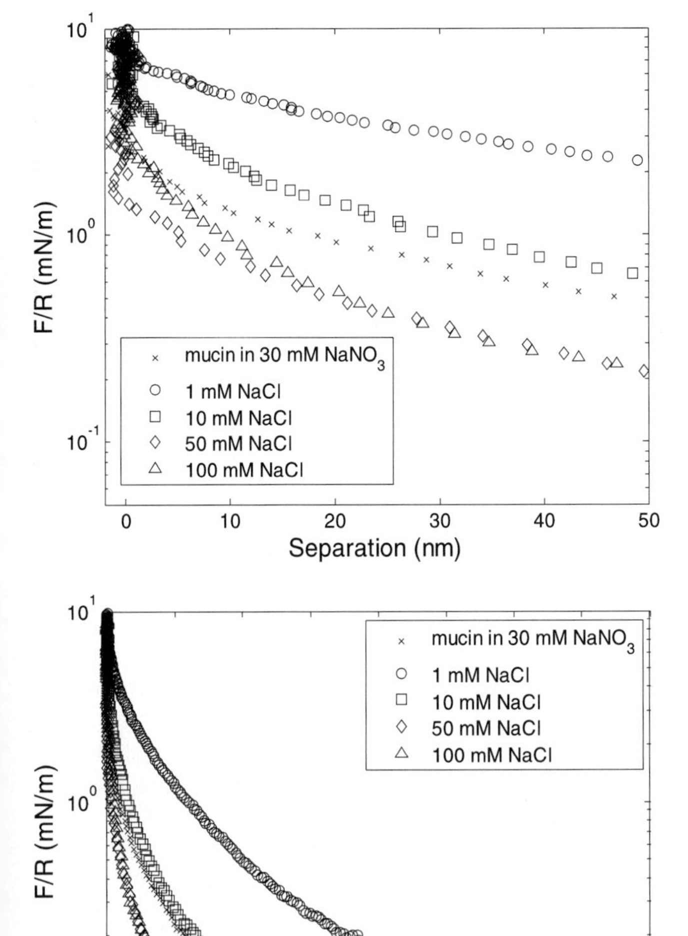

Fig. 6 Force normalized by radius between mercaptohexadecane-modified gold surfaces (a flat and a colloidal probe) carrying a preadsorbed mucin layer as a function of separation. Data are shown for interactions across 30 mM $NaNO_3$, and across a range of NaCl solutions. The *top graph* focuses on the intermediate-range interaction, whereas the *bottom graph* shows the same data on an expanded distance scale

increase in Debye screening length from 1.8 nm to 9.6 nm. By increasing the NaCl concentration to 10 mM a reduction in the range and strength of the interaction is achieved, signifying that the tails again contract and approach their initial state in 30 mM $NaNO_3$. A further increase in NaCl concentration to 50 mM results in further tail contraction. In 100 mM NaCl, a slight reswelling appears to occur, as evidenced by a slight increase in the repulsion observed below 10 nm. We conclude that the electrostatic repulsion between the anionic groups determines the mucin layer extension in 1 : 1 electrolyte. The slight reswelling observed when increasing the NaCl concentration from 50 to 100 mM remains unexplained.

Effect of Cation Valency and Concentration

In another set of experiments, the effects of adding $CaCl_2$ to the solution was explored. Addition of 1 mM $CaCl_2$ resulted in significant reduction in the range and strength of the force, as seen in Fig. 7. Thus, addition of $CaCl_2$ results in extensive contraction of the tails. We note that the Debye-length in 1 mM $CaCl_2$ is 5.6 nm, compared to 1.8 nm in 30 mM $NaNO_3$. Hence, by just considering the electrostatic screening length the layer extension is expected to be larger in 1 mM $CaCl_2$ than in 30 mM $NaNO_3$. However, the opposite is observed and this strongly suggests that calcium ions bind to the mucin layer and that this binding causes the layer to compact. An increase in $CaCl_2$ concentration to 10 mM results in further tail contraction, whereas further increasing the $CaCl_2$ concentration to 50 and 100 mM results in a limited reswelling of the tail region. This observation is suggested to be due to a slight recharging of the layer, i.e. the total positive charge of calcium ions incorporated in the mucin layer exceeds the total negative charge of the mucin.

The forces acting between the preadsorbed BSM layers across $LaCl_3$ solutions are shown in Fig. 8. The effects of $LaCl_3$ are similar to those of $CaCl_2$, but for $LaCl_3$ the most compact tail structure is found in the 1 mM solution. This suggests that recharging has already occurred at 10 mM $LaCl_3$.

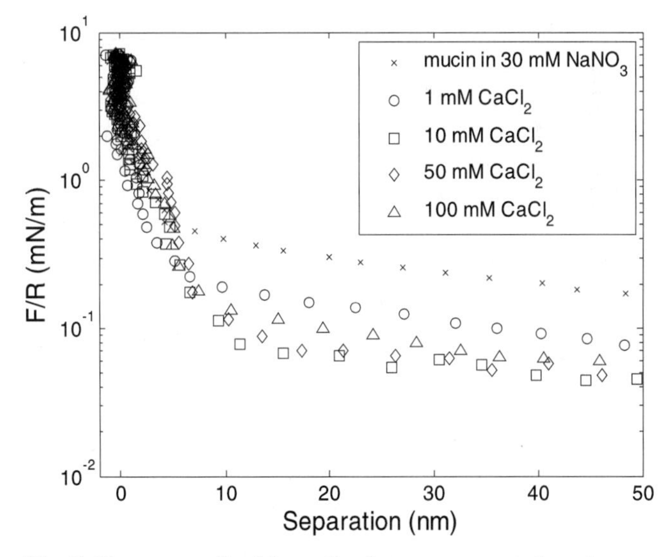

Fig. 7 Force normalized by radius between mercaptohexadecane-modified gold surfaces (a flat and a colloidal probe) carrying a preadsorbed mucin layer as a function of separation. Data are shown for interactions across 30 mM $NaNO_3$, and across a range of $CaCl_2$ solutions

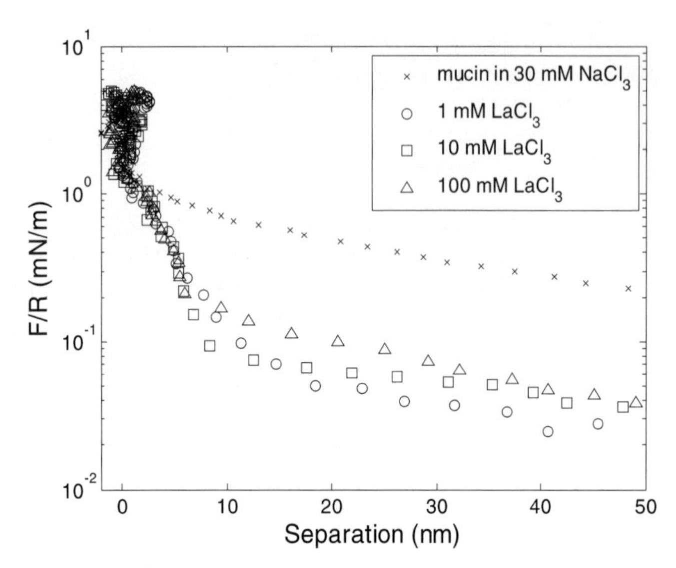

Fig. 8 Force normalized by radius between mercaptohexadecane-modified gold surfaces (a flat and a colloidal probe) carrying a preadsorbed mucin layer as a function of separation. Data are shown for interactions across 30 mM $NaNO_3$, and across a range of $LaCl_3$ solutions

Discussion

Structure of the Adsorbed Layer

The quartz crystal microbalance with dissipation, QCM-D, technique allows determination of two characteristics of an adsorbed layer. The first is the true sensed mass, which includes the mass of the adsorbing species and the change in mass of solvent oscillating with the crystal due to formation of the adsorption layer. The solvent can be either bound to the adsorbing species or mechanically trapped within the layer [25]. In the case of preadsorbed mucin on mercaptohexadecane-coated gold at different ionic strengths, both changes in bulk properties (viscosity and density) and effects due to the viscoelasticity of the layer have to be considered before the true sensed mass can be evaluated [11]. The second quantity that can be determined in a QCM-D experiment is the change in energy dissipation due to formation of the adsorbed layer, where a large positive value signifies formation of a viscoelastic layer.

In cases where both the adsorbed mass (Γ), e.g. obtained by reflectometry, and the true sensed mass (m_0), obtained with QCM-D, are known, then a QCM-thickness (d_{QCM}) can be evaluated using Eq. 2 [26]:

$$d_{QCM} = \frac{m_0}{\rho_{eff}} = \frac{m_0}{\rho_m \frac{\Gamma}{m_0} + \rho_s \left(1 - \frac{\Gamma}{m_0}\right)}, \tag{2}$$

where ρ_{eff} is the effective density of the layer, ρ_m the density of mucin (a typical mucin value [27] of $1.4\,g\,mL^{-1}$ was used in the calculation), and ρ_s the density of the solvent (reported by Feldötö et al. [11]). The true sensed mass measured under different conditions is provided in Table 1.

The equation above builds on a shear model that assumes that all material in the distance range 0–d_{QCM} oscillates with the crystal and thus contributes fully to the true sensed mass, whereas the material located further away from the surface does not oscillate with the crystal and does not contribute to the sensed mass. This is clearly a simplification, as are the optical models used for evaluating the ellipsometric thickness. It has been shown that the QCM thickness and the ellipsometric thickness are similar for relatively compact and homogeneous layers [28]. We do not expect this to be the case for more diffuse polymer layers since the ellipsometric thickness is directly influenced by the segment density profile [29], whereas the QCM thickness is influenced by the amount of water that oscillates with the crystal, and this quantity is at present an unknown function of the segment density profile.

For the extended mucin layers investigated here we will use the QCM thickness only as a qualitative indicator where a decrease in thickness means a reduction in water content of the adsorption layer, which in turn indicates a contraction of the layer. In the evaluation we used the reflectometry data of Dedinaite et al. [23] and the true sensed mass reported by Feldötö et al. [11], where both results were obtained for BSM adsorbing on hydrophobic surfaces from 25 ppm solutions in 30 mM $NaNO_3$. The surfaces used by Feldötö et al. are identical to ours, whereas Dedinaite et al. utilized hydrophobized silica. It is plausible that the adsorbed amounts on these two surfaces

Table 1 True sensed mass determined by QCM-D

Electrolyte	True sensed mass m_0 [$mg\,m^{-2}$] Initial in 30 mM $NaNO_3$	In 1 mM	In 10 mM	In 50 mM	In 100 mM	Final in 30 mM $NaNO_3$
NaCl	4.3	3.3[a]	4.0	4.1	4.2	4.3
$CaCl_2$	4.1	3.3	3.4	3.3	3.5	4.2
$LaCl_3$	3.9	2.7	2.9	3.4	4.0	2.9

[a] This value is not reliable due to limitations of the QCM technique, as discussed in the text

are similar, but it is not obviously that they are identical. Hence, to evaluate how critical it is to know the adsorbed mass accurately we have varied this quantity for two cases, keeping the other quantities constant. The result is shown in Fig. 9a. Clearly, for a constant sensed mass, the absolute value of d_{QCM} decreases with increasing adsorbed amount since the density of mucin is larger than that of the solvent. However, for qualitative comparison between two results it is not necessary to know the adsorbed mass to a very high precision. We emphasize that in the present case the absolute values presented in Fig. 9b have very limited significance, but the trends in the results provide useful information.

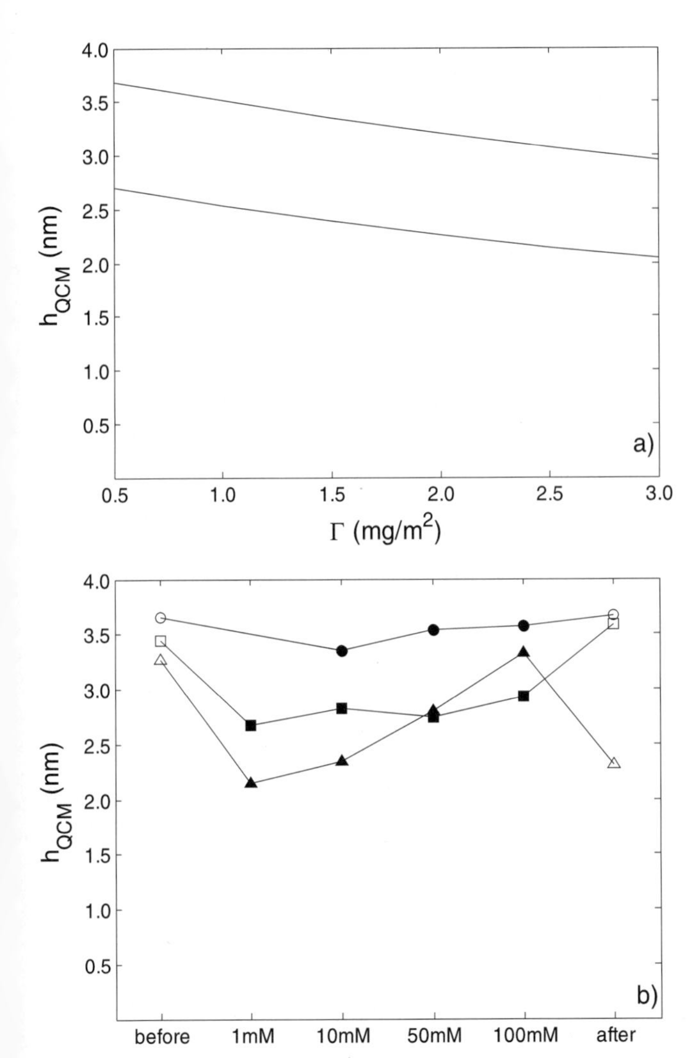

Fig. 9 **a** Variation of QCM thickness with the assumed adsorbed amount for two cases. The *top line* is for mucin preadsorbed to mercaptoethanol-coated gold before exposure to any other salt solution, and the *bottom line* is for the same mucin layer after exposure to 100 mM $LaCl_3$. **b** QCM thickness for different mucin–electrolyte combinations, using an adsorbed amount of 1.8 mg m^{-2}

In the evaluation of the QCM thickness we have ignored the mass of the ions associated with the mucin layer. We can estimate how severe this approximation is by considering the mass added to the layer by the ions at the charge neutralization point, i.e. when the number of charges from the ions equals the number of charges from sialic acid residues. For La^{3+} the added mass is about 0.04 mg m^{-2}, and for the other ions of this investigation it is lower by about a factor of two. For $LaCl_3$ the neglect of this mass does result in an overestimation of the QCM layer thickness by 0.01 nm, which is negligible.

The first feature to note is that the QCM thickness is orders of magnitude lower than the range of the steric force detected with AFM. This is not unexpected. To measure a thickness of an adsorbed polymer layer that is diffuse is by necessity dependent on the measuring technique. Surface force measurements, and measurements of hydrodynamic thicknesses using e.g. dynamic light scattering, are very sensitive to the longest tails and thus provide a large value for the thickness. On the other hand, the models used to evaluate the QCM thickness, and also the ellipsometric thickness, are insensitive to a small fraction of long tails. Thus, the range of the measured forces provides information related to the length of the longest tails, whereas the QCM thickness in our case provides complementary information related to the amount of water hydrodynamically coupled to the layer. We note that charged macromolecules adsorbing on an uncharged surface [29] are expected to have most of their mass close to the surface. The reason is that only the segments that bind directly to the surface contribute to a lowering of the free energy of the system. Negatively charged segments in the layer, but not directly in contact with the surface, on the other hand contribute with an increase in free energy of the system since they are brought a region with a negative potential (due to the accumulation of negative charges in the layer) relative to that in bulk solution. Thus, it seems likely that the majority of the adsorbed mucin mass is located close to the surface, that the trapped water in the layer is present in this region, and that the number of extending tails is relatively small. Such a structure has previously been observed for highly charged pig gastric mucine on hydrophobized mica [30].

That the tail region indeed is dilute is emphasized by the low force required to compress the layers and by the exponential decay of the long-range part of the force curve. We note that the segment density in the tail region has been predicted to decay exponentially with distance from the surface [31] Thus an exponential decay of the force is predicted in the weak overlap model [29] where the force is only due to the increased osmotic pressure caused by the increased segment density (including counterions) at the midpoint between the surfaces. Of course, at small distances the weak overlap approximation breaks down and other contributions, e.g. loss of conformational entropy, become important and the measured force is expected and observed to deviate from the exponential distance dependence.

A closer inspection of the data presented in Fig. 9 allows some other conclusions to be drawn. First, the QCM layer thickness before and after exposing the preadsorbed BSM layer to NaCl and $CaCl_2$ solutions are very similar. Thus, these ions do not alter the layer structure irreversibly and these ions are largely exchanged when the electrolyte type or concentration is varied. This is discussed further by Feldötö et al. [11]. In contrast, exposing the BSM layer to $LaCl_3$ solution results in an irreversible (over the time scale of the experiment, 60 min, for the QCM data) decrease in QCM layer thickness, i.e. in the amount of solvent that is hydrodynamically coupled to the layer. Thus, La^{3+} ions are not easily removed from the layer.

For NaCl the QCM thickness is roughly independent of concentration (the data at 1 mM was omitted since QCM cannot monitor this layer with very extended tails accurately, see below). The same is true for $CaCl_2$, but the QCM thickness is consistently lower in $CaCl_2$ solutions than in NaCl solutions. Thus, the presence of Ca^{2+} ions results in expulsion of water, most likely from the inner part of the adsorbed layer where trapped water is expected to be found, as well as in a reduction in the range of the steric force monitored by AFM that reflects the length of the tails. The QCM thickness in 1 mM $LaCl_3$ is lower than found for any of the other salts, demonstrating the compacting effect of the trivalent cation. In contrast to what was found for the other electrolytes, the QCM thickness increases with $LaCl_3$ concentration, indicating incorporation of water due to swelling of the inner part. We attribute this to recharging of the layer, as also demonstrated for the tail region by the force measurements reported in Fig. 7.

Comparison with Previous Measurements on Interactions Between Mucin Layers

The first report describing interactions between mucin layers is that of Perez et al., who measured the interaction between BSM adsorbed on negatively charged mica surfaces [32]. They found that long-range steric forces predominated in the interaction, and that for a constant adsorbed amount the forces were less repulsive in 0.15 M NaCl than in 0.1 mM NaCl. This result is consistent with our data that show a reduction in the repulsive forces between BSM-coated hydrophobic surfaces with increasing NaCl concentration. Perez et al. interpreted their results as being due to a stiffening of the mucin molecule at higher ionic strength, but we suggest that a more plausible interpretation of their and our results is screening of electrostatic forces, leading to contraction of extended tails at higher ionic strengths. The fact that qualitatively the same effect, i.e. compaction with increasing NaCl concentration, is found with two techniques (SFA was used by Perez et al.) and on both hydrophilic anionic mica and on uncharged hydrophobic surfaces strongly suggests that the compaction is due to decreased repulsion within the mucin layer rather than to a change in mucin–surface affinity.

Forces between BSM layers preadsorbed on negatively charged mica surfaces have also been reported by Dedinaite et al. [33]. They also found long-range steric interactions to be predominant, but they did not explore the effect of ionic strength. However, they showed that the anionic surfactant sodium dodecyl sulfate, SDS, has the ability to remove mucin from the surface and that adsorption of chitosan, a cationic polysaccharide, on top of mucin can protect the mucin layer from being removed by the surfactant.

The interactions between hydrophobized mica surfaces coated with rat gastric mucin (RGM) was investigated by Malmsten et al. [30]. RGM has an unusually small glycosidation and charge density for belonging to the mucin family and they suggested that the low charge density population was preferentially adsorbed to the hydrophobic surface (adsorbed amount $\approx 3\,\mathrm{mg\,m^{-2}}$). The steric forces generated by RGM showed no significant dependence on salt concentration, in the investigated range 0.1–150 mM. This is expected considering the low charge density of RGM. Comparing this finding with our results for BSM suggests that the charges in the highly glycosidated regions indeed are responsible for the effect of NaCl concentration on long-range repulsion. A compaction with increasing salt concentration is only observed when the charge carried by the mucin layer is sufficiently high.

We conclude that all surface force studies to date show that mucin form layers with extended tails on both hydrophilic anionic surfaces and on hydrophobic uncharged surfaces, resulting in long-range steric forces. The exact range and magnitude is influenced by the nature of the surface and by the composition of the mucin used. Low charge density mucin layers are not strongly affected by salt, whereas more highly charged layers become more compact as the NaCl concentration is increased.

Effect of NaCl Concentration

The effects of NaCl concentration on preadsorbed mucin layers have also been elucidated using QCM-D [11]. One unexpected finding was that the true sensed mass decreased when 30 mM $NaNO_3$ was replaced with 1 mM NaCl. The force curves shown in Fig. 6 can be used to rationalize this finding. The typical decay length of the acoustic wave under the experimental conditions used by Feldötö et al. is about 200 nm [34]. This is considerably shorter than the range of the steric force observed in this study, 700 nm, demonstrating that the tail region of the layer on each surface is at least 350 nm thick. Thus, the QCM-D is not able to sense the complete adsorbed layer in low ionic strength NaCl solutions, and thus the data become unreliable. The QCM-D investigation also reports a decreased dissipation with increasing NaCl concentration, i.e. the layer sensed by QCM-D becomes less efficient in dissipating the energy of the oscillating crystal with increasing NaCl concentration. The force data displayed in Fig. 6 strongly suggest that this is caused by a contraction of the extended tails.

Effect of Cation Valency

We have not been able to find any previous report on the effects of di- and trivalent cations on the interactions between mucin layers. However, we can discuss our findings in relation to the QCM-D study of Feldötö et al. [11], where it was found that addition of $CaCl_2$ resulted in a decrease in true sensed mass and dissipation, findings that suggest compaction of the adsorbed layer and expulsion of water. The dissipation decreased with increasing $CaCl_2$ concentration, whereas a slight increase in true sensed mass was observed when the $CaCl_2$ concentration was increased from 50 to 100 mM. This latter feature indicates some swelling due to recharging. The force measurements displayed in Fig. 6 are consistent with this trend. Addition of $CaCl_2$ results in significant contraction of the tails that contributes to the reduced dissipation. Water is also expelled from the inner part of the layer, which results in a decreased QCM thickness as seen in Fig. 9b. The QCM-D results demonstrate compaction in $LaCl_3$ solutions [11], and this is also observed in the force measurements.

Conclusion

The forces acting between preadsorbed BSM layers have been investigated using the AFM colloidal probe technique. It has been demonstrated that the electrostatic repulsion, mainly between sialic acid groups, determines the response of the mucin layer to changes in salt concentration and cation valency. The range and magnitude of the steric forces encountered between mucin layers decreases with increasing NaCl concentration. This is due to the decrease in Debye-length, which allows the tails to adopt less extended conformations. Multivalent cations, such as Ca^{2+} and La^{3+}, cause a significant contraction of the tails even in dilute solutions (1 mM). This cannot be explained by a variation of the Debye-length, but strongly suggests that these ions are binding to anionic sites in the mucin layer. It seems plausible that multivalent ion bridges are formed and that these promote the structural change. The force measurements reported here aid the interpretation of some recent QCM-D data for the same systems. In dilute NaCl solutions (1 mM) the adsorbed layer is too extended to be fully sensed by the QCM, which rationalizes the decrease in sensed mass registered by this technique under such conditions. The steric force observed in this study is very long-range and attributed to the presence of some extending tails. On the other hand, the QCM thickness that in our case reflects the amount of hydrodynamically coupled water is small, only a few nanometers. Thus, the few extended tails do not cause the surrounding water to oscillate with the crystal but rather suggests that the hydrodynamically coupled water is trapped in the layer close to the surface. Both the tail length and the amount of hydrodynamically coupled water are reduced in the presence of electrolytes with multivalent cations. The strong influence of multivalent cations on the structure of mucin layers shows that the presence of these ions is likely to affect the lubricating and hydrating properties of mucin coatings. This is a topic that deserves further attention.

Acknowledgement The authors acknowledge the Swedish Research Council, VR, for financial support.

References

1. Carlstedt I (1988) PhD Thesis. University of Lund
2. Shogren RL, Jamieson AM, Blackwell J, Jentoft N (1986) Biopolymers 25:1505–1517
3. Carlstedt I, Lindgren H, Sheehan JK (1983) Biochem J 213:427–435
4. Scheinthal BM, Bettelheim FA (1968) Carbohyd Res 6:257
5. Ullner M, Woodward CE (2002) Macromolecules 35:1437–1445
6. Bastardo L, Iruthayaraj J, Lundin M, Dedinaite A, Vareikis A, Makuska R, Van der Wal A, Furó I, Garamus VM, Claesson PM (2007) J Colloid Interf Sci 312:21–33
7. Raynal BDE, Hardingham TE, Sheehan JK, Thornton D (2003) J Biol Chem 278:28703–28710
8. Exley C (1998) J Inorg Chem 70:195–206
9. Shrivastava HY, Nair BU (2003) J Inorg Biochem 20:575–587
10. Shrivastava HY, Dhathathreyan A, Nair BU (2003) Chem Phys Lett 367:49–54
11. Feldötö Z, Pettersson T, Dedinaite A (2008) Langmuir 24:3348–3357
12. Bastardo L, Claesson PM, Brown W (2002) Langmuir 18:3848–3853
13. Feiler A, Sahlholm A, Sandberg T, Caldwell KD (2007) J Colloid Interf Sci 315:475–481
14. Sader JE, Chon JWM, Mulvaney P (1999) Rev Sci Instrum 70:3967–3969
15. Raiteri R, Preuss M, Grattarola M, Butt H-J (1998) Colloid Surf A 136:191–197
16. Cabanillas ED, Pasqualini EE, López M, Cirilo D, Desimoni J, Mercader RC (2001) Hyperfine Interact 134:179–185
17. Ralston J, Larson I, Rutland MW, Feiler AA, Kleijn M (2005) Pure Appl Chem 77: 2149–2170
18. Ducker WA, Senden TJ, Pashley RM (1991) Nature 353:239–241
19. Ducker WA, Senden TJ, Pashley RM (1992) Langmuir 8:1831–1836
20. Derjaguin B (1934) Kolloid Zeits 69:155–164
21. Ederth T, Claesson P, Liedberg B (1998) Langmuir 14:4782–4789
22. Wallqvist V, Claesson PM, Swerin A, Schoelkopf J, Gane PAC (2007) Langmuir 23:4248–4256

23. Dedinaite A, Bastardo L (2002) Langmuir 18:9383–9392
24. Cohen Stuart MA, Fleer GJ (1996) Ann Rev Mater Sci 26:463–500
25. Macakova L, Blomberg E, Claesson PM (2007) Langmuir 23:12436–12444
26. Höök F, Kasemo B, Nylander T, Fant C, Sott K, Elwing H (2001) Anal Chem 73:5796–5804
27. Creeth JM, Bhaskar KR, Horton JR (1977) Biophys J 167:557–569
28. Lundin M, Macakova L, Dedinaite A, Claesson P (2008) Langmuir 24:3814–3827, doi: 10.1021/la702653m
29. Fleer GJ, Cohen Stuart MA, Scheutjens JMHM, Cosgrove T, Vincent B (1993) Polymers at interfaces. Chapman & Hall, London
30. Malmsten M, Blomberg E, Claesson PM, Carlstedt I, Ljusegren I (1992) J Colloid Interf Sci 151:579–590
31. Scheutjens JMHM, Fleer GJ, Cohen Stuart MA (1986) Colloid Surf 21:285–306
32. Perez E, Proust JE (1987) J Colloid Interf Sci 118:182–191
33. Dedinaite A, Lundin M, Macakova L, Auletta T (2005) Langmuir 21:9502–9509
34. Rodahl M, Kasemo B (1996) Sens Actuators B 37:111–116

Progr Colloid Polym Sci (2008) 134: 11–18
DOI 10.1007/2882_2008_077

Published online: 20 May 2008

M. Maas
H. Rehage
H. Nebel
M. Epple

Formation and Structure of Coherent, Ultra-thin Calcium Carbonate Films below Monolayers of Stearic Acid at the Oil/Water Interface

M. Maas (✉) · H. Rehage
Chair of Physical Chemistry II, Technical University of Dortmund, 44227 Dortmund, Germany
e-mail: michael.maas@uni-dortmund.de

H. Nebel · M. Epple
Institute of Inorganic Chemistry, University of Duisburg-Essen, 45117 Essen, Germany

Abstract Detailed investigations of interfacial crystallization procedures are important to understand the basic principles of biomineralization processes. These interfacial phenomena can also be used to form new types of biomimetic composite materials. In a series of experiments we studied the influence of Langmuir-monolayers on the formation of ultra-thin calcium carbonate films. We systematically compared experiments performed at the water surface with results obtained at the water/oil interface. For stearic acid monolayers formed at the pure water surface, we were able to observe densely packed dispersions of ultra-thin $CaCO_3$ crystals, which were adsorbed below the surfactant membranes. We analyzed details of these structures by means of Brewster-angle-microscopy and other microscopic techniques. At the oil/water interface, however, we observed the formation of coherent, ultra-thin calcium carbonate films. These extended, two-dimensional crystalline structures were characterized by scanning electron microscopy, X-ray diffraction, and other techniques like interfacial-shear-rheology.

Keywords Biomineralization · Calcium carbonate · Stearic acid · Thin film · Amorphous precursor · Monolayer · Interface

Introduction

Biomineralization is a natural process that ultimately leads to the formation of complex nano-structured materials. These special structures are assembled by highly controlled growth mechanisms at the organic/inorganic interface [1–5]. Extensive research in the field of biomineralization is not only important for the comprehension of the processes in nature but could also lead to the development of advanced techniques for the architecture of new materials.

Calcium carbonate is the most important biomineral, due to its presence in, e.g., mollusk shells, skeletons of foraminiferes, coccolithophores, or corals (see [6] for a recent review).

A straightforward attempt that can be used to study biomineral growth is based on the application of Langmuir films. These lipid monolayers provide a simple model system for the organic/inorganic interface present in biomineralization, e.g., membranes or protein surfaces. In most of our experiments we studied stearic acid monolayers that were spread on a calcium hydrogencarbonate ($Ca(HCO_3)_2$) subphase [7]. These films were studied by Brewster angle microscopy (BAM), scanning-electron-microscopy (SEM), and other techniques.

The air/water interface in combination with Langmuir films of lipids or fatty acids is generally considered as a good modeling system for biomineralization processes occurring in cells or vesicles. In nature, of course, the mineralization takes place within the organisms without contact to air. So our approach was to transfer the experiments we first performed at the air/water interface to the oil/water interface. This, again, is a simplification compared to the bilayer membranes of cells that separate two aqueous phases. The results that we observed at the oil/water interface were completely different from what we measured at the air/water surface. In the latter studies only transient precursor films were observed, which turned into isolated crystals after a short period of time. At the oil/water interface, however, closed, durable films formed spontaneously within 12 h. Solid films of calcium carbonate composite materials were already prepared by other groups with additives such as polyelectrolytes [8–19] (mainly polyacrylic acid (PAA)) or high ratios of magnesium salts [20–22]. The additives serve as crystallization inhibitors in the bulk phase and tend to stabilize the $CaCO_3$ amorphous precursor phase in proximity to the interface. After addition of these compounds, macroscopic films with micrometer to nanometer thickness were obtained. In our experiments we observed the formation of stable films without adding crystallization inhibitors. The shear rheological properties of these films were measured with a custom built 2d-Couette system in order that the growth time and the stability of the films were evaluated. The interpretation of the rheological data also allows suggestions about the composition and structure of the films. Further, scanning electron micrographs and X-ray powder diffraction analysis are provided for a more detailed characterization. In this publication we focus on a brief overview of our work and compare the two strategies of biomineralization studies.

Materials and Methods

Substances

All substances were purchased at analytical grade from VWR company and used without further purification. Millipore grade water was prepared by an Elsa Purelab Ultra at 18.2 MΩ.

Preparation of the $Ca(HCO_3)_2$ Solutions

For the preparation of the $Ca(HCO_3)_2$ solutions several grams of $CaCO_3$ were suspended in 0.5 L water and flushed with gaseous CO_2 for about 2 h. In order to remove excess $CaCO_3$ the solution was filtered. In addition, the solutions were filtered again before each application. The concentration of the $Ca(HCO_3)_2$ solutions prepared by this method was approximately 8 mM (determined by titration with EDTA). The pH of this solution was 7.2.

Experiments at the Air/Water Interface

BAM Experiments. For all experiments a Langmuir–Blodgett trough (NIMA-Technology, Coventry, England, Type 601 BAM) was filled with 200 mL of water (Elsa Purelab Ultra, 18.2 MΩ). Afterwards, stearic acid solution was spread at the water surface. After evaporation of the solvent chloroform (about 5 min) the lipid monolayer was compressed until a solid-condensed film occurred with Π_o (surface pressure) $\approx 25\,\mathrm{mN\,m^{-1}}$. In order to generate well-defined domain structures, the solid condensed film was expanded and compressed several times. Afterwards, the 8 mM solution of $Ca(HCO_3)_2$ was induced into the subphase by means of a perfusor pump, and the excess water was simultaneously removed in order to obtain a constant water level. The induction speed was $25\,\mathrm{mL\,h^{-1}}$, so that a homogeneous distribution of the ions in the subphase was reached by diffusion. This distribution was created in order to ensure a controlled monitoring of the crystallization processes. Experiments at which the stearic acid was directly spread onto the subphase were performed as well, but there it was difficult to observe the first crystallization steps. Overall the results of these simpler experiments were equivalent to the infusion technique. During these experiments, the water surface was monitored by Brewster angle microscopy.

Scanning Electron Microscopy. For scanning electron microscopy 150 µL stearic acid (1 mM) was spread onto 400 mL of a 6 mM $Ca(HCO_3)_2$ solution in a Langmuir–Blodgett trough. The films were compressed to surface pressures of about $30\,\mathrm{mN\,m^{-1}}$. At a specified time after the surface pressure was obtained (app. 30 min after the beginning of the experiment), the films were transferred by the same technique as described in the above paragraph. Remaining water droplets were quickly removed from the surface. A silicon wafer was applied as substrate, which was made hydrophobic. For this purpose silicon wafers were aligned next to a drop of dimethyldichlorsilane, left in an exsiccator over night and, afterwards, were washed with freshly distilled acetone. Different samples were prepared at temperatures between 5 and 40 °C. At 20 °C additional samples were prepared at different $Ca(HCO_3)_2$ concentrations. The growth time and the pH were varied. These samples were investigated by a FEI ESEM Quanta 400 scanning electron microscope that was equipped with energy-dispersive X-ray spectroscopy (EDAX EDS Genesis 4000).

Experiments at the Oil/Water Interface

General Experiment. 50 mL of the freshly filtered Ca $(HCO_3)_2$ solution was put into a 100 mL crystallization dish (diameter 70 mm, height 40 mm). 40 mL of a 1 mM solution of stearic acid, in either dodecane or toluene, was added to the original solution. Samples for SEM or XRD were taken after specific amounts of time. Control experiments were carried out by exchanging the aqueous phase with 4 mM $Na(HCO_3)_2$, 4 mM $CaCl_2$, and pure water while maintaining the organic phase (and exchanging the stearic acid with different lipids, while maintaining the aqueous phase). No films were obtained under these conditions.

Scanning Electron Microscopy. For scanning electron microscopy at a specified time after the start of the experiments, a silicon wafer was quickly dipped with the sharp edge through the film and carefully pulled out again in a way that the lower side of the film was attached to the surface of the wafer. Samples were prepared at different $Ca(HCO_3)_2$ concentrations and different growth times at room temperature. These samples were investigated by a FEI ESEM Quanta 400 scanning electron microscope which was equipped with energy-dispersive X-ray spectroscopy (EDAX EDS Genesis 4000).

X-ray Powder Diffractometry. The material was obtained by collecting film fragments out of the crystallization dishes with a spatula and then studied in transmission mode on a Kapton foil with a Siemens D5000 diffractometer using Cu K_α radiation ($\lambda = 1.54$ Å). Additionally, one sample was measured with pure $CaCO_3$ powder and stearic acid (the same as used for the preparation of the samples).

Rheology. The shear rheological properties of the films were determined by a Rheometrics fluid spectrometer (RFS II), which was equipped with a modified shear system [23]. The measuring cell consisted of a quartz dish (diameter 83.6 mm) and a thin biconical titanium plate (angle 2°, diameter 60 mm), which were placed exactly at the interface between oil and water. The dish was first filled with the aqueous phase (100 mL). The titanium plate was then positioned at the water surface and the stearic acid solution was added (40 mL). We measured the torque required to hold the plate stationary as the cylindrical dish was rotated with the sinusoidal angular frequency ω. In such experiments, the two-dimensional storage modulus $\mu'(\omega)$ and the loss modulus $\mu''(\omega)$ were evaluated from the amplitude and phase angle of the stress and deformation signals.

Results and Discussion

Crystallization at the Air/Water Interface

In analogy to experiments performed by Hacke [24, 25], we investigated the crystallization of $CaCO_3$ under a monolayer of stearic acid. For a more detailed presentation of these results see [7]. If stearic acid is quickly spread onto an aqueous subphase of calcium hydrogencarbonate, thin films of $CaCO_3$ emerge in direct proximity to the Langmuir film. In the aqueous subphase the following equilibrium reaction occurs:

$$CaCO_3(s) + H_2O(l) + CO_2(g) \rightleftarrows Ca(HCO_3)_2(aq)\,. \quad (1)$$

This reaction also takes place during natural mineralization processes. Therefore, this simple membrane system is well suited to model biomineralization.

In the following, only these phenomena that are linked to the formation of thin films or precursor aggregates are shown. In almost all cases, a large amount of calcite single crystals appeared sooner or later together with those structures and eventually sedimentate towards the ground of the reaction vessel. The formation of such calcite crystals is well documented in the literature [26–31] and shall only be discussed here if they contribute to understanding the observed processes.

X-ray diffraction analysis of the films showed calcite, vaterite, and a broad amorphous region due to the organic material and/or amorphous calcium carbonate.

Brewster-Angle-Microscopy

The thin $CaCO_3$ film which was formed under the Langmuir monolayer of stearic acid could easily be investigated by means of BAM. A typical BAM-image of such a film is shown in Fig. 1.

In a series of different experiments, we observed the formation of transient films only under densely packed monolayers of stearic acid ($\Pi = 25$ mN/m). In a typical experiment, 50 mL of an 8 mM aqueous solution of $Ca(HCO_3)_2$ were introduced into the subphase during two hours (starting concentration: 0 mM, final concentration: 2 mM). Immediately after the beginning of the experiment, first particles with a diameter of around 200 nm (the maximum lateral resolution) formed at defects of the monolayer. These defects appeared as bright dots that were present already before the $Ca(HCO_3)_2$ solution was induced. The calcium carbonate particles appeared as white dots and could only be differentiated by extensive blank tests with pure water or calcium chloride as subphase. Consecutively these particles started to form fractal aggregates. With increasing time (accordingly increasing the $Ca(HCO_3)_2$) concentration), the aggregates became more densely packed, until finally a thin closed film was present. Note that it was difficult to derive the size of the particles from these experiments. The maximum lateral resolution of BAM is of the order of 0.2 μm, therefore the occurrence of larger and smaller particles at the same time leads to a superposition of particles that can be regularly depicted and those which are smaller than the resolution of the microscope. The latter are observed as Newtonian rings. Therefore, the fine structure in Fig. 1 can be misleading.

Scanning Electron Microscopy

A more detailed inspection of the observed structures was made possible by means of scanning electron microscopy (SEM). The selected images provide a close look onto the crystal shapes and surfaces. However, it is more difficult to get information by SEM about the height or the thickness of the films. The structures observed here can be compared to those described above. Generally, the dark, plane regions represent the lipid monolayer with the head groups

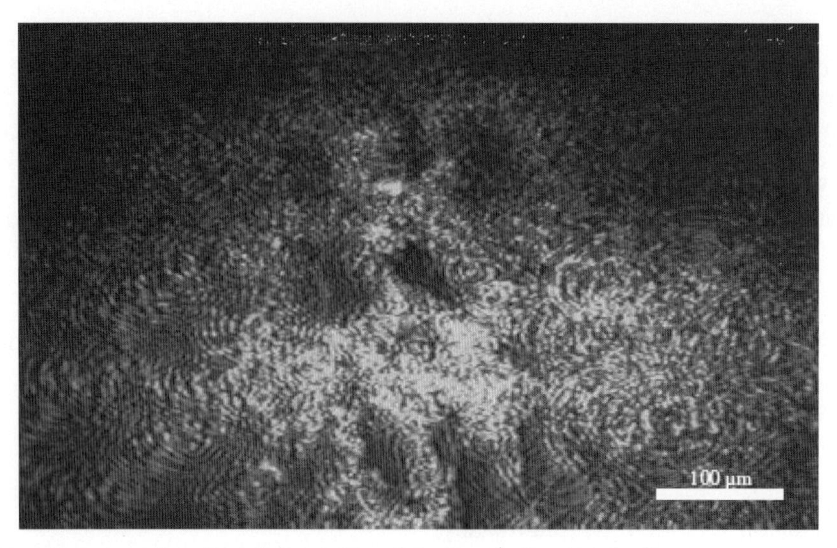

Fig. 1 Thin film of $CaCO_3$ below a stearic acid monolayer ($\Pi_0 = 25\,\mathrm{mN\,m^{-1}}$, $T = 22\,°\mathrm{C}$, pH = 7)

facing towards the observer. As reported earlier [32–35] calcite and vaterite crystals were observed, depending on the experimental conditions. Here attention was drawn to the more irregular, yet reproducible structures. All of the presented objects consisted of calcium carbonate as verified by quantitative EDX analysis and were rather seldomly observed but were reproducible. At 20 °C we observed large, layered structures in the vicinity of single crystals. A closer inspection of the surface of these objects revealed that these structures consisted of calcium carbonate aggregates, composed of small particles with a diameter of about 30–60 nm. The reason that only a few of these objects can be observed could be that they are metastable. Over time, they may develop into branched vaterite crystals, or they are forming clusters grown out calcite or vaterite crystals. This is concluded by comparison of the structures from BAM and SEM, as it is impossible to observe phase transitions under these conditions (in the given time frame) due to the lack of water.

At a lower temperature of 5 °C, a large number of leaf-like structures was observed. Again these objects consisted of smaller particles that typically had diameters between 50 and 100 nm. Similar structures were observed at very low subphase concentrations (see Fig. 2). The structures on the lower side of the lower "leaf" shown in Fig. 2 resemble small vaterite florets in an early growth state. So it may be assumed that these objects are metastable intermediates or precursor-aggregates that are trapped in this state because of the removal of water. The size of the particles from which these structures were formed was about 50–100 nm. From these observations it can also be concluded that the primary calcium carbonate particles are more stable at lower temperatures or very low subphase concentrations.

Fig. 2 Typical SEM-Images of leaf-like calcium carbonate films formed below a monolayer of stearic acid at $T = 5\,°\mathrm{C}$, $[Ca(HCO_3)_2] = 6\,\mathrm{mM}$, $t_{growth} = 1.5\,\mathrm{h}$, $\Pi_0 = 30\,\mathrm{mN\,m^{-1}}$, pH = 7. The films consisted of small particles with typical diameters between 50 and 100 nm

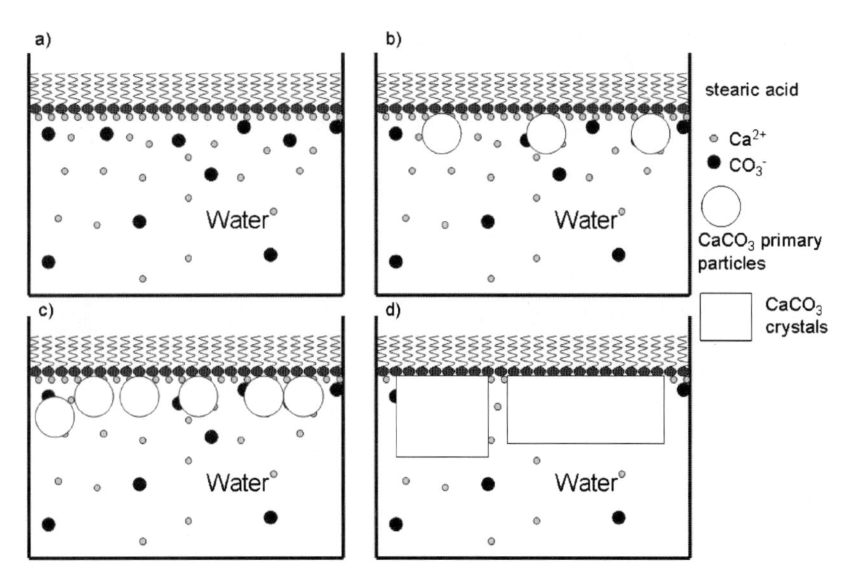

Fig. 3 Schematic picture of the formation of a $CaCO_3$ thin film under a monolayer of stearic acid

Under these assumptions these structures can be explained by mechanisms of colloidal growth processes. In other words, a self-organization process on the mesoscale through the assembly of calcium carbonate nanoparticles is assumed [36]. The polymorphic phase of these primary particles could not be determined in these experiments due to the small amount of substance available.

Conclusion

There are still some basic questions unresolved regarding the aggregation processes of $CaCO_3$ primary particles. As the particles have a net negative surface charge [37], adsorption and aggregation processes to the negatively charged stearic acid film appear to be energetically unfavorable. Here, the accumulation of the particles in proximity to the monolayer can only be induced by the compensation of the negative charges at the particle surfaces. This can be achieved by positively charged counter ions that adsorb on the surface of the nanoparticles. In monolayers of fatty acids protons can be supplied by the carboxylic head groups and cations from the subphase can be accumulated. Thus, the surface charge of the particles can be reduced or compensated so that the particles become able to absorb at the monolayer. Concerning the formation of $CaCO_3$ crystals, another electrostatic effect has to be taken into account. The importance of this effect becomes clear by considering the following reaction:

$$Ca^{2+}(aq) + 2HCO_3^-(aq) \rightleftarrows Ca^{2+}(aq) + 2CO_3^{2-}(aq) + 2H^+(aq) \,. \quad (2)$$

If the proton concentration is increased, the equilibrium shown in this equation shifts to the left. That means that the formation of $CaCO_3$ crystals is disfavored. It may be possible that this also means that the stability of the primary particles, which were formed before grown out $CaCO_3$ crystals occurred [38, 39], increases. Note that only the proton concentration in proximity to the head groups of the lipids is concerned. If the bulk pH is altered the crystallization is no longer controlled by the interface. The observed formation of the $CaCO_3$ thin films can be described as presented in Fig. 3: Below the stearic acid monolayer, a supersaturation of calcium and hydrogencarbonate ions occurred (Fig. 3a). From this supersaturated phase in proximity to the Langmuir film calcium carbonate particles formed (Fig. 3b). If the primary particles remained stable for a longer period of time, the particles could aggregate (Figure 3c) and form a thin film. This precursor film finally converted into more stable $CaCO_3$ modifications, like leaf-like structures, vaterite-flower-crystals, or clusters of well-developed calcite crystals and vaterite crystal agglomerations (Fig. 3d).

Crystallization at the Oil/Water Interface

After a growth time of 12 h, the films were visible to the unaided eye. They appeared as a translucent, slightly glossy membrane at the oil/water interface. After drying, the films appeared white with a crystalline glimmer. The drying of the films often led to breaking and sometimes to furling, probably due to different features of the upper and lower side. Derived from the experimental setup, the upper side of the films is the side that was exposed to the organic phase and the lower side the one exposed to the aqueous phase. Table 1 summarizes the different experiments. X-ray diffraction analysis of the films showed calcite, stearic acid, calcium stearate, and traces of vaterite.

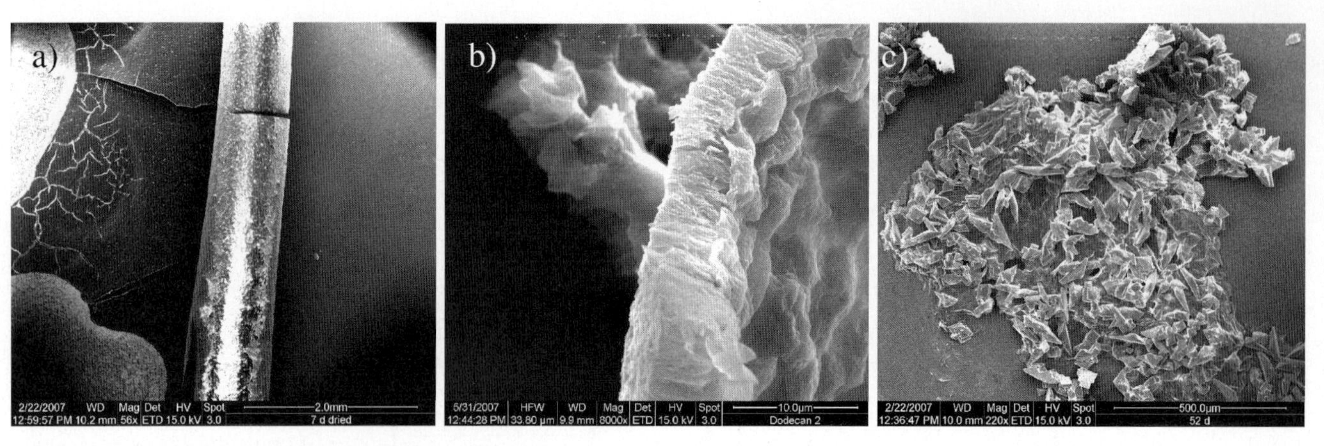

Fig. 4 4 mM $Ca(HCO_3)_2$, 1 mM stearic acid in toluene, growth time 7 days. **a** *Upper side* with furled section; **b** *Side view*; **c** 4 mM $Ca(HCO_3)_2$, 1 mM stearic acid in toluene, growth time 52 days

Table 1 Summary of experiments

Organic phase	Aqueous phase	Film formation
1 mM stearic acid in toluene	4 mM $Ca(HCO_3)_2$	Yes
Toluene	4 mM $Ca(HCO_3)_2$	No
1 mM stearic acid in toluene	Water	No
1 mM stearic acid in toluene	4 mM $Na(HCO_3)$	No
1 mM stearic acid in toluene	4 mM $CaCl_2$	No

Scanning Electron Microscopy

SEM provides a close look at dried versions of the films. Figure 4a depicts a large fragment of a film that furled on the right side so that the lower side became visible. Note the difference in roughness between both sides. Figure 4b allowed the estimation of the film thickness, which was roughly about 10 μm after 7 days of growth.

Figure 4c shows a film that was left in the reaction vessel for one and a half months. In contrast to the films that grew over one week, here, crystalline shapes were observed as components of the membrane. Due to the rectangular shapes of these crystals and compared to the XRD analysis they were mainly deformed calcite crystals.

2d-Shear-Rheology

At constant deformation and frequency, the two-dimensional viscoelastic properties of the films were measured as a function of time. With this kind of experiment the growth of the films was monitored. In case of the toluene films (Fig. 6a), 12–24 h after the beginning of the experiments, an elastic film formed. After the initial growth, the moduli decreased again, and the distance between storage and loss module increased in favor of the storage module. The local minimum and the increase in elasticity were probably due to crystallization processes inside the films. After approximately three days the moduli remained stable

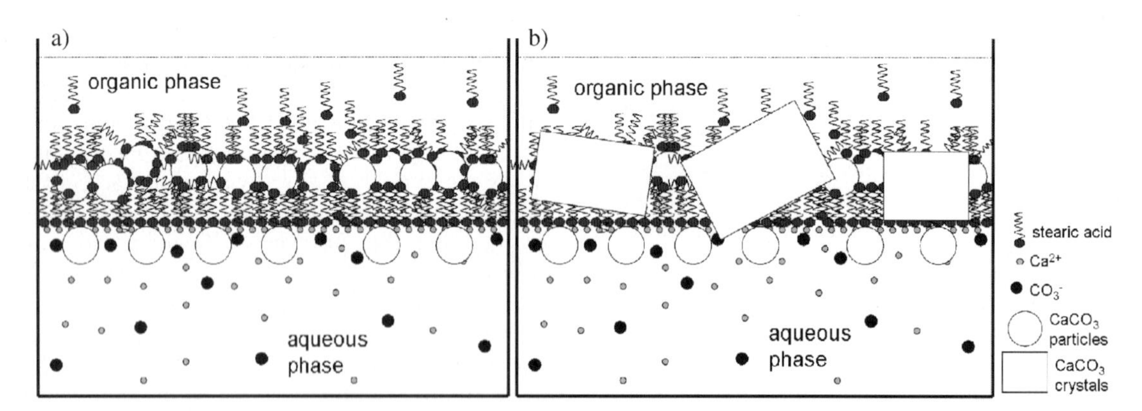

Fig. 5 Simplified model of the formation of the thin films: **a** Aggregation of precrystalline calcium carbonate particles at the oil/water interface, **b** Crystallization of the precrystalline calcium carbonate

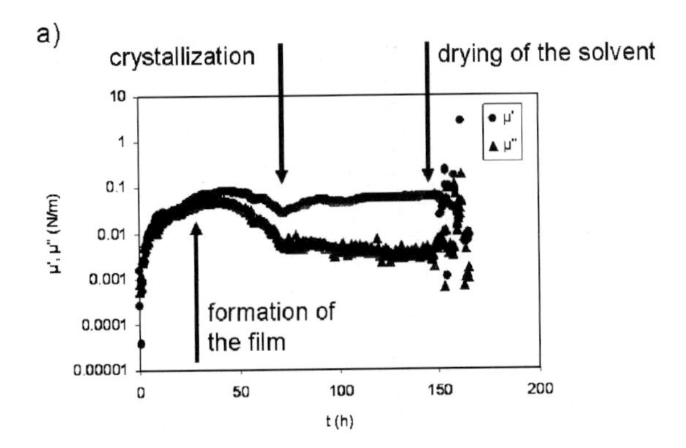

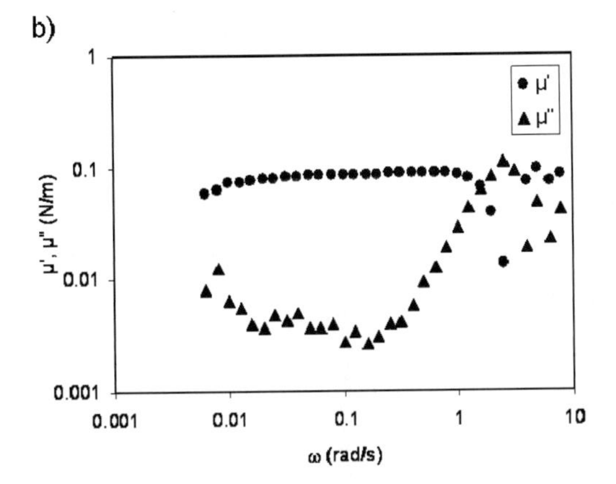

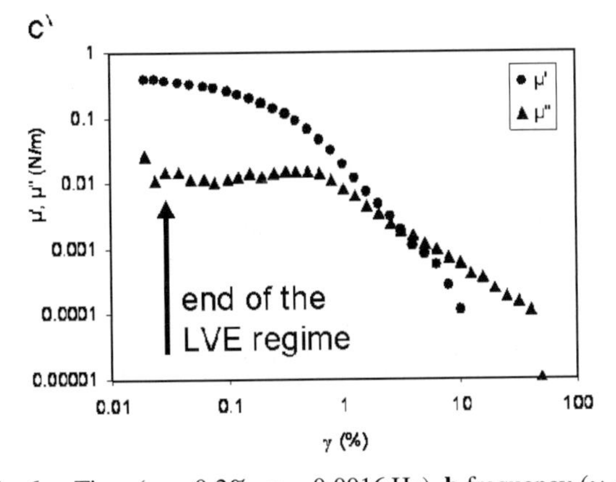

Fig. 6 **a** Time ($\gamma = 0.2\%$, $\omega = 0.0016$ Hz), **b** frequency ($\gamma = 0.1\%$) and **c** deformation ($\omega = 0.016$ Hz) tests for 4 mM $Ca(HCO_3)_2$, 1 mM stearic acid in toluene. The frequency test was performed after three days of growth time, the deformation test after 4 days

until the toluene was evaporated. The reason for the overall values of the moduli being very small was due to the smaller thickness of the films.

If the films were sufficiently stable (after 3 days) it was possible to measure μ' and μ'' at very low frequencies (Fig. 6b). Such measurements typically last from four to eight hours. As a result of these experiments, it can be concluded that the films behaved like viscoelastic, solid materials. This is because the moduli were parallel over the whole frequency spectrum with the storage modulus being the higher one. As an exception to this, the loss module increased at very high angular velocities because of the glass-analogues behavior of the films.

The brittleness and crystalline character of the films was emphasized by the deformation test. The linear viscoelastic regime (the deformation range at Fig. 6c which the films are deformed reversible) was limited to deformations of about 0.1%, which is an extremely low value. This points to the presence of energy elastic systems. As a compromise to the resolution of the measuring device, in the other test setups the applied strain was set to 0.1 or 0.2%.

Conclusion

From the experimental data it can be concluded that the films have a diameter of about 5–10 μm and consisted of stearic acid, calcium stearate, and calcium carbonate (mainly calcite, no experimental information could be given concerning amorphous calcium carbonate at this point). The viscoelastic data pointed to a glassy state with a crystalline hardness. It is important to note that the films did not consist of a stearate monolayer, cross-linked by calcium ions (chalk soaps) as control experiments with $CaCl_2$ observed. Derived from other control experiments it became clear that the growth of the films was due to a specific interaction of crystalline (or precrystalline) calcium carbonate and stearic acid.

In analogy to the studies at the air/water interface we suggest a growth mechanism involving calcium carbonate nanoparticles, which served as precursor particles. Osmotic draining of locally constrained emulsion droplets could also lead to crystallization processes. Further experiments shall involve high resolution X-ray diffraction and FTIR studies in order to evaluate the role of amorphous calcium carbonate in the crystallization processes.

References

1. Baeuerlein E (2000) Biomineralization. Wiley-VCH, Weinheim
2. Baeuerlein E (2003) Angew Chem Int Edn 42:614–641
3. Baeuerlein E (2004) Biomineralization. Progress in Biology, Molecular Biology and Application. Wiley-VCH, Weinheim, New York
4. Lowenstam HA, Weiner S (1989) On biomineralization. Oxford University Press, New York
5. Mann S (2001) Biomineralization; principles and concepts in bioinorganic materials chemistry. Oxford University Press, Oxford
6. Meldrum FC (2003) Int Mater Rev 48:187–224
7. Maas M, Rehage H, Nebel H, Epple M (2007) Colloid Polym Sci 285:1301–1311
8. Yu S-H, Cölfen H (2004) J Mater Chem 14:2124–2147
9. Cölfen H, Mann S (2003) Adv Mater Int Edit 42:2350–2365
10. Xu G, Aksay IA, Groves JT (2001) J Am Chem Soc 123:2196–2203
11. Xu X, Han JT, Cho K (2004) Chem Mater 16:1740–1746
12. DiMasi E, Patel VM, Sivakumar M, Olszta MJ, Yang YP, Gower LB (2002) Langmuir 18:8902–8909
13. Volkmer D, Harms M, Gower L, Ziegler A (2005) Adv Mater Int Edit 44:639–644
14. Xu G, Yao N, Aksay IA, Groves JT (1998) Langmuir 120:11977–11985
15. Hosoda N, Kato T (2001) Chem Mater 13:688–693
16. Zhang S, Gonsalves KE (1998) Langmuir 14:6761–6766
17. Kato T, Suzuki T, Irie T (2000) Chem Lett 29:186–187
18. Wada N, Suda S, Kanamura K, Umegaki T (2004) J Colloid Interf Sci 279:167–174
19. Kato T, Sugawara A, Hosoda N (2002) Adv Mater 14:869–877
20. Jiao Y, Feng Q, Li X (2006) Mater Sci Eng C 26:648–652
21. Kitamura M (2000) J Colloid Interf Sci 236:318–327
22. Aizenberg J, Lambert G, Weiner S, Addadi L (2002) J Am Chem Soc 124:32–39
23. Pieper G, Rehage H, Barthès-Biesel D (1998) J Colloid Interf Sci 202:293–300
24. Hacke S (2001) Brewsterwinkel-Mikroskopie zur Untersuchung der Kristallisation von Calciumcarbonaten an Modell-Monofilmen an der Grenzfläche Wasser/Luft. PhD thesis. University of Göttingen, Germany
25. Hacke S, Möbius D (2004) Colloid Polym Sci 282:1242–1246
26. Backov R, Lee CM, Khan SR, Mingotaud C, Fanucci GE, Talham DR (2000) Langmuir 16:6013–6019
27. Benitez IO, Talham DR (2004) Langmuir 20:8287–8293
28. Buijnsters PJJA, Donners JJJM, Hill SJ, Heywood BR, Nolte RJM, Zwanenburg B, Sommerdijk NAJM (2001) Langmuir 17:3623–3628
29. Heywood BR, Mann S (1994) Chem Mater 6:311–318
30. Mann S, Heywood BR, Rajam S, Walker JBA (1991) J Phys D 24:154–164
31. Mann S, Archibald DD, Didymus JM, Douglas T, Heywood BR, Meldrum FC, Reeves NJ (1993) Science 261:1286–1292
32. Buijnsters PJJA, Donners JJJM, Hill SJ, Heywood BR, Nolte RJM, Zwanenburg B, Sommerdijk NAJM (2001) Langmuir 17:3623–3628
33. Heywood BR, Mann S (1994) Chem Mater 6:311–318
34. Mann S, Heywood BR, Rajam S, Walker JBA (1991) J Phys D 24:154–164
35. Mann S, Archibald DD, Didymus JM, Douglas T, Heywood BR, Meldrum FC, Reeves NJ (1993) Science 261:1286–1292
36. Cölfen H, Mann S (2003) Angew Chem Int Edn 42:2350–2365
37. Moulin P, Roques H (2003) J Colloid Interf Sci 261:115–126
38. Ogino T, Suzuki T, Sawada K (1987) Geochim Cosmochim Acta 51:2757–2767
39. Olszta MJ, Odom DJ, Douglas EP, Gower LB (2003) Connect Tissue Res 44:326–334

Progr Colloid Polym Sci (2008) 134: 19–29
DOI 10.1007/2882_2008_081

Published online: 2 August 2008

Helmut Durchschlag
Peter Zipper

Volume, Surface and Hydration Properties of Proteins

Helmut Durchschlag (✉)
Institute of Biophysics and Physical Biochemistry, University of Regensburg, Universitätsstrasse 31, 93040 Regensburg, Germany
e-mail: helmut.durchschlag@biologie.uni-regensburg.de

Peter Zipper
Physical Chemistry, Institute of Chemistry, University of Graz, Heinrichstrasse 28, 8010 Graz, Austria

Abstract The present knowledge of protein science includes information on amino acid sequence and 3D structure in terms of precise models on the atomic level. Recourse to the respective databanks and advanced computer programs allows a series of molecular features to be calculated. Application of analytical surface calculation programs (SIMS, MSRoll) based on atomic coordinates or the coordinates of gravity centers of amino acids allows precise molecular dot surfaces to be calculated, in addition to numerical data for anhydrous molecular surface and anhydrous molecular volume. Usage of in-house hydration programs (HYDCRYST, HYDMODEL) permits the putative localization of individual water molecules on the protein envelope to be addressed explicitly. To estimate the overall values of protein volume and hydration, simple approximations based on the amino acid composition and characteristic numbers for the constituents can be used. Derivation of secondary parameters such as volume-to-mass ratios or water occupancies of the surface additionally helps to differentiate between anhydrous and hydrated molecular properties and to establish the extent of hydration. In this context, a variety of proteins, ranging from small monomeric proteins to huge multisubunit protein complexes, have been tested. In all cases, biologically realistic models have been observed. The obtained results are of importance for manifold future applications including characterization of hydration contributions to biotechnologically relevant proteins, drug design projects, and surface physics and surface chemistry topics such as development of functionalized surfaces and colloids as well.

Keywords Individual water molecules · Molecular volume · Overall hydration · Proteins · Surface topography · Visualization and Modeling

Introduction

Volume, surface and hydration properties of low-molecular compounds and macromolecules in solution are required in the biosciences for numerous purposes: (i) correct application of many physicochemical techniques (e.g., analytical ultracentrifugation, solution scattering and diffraction techniques, electron microscopy, modeling approaches), (ii) understanding the behavior and the interactions of native and denatured biopolymers in the solvent water or aqueous solutions, (iii) interpretation of parameter changes occurring in multicomponent solutions (e.g. proteins or nucleic acids at elevated concentrations of cosolvents, particularly electrolytes and denaturing agents), (iv) insight into the peculiarities of hydrated biopolymers, as crucial prerequisites for flexibility, dynamics and functionality (binding characteristics) of these molecules, (v) prediction of biopolymer parameters in aqueous environment, (vi) construction of tailor-made biomacromolecules and macromolecule-ligand complexes in context with drug-design and other pivotal projects, etc.

Volume, surface and hydration parameters can be obtained by various experimental techniques and, nowadays, by calculative approaches as well [1–3]. Among the experimental techniques available, the small-angle X-ray scattering (SAXS) technique may provide information on volume, surface and hydration quantities of biopolymers. Partial and isopotential specific volumes of macromolecules in two- and multicomponent solutions may be derived from a thermodynamic analysis of density increments. Partial specific volumes of molecules in water may also be obtained from simple ab initio calculations or a calculus of differences; adequate knowledge of interaction parameters (for salt binding and hydration) allows the isopotential specific volumes to be calculated as well. If the precise sequence and the 3D structure of biopolymers are known (e.g. protein data stored in data banks such as SWISS-PROT [4] and Protein Data Bank PDB [5]), these data may be applied for computing analytically the exact surface topography of biopolymers; accurate surface calculations can be complemented by surface rendering techniques. The hydration of proteins can be calculated on the basis of the hydration numbers found for amino acid (AA) residues [6]; the presumable position of individual water molecules on the protein surface may be localized by our hydration algorithms for atomic or AA coordinates [1, 7–12] using the information about the surface topography.

The present paper summarizes typical examples of precise calculations and realistic approximations. In particular, volume, surface and hydration properties of several proteins of different characteristics are highlighted, including visualization of their atoms, AAs, ligands, preferentially bound water molecules (for short: "waters") and water clusters, position of water-coated crevices and channels.

Sources of Data and Methods

Detailed definitions concerning volume, surface and hydration properties of biopolymers have been given elsewhere [2, 13]. Therefore only a few particulars pertinent to the study of the proteins in this paper are given. For the following studies, only proteins were chosen where we had a wide experience in data handling, primarily pertaining to SAXS and advanced modeling investigations (e.g. [14–17]).

Proteins

The atomic coordinates and the masses of the proteins under analysis were obtained from the PDB and/or SWISS-PROT data banks. The proteins under investigation are listed in Table 1, together with their major characteristics required for the following calculations. The proteins selected span a wide variety of proteins and features concerning number of chains and AA residues ($N_{\text{chain}} = 1$–180, $N_{\text{AA}} = 124$–29808), molar mass ($M \approx$ 14–3500 kg mol^{-1}), and quite different shapes. In several cases, the values for N_{AA} and M obtained from the PDB and SWISS-PROT files, respectively, differ significantly. Apart from contributions of any ligands, this is caused by the fact that in crystallographic work very often a certain number of AA residues are not resolved.

To obtain fairly comparable results, for the calculations based on crystallographic data, only the atomic coordinates of the protein moieties of the available crystal structures are taken. By contrast, complementary volume and hydration estimations based on AA composition data only made use of SWISS-PROT data.

All protein models were visualized by the program RASMOL [18]. Protein atoms were displayed in *CPK colors* (C in *light grey*, O in *red*, N in *light blue*, and S in *yellow*), while AA residues, ligands and water molecules are highlighted in an appropriate manner.

Calculation of Volumes

The simplest procedure for calculating the partial molar volume, $\overline{V}$, and partial specific volume, $\overline{v}$, of small compounds or macromolecules goes back to Traube [19] and utilizes an additivity principle, quite simply by summing up the volume increments for the basic atoms and considering a few special increments. For the case of native, nonconjugated proteins in two-component solutions (protein in water or dilute buffer), a somewhat modified procedure, the method by Cohn and Edsall [20], is usually the method of choice for predicting the partial specific volume (by adding the contributions of the constituent AA residues).

The partial volume quantities, $\overline{V}$ (usually in cm^3 mol^{-1}) and $\overline{v}$ (in cm^3 g^{-1}), are related by the molar mass, M (in g mol^{-1}): $\overline{V} = M\overline{v}$. If individual molecules are considered, the corresponding molecular volumes, V (usually given in Å^3), are additionally related by Avogadro's number, N_{A}: $V = 10^{24}\, M\overline{v}/N_{\text{A}}$. The molecular volumes obtained by the Traube or Cohn and Edsall procedure are anhydrous ("dry") volumes.

Modern analytical procedures and programs for the visualization and analysis of protein molecules and models (see below) are based on crystallographic data sets as those in the PDB. These programs allow, among other things, the calculation of molecular volumes, V, and further volume and surface characteristics as well. The molecular volumes thus obtained are either anhydrous volumes, neglecting all hydration contributions, or consider the influence of the water molecules found. Even if the few waters found in crystallographic work (cf. [11]) are taken into account,

Table 1 Comparison of calculated parameters (N_{dot}, S, V) for anhydrous proteins as obtained by surface calculation approaches (SIMS, AA-SIMS, MSRoll) and different input parameters; the proteins are arranged according to increasing number of chains

Protein (source)	N_{chain} (N_{AA}) and M (kg mol^{-1}) [a]	Codes for PDB and SWISS-PROT	Surface calculation approach	N_a	r_{probe} (Å)	d_{dot} (Å^{-2})	N_{dot}	S (Å^2)	V (Å^3)
Ribonuclease A (bovine pancreas)	1 (124) 13.69	1RBX P61823	SIMS	951	1.525	3.0	17975	5529	15607
Lysozyme (hen egg white)	1 (129) 14.31	2LYZ P00698	SIMS	1001	1.475	5.0	29196	5772	16062
Malate synthase (*E. coli*)	1 (722) 80.36	1P7T, mol. A [b] P37330	SIMS	5414	1.45	5.0	130974	24699	90987
			AA-SIMS [c]	706	1.45	0.5	12008	23353	96277
					1.45	5.0	117434	23281	92743
		1P7T, mol. B [b] P37330	SIMS	5358	1.45	5.0	133355	25081	90200
			AA-SIMS [c]	705	1.45	0.5	11987	23389	95557
					1.45	5.0	117769	23350	92310
Citrate synthase (pig heart)	2 (874) 97.84	1CTS P00889	SIMS	6888	1.45	3.0	104595	32119	118359
					1.50	3.0	106403	31354	119308
Phosphorylase (*E. coli*)	2 (1592) 180.8	1AHP P00490	SIMS	12732	1.45	3.0	184141	56603	214778
Phosphorylase (rabbit muscle)	2 (1684) 194.3	3GPB [b] P00489	SIMS	13558	1.45	3.0	207950	63464	230999
Lactate dehydrogen- ase (dogfish)	4 (1328) 146.3	6LDH [b] P00341	SIMS	10168	1.40	3.0	166007	49547	176037
Catalase (bovine liver)	4 (2104) 239.1, 244.6 [d]	8CAT [b] P00432	MSRoll	16396	1.45		94282	80485	
					1.50		89564	78316	
					1.525		87466	77415	
Haemoglobin (*L. terrestris*), dodecameric subunit	12 (1767) 201.1, 208.5 [e]	1X9F [b] P13579 (A,E,I) P02218 (B,F,J) P11069 (C,G,K) O61233 (D,H,L)	SIMS	14566	1.45	1.0	96629	70841	251854
			AA-SIMS [c]	1755	1.45	2.0	125871	64532	249337
Bacteriophage fr capsid (*E. coli*)	180 (23220) 2472	1FRS P03614	AA-SIMS [c]	23220	1.45	0.5	406388	756803	3119751
Haemoglobin (*L. terrestris*), HBL complex	180 (29808) 3386 3475 [f]	2GTL, original [b,g] Q9GV76 (M) Q2I743 (N) Q2I742 (O)	AA-SIMS [c]	28884	1.45	1.0	1056455	1066232	4065634
		2GTL, modified [h]	AA-SIMS [c]	28908	1.45	1.0	1078871	1088497	4162778

[a] The numbers of AA residues (N_{AA}) and the molar masses (M) of unliganded proteins were obtained from SWISS-PROT.
[b] Some AA residues are missing in the crystal structure; therefore the number of groups (N_a) is lower than to be expected from the SWISS-PROT data.
[c] The program SIMS was applied to the crystal structure reduced to AA residues.
[d] The second value includes the mass of 4 haem and 4 NADPH groups.
[e] The second value includes the mass of 12 haem groups.
[f] The second value includes the mass of 144 haem groups.
[g] The crystal structure of the HBL complex comprises the haem-liganded globin chains A–L (see dodecameric subunit) and the haem-deficient linker chains M–O.
[h] The crystal structure of the complex was modified by addition of 24 large beads representing the mass of the AA residues missing in the N-terminal region of linker chains M–O and at the C-terminus of linker chain N.

the obtained volumes represent by no means the behavior of a properly hydrated protein in aqueous solution. Only consideration of all or at least the majority of water molecules, which can be bound preferentially to the protein surface, yields a biologically relevant hydrated protein volume.

Calculation of Surfaces

Because of the intricate definition of a smooth protein surface, its determination is a bit arbitrary and mathematically highly sophisticated. Usually the concept of a "molecular surface" and of a "solvent-accessible surface area" is applied. Both types of surfaces are obtained by a "rolling ball strategy", allowing a spherical solvent probe (water) of radius r_{probe} to roll over the outside of the molecule. The continuous sheet traced out by the ball surface is called the molecular surface, S (in Å^2), while the sheet produced by the locus of the probe center defines the solvent-accessible surface area. While the former quantity obviously characterizes an anhydrous protein surface area, the latter quantity is representative of a hydrated protein.

Among the surface calculation approaches tested (cf. [2]), the analytically accurate programs MSRoll [21] and, in particular, SIMS [22] turned out to yield the most reasonable results for protein envelopes in terms of molecular and accessible surfaces, in addition to the possibility to create a smooth "molecular dot surface" required for application of advanced hydration modeling strategies and values for the solvent-excluded volume. The input to both programs can be influenced by the values used for r_{probe} (usually approx. 1.45 Å), and in the case of SIMS also by the dot density, d_{dot} (in Å^{-2}), chosen.

Starting points for both surface calculation programs are the coordinates of the molecules/models to be analyzed. In the case of proteins, the atomic coordinates in the PDB may be used. In the present study, the program SIMS was applied either to the atomic coordinates ("SIMS") or to coordinates of the AA residues (i.e. the gravity centers of the AAs; "AA-SIMS"). Of course, the latter procedure is a necessary requirement if the surface of large molecules should be calculated.

Calculation of Hydration

As a consequence of preferential water binding to certain (primarily polar) AA residues on the protein surface, there exists some kind of water layer on the protein surface. The bound water molecules have characteristics different from those of the bulk water; in particular they are characterized by a higher order and lower mobility of the bound waters and a higher density as well [11].

Approximate trials to predict the amount of water bound around proteins simply utilize the hydration numbers found for individual AA residues [6]. Though these computations do not differentiate between AAs in the protein interior or its outside, they turned out to be good estimates for the overall protein hydration δ_1 (in g of water per g of protein) [1]. For these calculations, only the AA composition and the respective (pH-dependent) hydration numbers are required.

Knowledge of the exact anhydrous surface topography of protein envelopes in terms of numerous dot surface points, as obtained by the above surface calculation programs MSRoll or SIMS, can be used for application of advanced hydration algorithms. The huge number of dot surface points, N_{dot}, and the appendent normal vectors can be exploited as a pool of hypothetical points for potential positions of water molecules at a constant normal distance of size r_w (the radius of a water molecule, usually identical with r_{probe}) above the protein surface. Our in-house programs HYDCRYST and HYDMODEL [1, 7, 11] for the creation of models based on the initial atomic coordinates of the protein moiety or reduced models derived from AA coordinates, respectively, are applied for selecting preferential positions for bound water molecules. The selected waters, represented by spheres of volume V_w, and assigned to each accessible AA residue, were attached to the dry protein models. The relation between water volume, V_w, and the nominal radius of the waters, r_w, is rather intricate: $V_w = (2r_w)^3$ (i.e. V_w is the volume of the cube circumscribing the water sphere of radius r_w).

The extent of hydration can be modulated by a fine-tuning of input parameters, in particular by variation of V_w and the scaling factor f_K acting on the hydration numbers given by Kuntz [6]. The water molecules assigned by HYDCRYST may be applied either to the initial or the reduced anhydrous protein models.

Results and Discussion

Firstly, analytical surface calculations were performed with the whole set of proteins, in order to obtain precise protein contours and molecular parameters for the anhydrous proteins (Table 1). In the following, the data obtained were used for application of our advanced hydration algorithms to obtain protein models provided with an appropriate number of water molecules (Table 2). Finally, the obtained data were compared with simple approximations and among each other (Table 3). Selected images explain the performance of steps, starting from the anhydrous space-filling protein models and eventually leading to realistically hydrated protein models (Figs. 1–5).

Analytical Surface Calculations

The surface calculation approaches (SIMS, AA-SIMS, MSRoll) applied to the atomic or AA coordinates of

Table 2 Comparison of calculated parameters (N_w, δ_1, N_b, V) for hydrated proteins as obtained by special hydration approaches (HYDCRYST, HYDMODEL) and different input parameters

Protein (source)	Surface calculation approach	Hydration approach	r_w (Å)	V_w (Å^3)	f_K	N_w	δ_1 (g g^{-1}) [a]	N_b	V ($\times 10^3$ Å^3)
Ribonuclease A (bovine pancreas)	SIMS	HYDCRYST	1.525	28.4	1.5	305	0.401	1256	24.76
Lysozyme (hen egg white)	SIMS	HYDCRYST	1.475	25.7	1.0	225	0.283	1226	22.74
					1.25	272	0.342	1273	23.94
					1.5	303	0.381	1304	24.74
					1.75	329	0.414	1330	25.41
					2.0	340	0.428	1341	25.69
Malate synthase (*E. coli*), mol. A	SIMS ($d_{dot} = 5$ Å^{-2})	HYDCRYST	1.45	24.4	1.0	1110	0.260	6524	120.5
					2.0	1509	0.353	6923	130.2
					3.0	1691	0.396	7105	134.7
					6.0	1814	0.425	7228	137.7
	AA-SIMS ($d_{dot} = 5$ Å^{-2})	HYDMODEL	1.45	24.4	1.0	1135	0.266	1841	123.0
					2.0	1483	0.347	2189	131.5
					3.0	1610	0.377	2316	134.6
					6.0	1677	0.393	2383	136.2
–, mol. B	SIMS ($d_{dot} = 5$ Å^{-2})	HYDCRYST	1.45	24.4	1.0	1149	0.272	6507	120.5
					2.0	1547	0.366	6905	130.2
					3.0	1729	0.409	7087	134.6
					6.0	1840	0.435	7198	137.3
	AA-SIMS ($d_{dot} = 5$ Å^{-2})	HYDMODEL	1.45	24.4	1.0	1152	0.273	1857	123.2
					2.0	1501	0.355	2206	131.7
					3.0	1633	0.386	2338	135.0
					6.0	1708	0.404	2413	136.8
Citrate synthase (pig heart)	SIMS	HYDCRYST	1.45	24.4	1.0	1406	0.259	8294	153.7
					1.5	1736	0.320	8624	161.8
					2.0	1879	0.346	8767	165.2
			1.50	27.0	1.0	1340	0.247	8228	155.6
					1.5	1631	0.300	8519	163.4
					2.0	1759	0.324	8647	166.9
Phosphorylase (*E. coli*)	SIMS	HYDCRYST	1.45	24.5	1.0	2569	0.256	15 301	282.6
					1.5	3099	0.310	15 831	295.5
					2.0	3348	0.335	16 080	301.6
					2.5	3556	0.356	16 288	306.6
		HYDMODEL [b]	1.45	24.5	1.0	2584	0.258	4176	283.0
					1.5	3174	0.317	4766	297.4
					2.0	3477	0.348	5069	304.8
Phosphorylase (rabbit muscle)	SIMS	HYDCRYST	1.45	24.5	1.0	2792	0.262	16 350	302.2
					1.5	3350	0.314	16 908	315.8
					2.0	3558	0.333	17 116	320.9
					2.5	3778	0.354	17 336	326.2
		HYDMODEL [b]	1.45	24.5	1.0	2835	0.266	4501	303.2
					1.5	3469	0.325	5135	318.7
					2.0	3721	0.349	5387	324.8
Lactate dehydrogenase (dogfish)	SIMS	HYDCRYST	1.40	21.95	1.0	2229	0.277	12 397	227.4
					2.0	2976	0.370	13 144	243.8
					3.0	3205	0.398	13 374	248.8
					4.0	3302	0.410	13 470	250.9

the proteins under analysis revealed data for the anhydrous proteins summarized in Table 1. Initially, for the proteins studied (composed of large numbers of atoms or AA residues, $N_a \approx 700$–29 000), huge numbers of dot surface points were created on the protein surface ($N_{dot} \approx 12\,000$–1 080 000). When using the SIMS approaches (SIMS or AA-SIMS), N_{dot} is governed by the choice of d_{dot}. Usage of the surface programs also fur-

Table 2 (continued)

Protein (source)	Surface calculation approach	Hydration approach	r_w (Å)	V_w (Å^3)	f_K	N_w	δ_1 (g g^{-1}) [a]	N_b	V ($\times 10^3$ Å^3)
		HYDMODEL [b]	1.40	21.95	1.0	2172	0.270	3488	226.2
					2.0	3084	0.383	4400	246.2
					3.0	3470	0.431	4786	254.7
					4.0	3567	0.443	4883	256.8
Catalase (bovine liver)	MSRoll	HYDMODEL [c]	1.45	24.5	1.0	2746	0.213	4746	345.4
					2.0	3269	0.254	5269	358.2
					10.0	3577	0.278	5577	365.7
			1.50	26.9	1.0	2586	0.201	4586	347.6
					2.0	3072	0.238	5072	360.7
					10.0	3342	0.259	5342	368.0
			1.525	28.4	1.0	2539	0.197	4539	350.2
					2.0	2974	0.231	4974	362.5
					10.0	3217	0.250	5217	369.4
Haemoglobin (*L. terrestris*), dodecameric subunit	SIMS	HYDCRYST	1.45	24.4	1.0	3125	0.273	17 165	326.3
					1.5	3706	0.324	17 746	340.4
					2.0	3952	0.345	17 992	346.4
	AA-SIMS	HYDMODEL	1.45	24.4	1.0	3148	0.275	4903	326.8
					1.5	3696	0.323	5451	340.1
					2.0	3915	0.342	5670	345.5
Bacteriophage fr capsid (*E. coli*)	AA-SIMS	HYDMODEL	1.45	24.4	1.0	31 904	0.232	55 124	3774
					1.5	36 773	0.268	59 993	3893
					2.0	38 892	0.283	62 112	3944
					3.0	41 778	0.304	64 998	4015
					7.0	42 831	0.312	66 051	4040
Haemoglobin (*L. terrestris*), HBL complex, original	AA-SIMS	HYDMODEL	1.45	24.4	1.0	50 862	0.272	79 746	5297
					3.0	63 608	0.340	92 492	5608
					5.0	64 965	0.348	93 849	5641
					7.0	65 235	0.349	94 119	5648
–, modified [d]	AA-SIMS	HYDMODEL	1.45	24.4	1.0	52 636	0.275	81 544	5435

[a] For calculating the hydration values δ_1, the molar masses M_{cryst} corresponding to the crystal structures were used. These masses deviate from the M values listed in Table 1 in the following cases: malate synthase: $M_{cryst} = 76.97$ kg mol^{-1} (mol. A) and 76.13 kg mol^{-1} (mol. B); phosphorylase: $M_{cryst} = 180.2$ kg mol^{-1} (*E. coli*) and 192.2 kg mol^{-1} (rabbit muscle); lactate dehydrogenase: $M_{cryst} = 145.0$ kg mol^{-1}; catalase: $M_{cryst} = 232.1$ kg mol^{-1} (including haem and NADPH); haemoglobin: $M_{cryst} = 206.3$ kg mol^{-1} (dodecameric subunit), 3367 kg mol^{-1} (HBL complex, original), and 3449 kg mol^{-1} (HBL complex, modified), all three values including CO-liganded haem.

[b] The reduced crystal structure (represented by AA residues) was hydrated by using the potential water points obtained by application of SIMS to the unreduced crystal structure.

[c] The reduced crystal structure (represented by AA residues) was hydrated by using the potential water points obtained by application of MSRoll to the unreduced crystal structure.

[d] The crystal structure of the complex was modified by addition of 24 large beads.

nishes values for the anhydrous molecular surface, S, and volume, V. As may be expected, the values for S and V grow in a systematic manner with increasing protein masses.

Figures 1A,B and 2A show the initial space-filling models of ribonuclease A and malate synthase, respectively, in *CPK colors* (atomic representation) or in *grey* (AA representation) plus the obtained surface dots in *black*. The surface dots represent the anhydrous protein contour as obtained by SIMS or AA-SIMS, respectively; the dots cover the protein model continuously (in a few cases the dots are hidden slightly underneath the surface of the model beads).

Advanced Hydration Calculations

Application of our special hydration algorithms to the selected protein set, after preceding surface calculations, yielded several molecular characteristics for the hydrated proteins analyzed (Table 2). The number of bound water molecules, N_w, increases continuously with increasing

Table 3 Further hydration and volume data (δ_1, V) for anhydrous and hydrated proteins as obtained by simple calculation procedures, together with some secondary parameters (V/M, N_w/S, $N_w(2r_w)^2/S$). Calculation procedures: hydration δ_1 according to Kuntz (K); volumes V according to Traube (T) or Cohn and Edsall (CE); these calculations were performed without ligands, using the molar masses M from the SWISS-PROT database (Table 1). Volumes V obtained by the programs SIMS (Table 1) or HYDCRYST (HC) and/or HYDMODEL (HM) (Table 2) are inclusive of the contributions of ligands; for calculating the V/M ratios, the molar masses from the crystal structure (M_{cryst}, Table 2) were used if these were different from the SWISS-PROT data

Protein (source)	δ_1 ($g\,g^{-1}$) [a]	V ($\times 10^3$ Å^3)	V ($\times 10^3$ Å^3)	V/M (Å^3 kg^{-1} mol)	V/M (Å^3 kg^{-1} mol)	V/M (Å^3 kg^{-1} mol) [b]	V/M (Å^3 kg^{-1} mol) [b]	N_w/S (Å^{-2}) [b]	$N_w(2r_w)^2/S$ [b]
	K	T	CE	T	CE	SIMS	HC/HM	HC/HM	HC/HM
Ribonuclease A (bovine pancreas)	0.397	16.10	16.15	1.18	1.18	1.14	1.81	0.055	0.51
Lysozyme (hen egg white)	0.348	16.95	17.03	1.18	1.19	1.12	1.59–1.79	0.039–0.059	0.34–0.51
Malate synthase (*E. coli*), mol. A	0.395	97.31	98.11	1.21	1.22	1.18–1.20	1.57–1.79	0.045–0.073	0.38–0.62
–, mol. B	0.395	97.31	98.11	1.21	1.22	1.18–1.21	1.58–1.80	0.046–0.073	0.39–0.62
Citrate synthase (pig heart)	0.386	119.4	120.4	1.22	1.23	1.21–1.22	1.57–1.71	0.043–0.059	0.37–0.50
Phosphorylase (*E. coli*)	0.394	220.4	222.2	1.22	1.23	1.19	1.57–1.70	0.045–0.063	0.38–0.53
Phosphorylase (rabbit muscle)	0.401	236.5	238.4	1.22	1.23	1.20	1.57–1.70	0.044–0.060	0.37–0.50
Lactate dehydrogenase (dogfish)	0.396	179.7	181.7	1.23	1.24	1.21	1.56–1.77	0.044–0.072	0.35–0.56
Catalase (bovine liver)	0.401	286.8	289.0	1.20	1.21	–	1.49–1.59	0.033–0.044	0.29–0.39
Haemoglobin (*L. terrestris*), dodecameric subunit	0.420	243.1	246.3	1.21	1.22	1.21–1.22	1.58–1.68	0.044–0.061	0.37–0.51
Bacteriophage fr capsid (*E. coli*)	0.367	2995	3008	1.21	1.22	1.26	1.53–1.63	0.042–0.057	0.35–0.48
Haemoglobin (*L. terrestris*), HBL complex, original	0.432	4063	4117	1.20	1.22	1.21	1.57–1.68	0.048–0.061	0.40–0.51

[a] The Kuntz hydration was calculated for neutral pH.
[b] Calculations were performed for the whole range of data given in Tables 1 and 2; for clarity only the resulting minimum and maximum values are given.

M ($N_w \approx 200$–$65\,000$); and the same is true for the hydrated molecular volume. The values found for hydrated V exceed those given for the anhydrous protein models given in Table 1 by far. The amount of water bound to different proteins is relatively constant ($\delta_1 \approx 0.2$–$0.4\,g\,g^{-1}$). The number of beads, N_b, obtained for the hydrated models is composed of protein and water beads ($N_b = N_a + N_w$) and varies between ≈ 1200 and $94\,000$. The values obtained for water binding (N_w, δ_1) depend to some degree on the input variables, in particular on f_K. This scaling parameter, however, can be used advantageously for modulating the extent of hydration (low, medium, high hydration). In general, the absence of scaling ($f_K = 1$) produces only an imperfect water sheet around a protein, while use of enhanced scaling factors leads to a more continuous water shell. In no case a complete water shell will be produced. This result, however, is in agreement with the finding of several hydrophobic AA patches on the protein surface.

As shown by the examples where HYDCRYST as well as HYDMODEL were applied, both approaches are capable of generating similar hydration values. Similar results are also found if SIMS and AA-SIMS are used concomitantly; this proves the usability of both approaches for the creation of surface dots.

Figures 1 and 2 illustrate the models resulting after applying the initial surface calculations and the ensuing

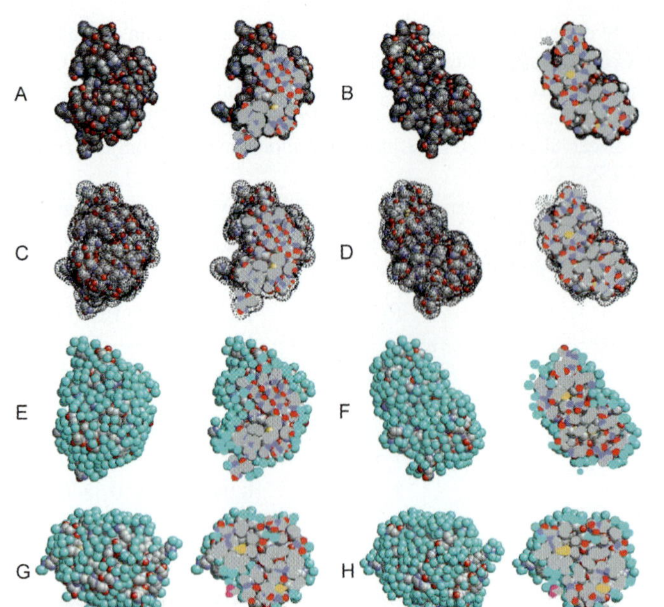

Fig. 1 **A–F** The various steps in modeling the hydration of the crystal structure of ribonuclease A (1RBX). **A,B** Two different views together with the corresponding slabs, showing the surface dots (*black*) obtained by application of SIMS ($r_{probe} = 1.525$ Å, $d_{dot} = 3.0$ Å^{-2}; $N_{dot} = 17\,975$) to the crystal structure; **C,D** two views together with slabs, displaying the potential water positions (*black*) located at a distance of $r_w = 1.525$ Å above the surface dots; **E,F** two views together with slabs, presenting the hydration waters selected by HYDCRYST from the pool of potential positions ($V_w = 28.4$ Å^3, $f_K = 1.5$; $N_w = 305$). **G,H** Views together with slabs of the crystal structure of lysozyme (2LYZ) hydrated by application of SIMS and HYDCRYST ($d_{dot} = 5.0$ Å^{-2}, $r_{probe} = r_w = 1.475$ Å, $V_w = 25.7$ Å^3), using $f_K = 1.0$ (**G** low hydration, $N_w = 225$) and $f_K = 2.0$; (**H** high hydration, $N_w = 340$). Atoms are displayed in *CPK colors* and hydration waters in *cyan*

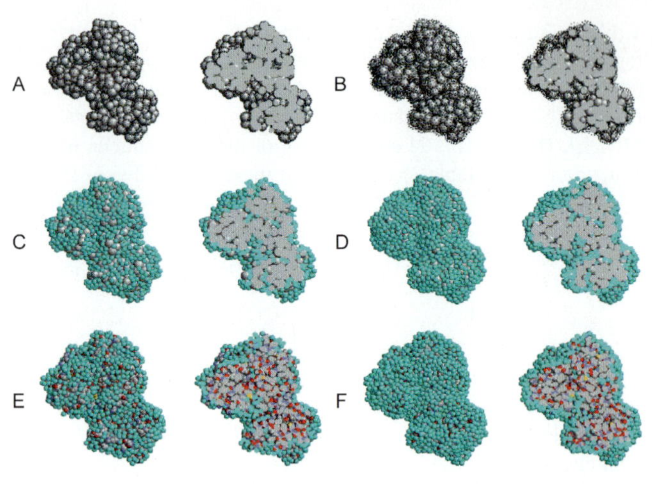

Fig. 2 **A–D** Illustration of the hydration modeling procedure when starting with a reduced cystal structure. **A** View together with slab of the AA representation of the crystal structure of malate synthase (1P7T, mol. B), showing the surface dots (*black*) obtained by AA-SIMS ($r_{probe} = 1.45$ Å, $d_{dot} = 0.5$ Å^{-2}; $N_{dot} = 11\,987$); **B** the analogous view and slab, displaying the corresponding potential water positions (*black*); **C,D** views with slabs of the hydrated AA representation of the protein, as obtained by AA-SIMS and HYDMODEL ($d_{dot} = 5.0$ Å^{-2}, $r_{probe} = r_w = 1.45$ Å, $V_w = 24.4$ Å) with $f_K = 1.0$ (**C** $N_w = 1152$) and $f_K = 6.0$ (**D** $N_w = 1708$). **E,F** views with slabs of the hydrated crystal structure of malate synthase, as obtained by SIMS and HYDCRYST ($d_{dot} = 5.0$ Å^{-2}, $r_{probe} = r_w = 1.45$ Å, $V_w = 24.4$ Å^3) with $f_K = 1.0$ (**E** $N_w = 1149$) and $f_K = 6.0$ (**F** $N_w = 1840$). Atoms are displayed in *CPK colors*, AA residues in *grey*, and hydration waters in *cyan*

hydration approaches, shown for simple (= nonconjugated and unliganded) monomeric proteins, i.e. the small proteins ribonuclease A and lysozyme and application of SIMS, and the essentially larger protein malate synthase and usage of the SIMS or AA-SIMS approach. After preceding construction of the anhydrous models (Figs. 1A,B; 2A) potential water sites (*black dots*) are positioned at appropriate positions (Figs. 1C,D; 2B). Owing to the different size of the proteins addressed, the distance between anhydrous protein surface and hypothetical water points is better visible with the smaller protein ribonuclease A. The water molecules (in *cyan*) bound to protein sites selected by HYDCRYST or HYDMODEL are illustrated in Fig. 1E–H and in Fig. 2C–F. Irrespective of the level of hydration (obtained by varying f_K), in no case a complete coverage of the protein surface by water molecules is achieved. On the other hand, except for some crevices, no waters are located in the protein interior (cf. *slabs*).

Figure 3 summarizes some representative hydrated models for simple oligomeric proteins, i.e. dimeric citrate synthase (A,B), two dimeric phosphorylases from different sources (C–F), and tetrameric lactate dehydrogenase (G,H), after applying quite different input variables and calculation approaches. An inspection of the slabs created reveals that proteins composed of several chains may exhibit water accumulations between the constituent chains, unless compact units are formed.

Calculation and visualization of oligomeric, conjugated proteins requires consideration of the respective ligands. This is shown in space-filling format for two haemoproteins, i.e. tetrameric catalase and the dodecameric domed-up subunit of the earthworm haemoglobin and their ligands (in *black* and *blue*) (Fig. 4). As shown impressively in the central slabs, water covers the protein surface. Again, all protein chains seem to be slightly separated and well coated by waters. Usage of SIMS plus HYDRCRYST or AA-SIMS plus HYDMODEL renders very similar water envelopes (cf. C,D with E,F), proving their effectiveness for hydration calculations.

The analysis of giant, multisubunit protein complexes (Fig. 5) poses a particular challenge for modelers, in particular because of the multitude of beads to be handled. The 180-meric bacteriophage capsid and the 180-meric

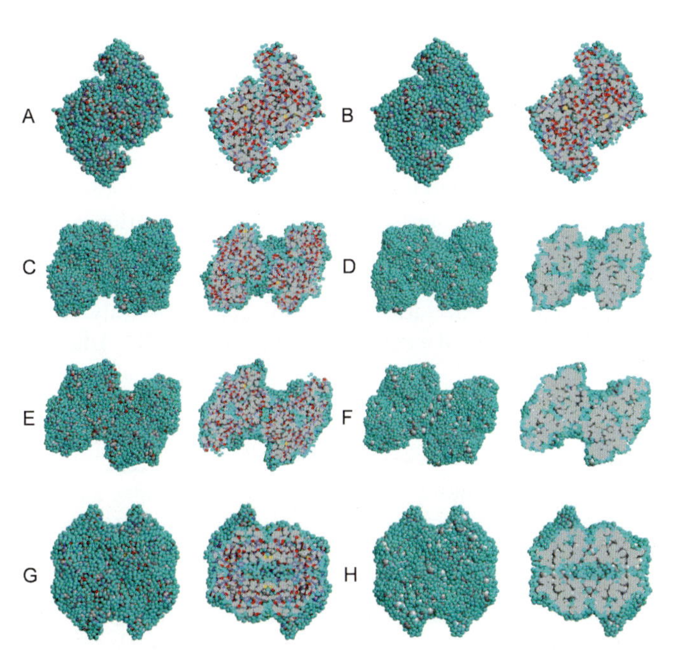

Fig. 3 Hydrated models of selected oligomeric proteins. **A**,**B** Views and slabs of the hydrated crystal structure of citrate synthase (1CTS), obtained by application of SIMS and HYDCRYST ($d_{dot} = 3.0\,Å^{-2}$, $r_{probe} = r_w = 1.50\,Å$, $V_w = 27.0\,Å^3$) using $f_K = 1.0$ (**A** $N_w = 1340$) and $f_K = 2.0$ (**B** $N_w = 1759$). **C**–**F** Views with slabs of the hydrated crystal structure (**C**,**E**) and the hydrated AA representation (**D**,**F**) of phosphorylase from *E. coli* (1AHP; **C**,**D**) and from rabbit muscle (3GPB; **E**,**F**), as obtained by application of SIMS and HYDCRYST (**C**,**E**) or HYDMODEL (**D**,**F**) using $d_{dot} = 3.0\,Å^{-2}$, $r_{probe} = r_w = 1.45\,Å$, $V_w = 24.5\,Å^3$, and $f_K = 2.0$ (**C** $N_w = 3348$; **D** $N_w = 3477$; **E** $N_w = 3558$; **F** $N_w = 3721$). **G**,**H** Views with slabs of the hydrated crystal structure (**G**) and the hydrated AA representation (**H**) of lactate dehydrogenase (6LDH), as obtained by application of SIMS and HYDCRYST ((**G**) or HYDMODEL (**H**), using $d_{dot} = 3.0\,Å^{-2}$, $r_{probe} = r_w = 1.40\,Å$, $V_w = 21.95\,Å^3$), and $f_K = 2.0$ (**G** $N_w = 2976$; **H** $N_w = 3084$). Atoms are displayed in *CPK colors*, AA residues in *grey*, and hydration waters in *cyan*

HBL complex of earthworm haemoglobin represent MDa proteins which are composed of thousands of AAs (approx. 23 000–30 000). By sophisticated reduction procedures for these multibead models [23], however, also this Sisyphean task can be solved and finally manageable entities are obtained. Visualization of the hydrated bacteriophage capsid (A,B) shows water coverage of both the outer and inner side of the protein shell. Even at the highest hydration level obtained no complete water shell can be achieved. Interestingly, the respective slabs unveil fine water channels in the protein shell, presumably required for its biological function. The hydrated models for the HBL complex of earthworm haemoglobin including haem groups (in *black*) (C–F) exhibit the hexagonal bilayer appearance of the complex and an internal cavity, and reveal the water-coated arrangement of the building blocks.

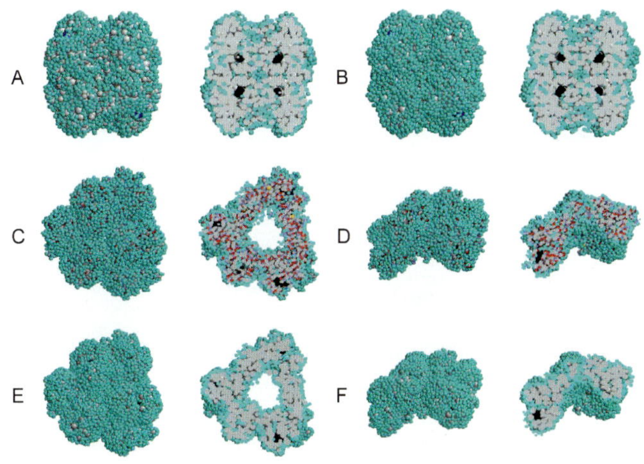

Fig. 4 Hydrated models of selected liganded proteins. **A**, **B** Views with slabs of the hydrated AA representation of catalase (8CAT) with ligands haem (*black*) and NADPH (*blue*), obtained by application of MSRoll ($r_{probe} = 1.45\,Å$) together with HYDMODEL ($r_w = 1.45\,Å$, $V_w = 24.5\,Å^3$) using $f_K = 1.0$ (**A** $N_w = 2746$) and $f_K = 10.0$ (**B** $N_w = 3577$). **C**–**F** Two different views with slabs of the crystal structure (**C**, **D**) or the AA representation (**E**, **F**) of the dodecameric subunit of *L. terrestris* haemoglobin (1X9F) including haem ligands (in *black*), as obtained by application of SIMS and HYDCRYST (**C**, **D** $d_{dot} = 1.0\,Å^{-2}$, $r_{probe} = r_w = 1.45\,Å$, $V_w = 24.4\,Å^3$, $f_K = 2.0$; $N_w = 3952$) or AA-SIMS and HYDMODEL (**E**, **F** $d_{dot} = 2.0\,Å^{-2}$, $r_{probe} = r_w = 1.45\,Å$, $V_w = 24.4\,Å^3$, $f_K = 2.0$; $N_w = 3915$). Atoms are displayed in *CPK colors*, AA residues in *grey*, and hydration waters in *cyan*

Simple Hydration and Volume Estimates and Parameter Correlations

The simple hydration approach by Kuntz, based on the AA composition and hydration numbers for AA residues only, allows a reasonable assessment of water binding (Table 3). The δ_1 values vary in a relatively narrow range ($\delta_1 \approx 0.35$–$0.43\,g\,g^{-1}$), they coincide largely with the values listed in Table 2 which have been found by the above advanced hydration approaches.

The simple volume estimates, based on the AA composition and the volume increments given by Traube or Cohn and Edsall, lead to dry volumes (Table 3) very similar to the anhydrous volumes found by SIMS (Table 1).

A critical comparison of all volume estimates in terms of V/M values (volumes normalized by the masses) clearly shows (Table 3) that the approaches by Traube, Cohn and Edsall and SIMS give similar values for all proteins studied (V/M about 1.2); these values are characteristic of anhydrous (dry) proteins. By contrast, the V/M data observed for hydrated proteins exceed those of the anhydrous proteins (V/M about 1.6–1.7 for different levels of hydration).

The quotients N_w/S or $N_w(2r_w)^2/S$ may be considered as measures for the water occupancy of the protein surface. In agreement with the relation given between V_w

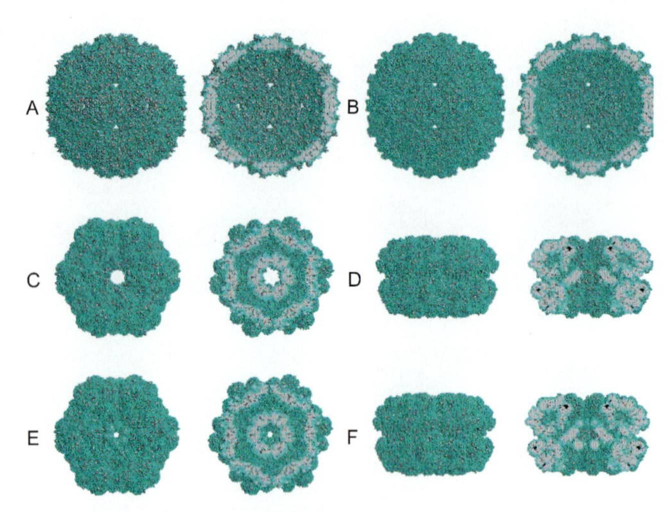

Fig. 5 Hydrated models of giant protein complexes. **A,B** Views with slabs of the hydrated AA representation of the capsid of bacteriophage fr (1FRS), as obtained by AA-SIMS and HYDMODEL ($d_{\mathrm{dot}} = 0.5\,\text{Å}^{-2}$, $r_{\mathrm{probe}} = r_{\mathrm{w}} = 1.45\,\text{Å}$, $V_{\mathrm{w}} = 24.4\,\text{Å}^3$) using $f_{\mathrm{K}} = 1.0$ (**A** $N_{\mathrm{w}} = 31\,904$) and $f_{\mathrm{K}} = 7.0$ (**B** $N_{\mathrm{w}} = 42\,831$). **C–F** Two different views with slabs of the hydrated AA representation of the unmodified (**C,D**) and modified (**E,F**) HBL complex of *L. terrestris* haemoglobin (2GTL), with haem groups displayed in *black*, as obtained by AA-SIMS and HYDMODEL ($d_{\mathrm{dot}} = 1.0\,\text{Å}^{-2}$, $r_{\mathrm{probe}} = r_{\mathrm{w}} = 1.45\,\text{Å}$, $V_{\mathrm{w}} = 24.4\,\text{Å}^3$, $f_{\mathrm{K}} = 1.0$); **C,D** $N_{\mathrm{w}} = 50\,862$; **E,F** $N_{\mathrm{w}} = 52\,636$. AA residues are displayed in *grey* and hydration waters in *cyan*

and r_{w}, the ratio $N_{\mathrm{w}}(2r_{\mathrm{w}})^2/S$ yields directly the fraction of the surface covered by waters. As may be taken from Table 3, again, the quotients found for different proteins remain nearly constant ($N_{\mathrm{w}}/S \approx 0.043$–$0.061\,\text{Å}^{-2}$ and $N_{\mathrm{w}}(2r_{\mathrm{w}})^2/S \approx 0.36$–$0.52$ for different hydration levels).

Conclusions

It was shown previously [11] that the positions of preferentially bound water molecules in lysozyme crystals and in solution are similar, as may be tested from the few water molecules fixed in crystallographic work, on the one hand, and our hydration algorithms, on the other. In this study, the comparative analysis of a variety of proteins in terms of volume, surface and hydration parameters and the respective models shows that the advanced surface and hydration calculations are able to obtain biologically realistic models for the hydrated proteins, irrespective of their nature, size and composition. As first approximations, even simple estimations based on the AA composition may supply reasonable results concerning volume and hydration.

The obtained hydrated models may be used as starting points for several further modeling purposes, including scattering and hydrodynamic modeling, prediction of parameters for hydrated proteins, visualization and analysis of structural details such as existence and localization of water channels and water compartments, sites of preferential water binding or hydrophobic patches etc. This may be also of interest in the future in context with several new applications, such as development of engineered proteins for several purposes, characterization of the solubility of all classes of proteins, and structure-based drug-design development [24].

Our approach to predict the existence and presumable localization of individual water molecules is mainly based on the existence of crystallographic (or NMR) data and hydration numbers for the individual components (AAs in the case of proteins). In principle, however, the procedure developed for proteins may also be applicable to other types of molecules, comprising biological and nonbiological macromolecules and molecules of smaller size as well. A gamut of possible (bio)technological and biomedical applications come to mind, including estimations of the flow behavior of biological material, prediction of the characteristics of functionalized surfaces and colloids (e.g. in context with the development of (bio)sensors, (bio)films, microelectronic, optoelectronic and lithographic material), diverse surface modifications of polymers and nanoparticles, design of biocidal and antifouling surfaces and materials, adsorption, adhesion and binding aspects (protein adsorption, antibody immobilization etc.). Due to the peculiarities of interfaces (such as charge and dipole characteristics, energies, balance between hydrophobic and hydrophilic groups, wettability, surface roughness, mobility of chains) [25–27], on the basis of our concepts and strategies, a variety of novel applications could be possible henceforth in the fields of surface and nanostructure engineering. This might include tailoring of polymers (e.g. block copolymers), polyelectrolytes, amphiphilic compounds, and antifreeze materials. In this context a lot may be learned from understanding and mimicking biological material such as proteins. In this respect, the possible realization of our approaches might be of particular importance, not only for the biosciences but also for surface physics and surface chemistry.

Future applications of the principles of our approaches to functionalized surfaces and polymers, however, should keep clearly in mind that this requires a substantial development of the present approaches. In the case of the proteins studied in this paper, crystallographic datasets and hydration numbers for all constituent AAs were available. In principle, appropriate adaptation of the surface calculation program SIMS, required for calculation of surface dots and potential water sites, should also allow this program to be applied to artificially generated models (by modification of the file ATRAD.SIZ); knowledge of the crystal structure of the molecules under analysis is not necessarily required. Similarly, further development of our recent version of the hydration approach HYDCRYST should enable adding the required information concerning atoms, groups, hydration numbers by means of parameter files. Accepting the efforts to be performed, this should provide various novel and interesting applications in the future.

Acknowledgement The authors are much obliged to Y.N. Vorobjev for use of the program SIMS and to R.A Sayle for RASMOL. H.D. thanks the Yukawa Institute for Theoretical Physics at Kyoto University, where discussions during the workshop YITP-W-06-04 on "Structures and Dynamics in Soft Matter – Beyond Self-Organization and Hierarchical Structures" initiated the basic concept of this work.

References

1. Durchschlag H, Zipper P (2001) Biophys Chem 93:141–157
2. Durchschlag H, Zipper P (2005) Calculation of Volume, Surface, and Hydration Properties of Biopolymers. In: Scott DJ, Harding SE, Rowe AJ (eds) Analytical Ultracentrifugation: Techniques and Methods. Royal Society of Chemistry, Cambridge UK, pp 389–431
3. Durchschlag H, Zipper P (2006) Bussei Kenkyu (Japan) 87:68–69
4. Boeckmann B, Bairoch A, Apweiler R, Blatter M-C, Estreicher A, Gasteiger E, Martin MJ, Michoud K, O'Donovan C, Phan I, Pilbout S, Schneider M (2003) Nucleic Acids Res 31:365–370
5. Berman HM, Westbrook J, Feng Z, Gilliland G, Bhat TN, Weissig H, Shindyalov IN, Bourne PE (2000) Nucleic Acids Res 28:235–242
6. Kuntz ID (1971) J Am Chem Soc 93:514–516
7. Durchschlag H, Zipper P (2002) J Phys Condens Matter 14:2439–2452
8. Zipper P, Durchschlag H (2002) Physica A 304:283–293
9. Zipper P, Durchschlag H (2002) Physica A 314:613–622
10. Durchschlag H, Zipper P (2002) Progr Colloid Polym Sci 119:131–140
11. Durchschlag H, Zipper P (2003) Eur Biophys J 32:487–502
12. Durchschlag H, Zipper P (2004) Progr Colloid Polym Sci 127:98–112
13. Durchschlag H (1986) Specific Volumes of Biological Macromolecules and Some Other Molecules of Biological Interest. In: Hinz H-J (ed) Thermodynamic Data for Biochemistry and Biotechnology. Springer, Berlin, pp 45–128
14. Zipper P, Durchschlag H (2003) J Appl Crystallogr 36:509–514
15. Zipper P, Durchschlag H, Krebs A (2005) Modelling of Biopolymers. In: Scott DJ, Harding SE, Rowe AJ (eds) Analytical Ultracentrifugation: Techniques and Methods. Royal Society of Chemistry, Cambridge UK, pp 320–371
16. Zipper P, Durchschlag H (2007) J Appl Crystallogr 40:s153–s158
17. Durchschlag H, Zipper P, Krebs A (2007) J Appl Crystallogr 40:1123–1134
18. Sayle RA, Milner-White EJ (1995) Trends Biochem Sci 20:374–376
19. Traube J (1899) Samml chem chem-tech Vortr 4:255–332
20. Cohn EJ, Edsall JT (eds) (1943) Proteins, Amino Acids and Peptides as Ions and Dipolar Ions. Reinhold, New York
21. Connolly ML (1993) J Mol Graph 11:139–141
22. Vorobjev YN, Hermans J (1997) Biophys J 73:722–732
23. Zipper P, Durchschlag H (2008) J Biol Phys, in press; DOI: 10.1007/s10867-008-9063-6
24. Durchschlag H, Zipper P (2008) 1st International Conference on Drug Design and Discovery, Dubai UAE, Abstract 35-jc-mi
25. Israelachvili JN (1992) Intermolecular and Surface Forces, 2nd ed. Academic Press, London
26. Butt H-J, Graf K, Kappl M (2006) Physics and Chemistry of Interfaces, 2nd ed. Wiley-VCH, Weinheim
27. Morra M (ed) (2001) Water in Biomaterials Surface Science. John Wiley & Sons, Chichester

Progr Colloid Polym Sci (2008) 134: 30–38
DOI 10.1007/2882_2008_088

Published online: 15 August 2008

M. Kolasińska
R. Krastev
T. Gutberlet
P. Warszyński

Swelling and Water Uptake of PAH/PSS Polyelectrolyte Multilayers

M. Kolasińska (✉) · R. Krastev (✉)
Max-Planck Institute of Colloids and Interfaces, 14424 Potsdam, Germany
e-mail: marta.kolasinska@mpikg.mpg.de, rumen.krastev@mpikg.mpg.de

T. Gutberlet
Laboratory for Neutron Scattering, ETH Zürich & Paul Scherrer Institute, 5232 Villigen PSI, Switzerland

T. Gutberlet
Present address:
Forschungszentrum Jülich GmbH, Jülich Centre for Neutron Science, Lichtenbergstr. 1, 85747 Garching, Germany

P. Warszyński
Institute of Catalysis and Surface Chemistry, Polish Academy of Sciences, Cracow, Poland

Abstract We studied the uptake of liquid water and water vapours in polyelectrolyte multilayer films (PEM) prepared from polyallylamine hydrochloride (PAH) and polysodium 4-styrenesulfonate (PSS). The experiments were performed using neutron reflectometry. The amount of water uptake was characterized by swelling of the samples and the change in their water content – quantities which are independently accessible in a neutron reflectometry experiment. The influence of the first branched polyethylene imine (PEI) anchoring layer on the structure and the properties of obtained PEMs was also studied. The first PEI layer acts as a uniform anchoring network for the consecutive layers formation and the resulting PE film is more homogeneous. The samples prepared with first anchoring PEI layer were slightly thicker than those without PEI layer. A small difference in the film thickness between PSS and PAH terminated layers was observed (ca. 1 nm).

When the samples were exposed to D_2O vapors or liquid D_2O they swelled with ca. 30 vol. %. The swelling was complemented with an increase of the D_2O content in the sample. The striking difference is that if the water content is ca. 30 vol. % when the samples were exposed to D_2O vapors it increases to above 60 vol. % when exposed to the liquid D_2O even though the thickness of the samples did not change.

PAH/PSS samples were treated with acid/base solutions with pH of 3/11 and respective ionic strength of 0.001 M. Changes in the PEM structure were found but they were not pH specific. This means that the observed changes were related to the ionic strength of the solutions. The water amount decreases from ca. 60 vol. % in the non-treated samples to ca. 30 vol. % in the treated samples. The decrease of the amount of D_2O in the films means that PEM are sensitive to any exposure to electrolyte solutions which affects their internal structure.

Keywords Neutron reflectometry · Polyelectrolytes · Polyelectrolyte multilayers · Swelling · Water uptake

Introduction

Due to their specific features, polyelectrolytes (PE) are used to modify surfaces of various materials improving their interfacial properties ([1, 2] and references cited therein). Polyelectrolyte thin multilayer (PEM) films are easily formed by sequential adsorption of polycation and polyanion layers using the layer-by-layer (LbL) deposition procedure [1], which is an efficient method for obtaining a variety of materials of exactly defined properties. The possibility to combine different polyelectrolytes extends number of applications, where such materials can be used.

Studies on PEM are of interest because of the versatility of multilayer formation process with respect to variety of support materials, combination with other assembly techniques and possibility of incorporation of different functional species [1–4]. PEM are applied in chemical and biochemical sensing or preparation of new biomaterials. When colloidal particles or emulsion droplets are used as cores for polyelectrolyte deposition one can obtain hollow micro- and nanocapsules. Such capsules have potential as drug delivery systems [5] capable of sustained release [6, 7], microreactors [8, 9] or catalytic systems [8–10]. Detailed summaries on the properties and application of PEM prepared by LbL technique can be found in [1, 11].

Neutron reflectometry (NR) is a powerful technique for studying the structure and the properties of thin layers on solid supports. The method was extensively used to characterise different PEM [12–15]. It allows measuring the film thickness and composition profiles along the z direction normal to the interface over a length scale up to 500 nm, with a resolution down to about 0.2 nm. As neutrons only weakly interact with most materials, in situ measurements at solid/liquid interfaces can be performed. An important advantage of the NR method is the absence of sample damage, even upon prolonged exposure to the neutron beam. The use of compounds with different isotopes of a same chemical element allows NR measurements to be focused on particular parts of a thin layer without perturbing the whole chemical structure. The specular neutron reflectivity method is based on the variation of the specular reflection R at the interface between two phases with the variation of the wave vector transfer $Q = (4\pi/\lambda)\sin\theta$, where θ is the angle of incidence of the neutron beam and λ is the neutron wavelength. The variation depends on the interfacial composition perpendicular to the layer interface characterized by the neutron scattering length density (SLD) $\rho(z) = \sum_i n_i b_i$. Here n_i is the number density of the element i, and b_i is its scattering length. Different isotopes are characterized by different values of the scattering length b_i. The SLD can be easily transformed into material density or composition of the PEM. Thus, the method is suitable to estimate the changes in the PEM thickness (swelling) and the amount of absorbed water (water uptake) in a single experiment.

PEM were extensively studied in the last decade [1, 4, 15] but the knowledge about their structure and properties as a function of the preparation conditions or post preparation treatment still needs to be completed and systematized with precise studies on their stability and response to various external stimuli. The aim of the present work was to summarise data about film swelling and water uptake of PEM prepared from polyallylamine hydrochloride/polysodium 4-styrenesulfonate (PAH/PSS) using different preparation procedures or post preparation treatment. A layer of highly branched poly(ethylene imine) (PEI) is often used as a first layer to anchor subsequent layers to the support material [1] when PEM are formed by LbL technique. We aimed to understand the influence of this first anchoring PEI layer on the multilayer systems. The swelling and the amount of water penetrating into the films in dependence of PEI anchoring layer and type of the terminating layer (PAH or PSS) so-called odd-even-effect [4, 13] were estimated. Changes in the PEM structure after treatment with solutions of different pH were followed as a function of the first PEI anchoring layer and the termination layer of the PEM.

Experimental

Polyelectrolytes used were: poly(allylamine hydrochloride) (PAH) with mean molecular weight of 70 kDa, branched poly(ethylene imine) (PEI) of molecular weight 750 kDa and poly(sodium 4-styrenesulfonate) (hPSS) of 70 kDa and perdeuteurated PSS (dPSS) of molecular weight ca. 80 kDa. PAH, PEI and hPSS were purchased from Sigma-Aldrich (Germany). The perdeuterated dPSS was from Polymer Standards Service (Mainz, Germany). The molecular structures of polyions used are depicted in Fig. 1. NaCl (99.5%) was obtained from Fluka. HCl, H_2SO_4 (96%) and H_2O_2 (32%) were from P.O.Ch (Gliwice, Poland). Deuterium oxide (D_2O) with purity of 98.9% used in the neutron reflectometry experiments was from Aldrich (Germany).

Silicon blocks (Siliciumbearbeitung Andrea Holm, Tann/Ndb., Germany) of dimensions $80 \times 50 \times 15\ mm^3$ and orientation $\langle 100 \rangle$ with two sides polished were used as supports for polyelectrolyte deposition in the neutron reflectometry experiments. The blocks were cleaned with piranha solution (H_2SO_4/H_2O_2 1 : 1). (*Caution! – The piranha solution is strong oxidizing agent and should be handled carefully*) for 30 min and carefully rinsed with distilled water. Finally, the Si substrates were dipped for 30 min in hot water (ca. 70 °C).

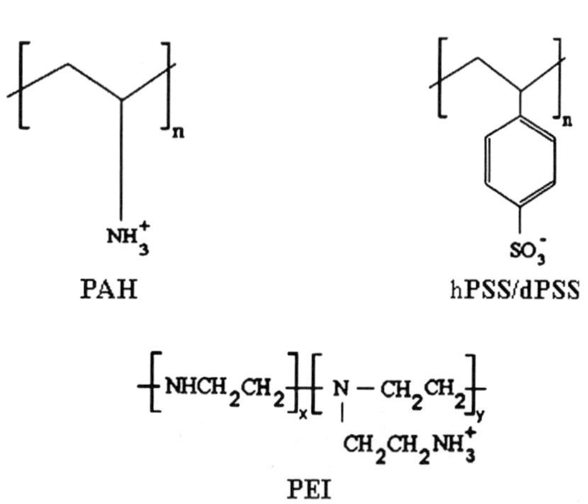

PAH

hPSS/dPSS

PEI

Fig. 1 Structural formulas of the used polyions

The polyelectrolyte multilayer films were prepared using the layer-by-layer (LbL) technique of electrostatically driven sequential adsorption of polyions from their solutions. The technique is described in details elsewhere [1, 17]. Adsorption of polyelectrolytes was performed from solutions of NaCl in H_2O of concentration 0.15 M at PE concentration of 0.5 g/l. Each deposition step lasted 20 min and rinsing in between the steps was done 3 times for 2 min in water. Always ultra pure water with Milli-Q quality was used. Ready samples were rinsed with D_2O before the NR experiments.

The samples were treated after the preparation by exposing them to electrolyte solutions (HCl or NaOH) having the same ionic strength (10^{-3} M) and pH values 3 or 11, respectively. Duration of that post-treatment was 48 h. Afterwards films were rinsed with H_2O and then with D_2O. Further handling of samples was carried out according to the procedures described below.

The experiments were performed with four types of samples – without and with anchoring PEI layer and different charge of the outermost layer – either negatively charged PSS terminated layer or positively charged PAH terminated layer. The same number of PE layers was used independent on the existence of PEI anchoring layer. Thus, effects due to different film thickness were excluded. The expected film thickness was chosen to be in the range accessible with the used neutron reflectometry instruments with high resolution. The PEM structures used in this study are summarized in Table 1. Lack of contrast between the fully protonated polyelectrolyte films and the Si support was expected for measurements against dry N_2 because of the similar values of the SLD of the two phases [12, 18]. A similar problem could have been present for experiments carried out in liquid D_2O using fully deuterated polymers. Therefore, to obtain as high contrast as possible for all experiments and to carry out all these measurements using very same samples, films were prepared with moderately deuterated polyelectrolytes using equimolar mixture of hPSS and dPSS as polyanion.

The experiments with not acid–base treated samples were carried out in the order: liquid D_2O, dry N_2, D_2O vapors, liquid D_2O. The samples were freshly prepared and without any drying the reflectivity curves in liquid D_2O were collected. Then the samples were dried and measurements were performed in dry nitrogen atmosphere (expected 0% relative humidity) followed by saturated D_2O vapors atmosphere (expected 100% relative humidity). Finally, the experiment was carried out again in liquid D_2O to check reversibility of the film structure after drying (swelling). The respective samples were allowed ca. 2 h to equilibrate before neutron reflectivity experiment after change of external conditions. The experiments with post preparation treated samples were performed only in liquid D_2O.

Table 1 Structures of the used PEM

Sample	Charge*	without PEI	with PEI
13 layers	(+)	$Si/(PAH/PSS)_6PAH$	$Si/PEI(PSS/PAH)_6$
14 layers	(−)	$Si/(PAH/PSS)_7$	$Si/PEI(PSS/PAH)_6PSS$

* The column shows the charge of the outermost PEM layer.

The NR experiments were carried out at two reflectometers. The V6 monochromatic (wavelength $\lambda = 0.47$ nm) reflectometer [19] at the Hahn-Meitner Institute, Berlin, Germany, was used in $\theta/2\theta$ geometry. The neutron reflectometer AMOR at the Paul Scherrer Institute, Villigen, Switzerland, [20] was used in time-of-flight (ToF) mode at three angles of incidence (0.4, 0.9, and 1.5°), covering the whole necessary Q range. Every single experiment took ca. 8 h (including necessary alignment before). The background signal was directly subtracted from the specular signal to obtain the corrected intensity. The reflectivity data were footprint-corrected for the varying flux on the sample as Q increased. Home made solid–liquid and solid–gas tight experimental chambers were used. Both chambers were described earlier [18, 21]. The geometries of the used chambers allows always the phase with higher SLD (either D_2O or Si) to be used as a lower phase for the incoming neutron beam. This assures observation of very well pronounced critical edge in the reflectivity curves.

The information that can be extracted in a single NR experiment includes the film thickness d, the scattering length density profile $\rho(z)$ across the film and the surface roughness σ between the different layers. The experimentally obtained reflectivity curves were analyzed by applying the standard fitting routine Parratt 32 [22]. It determines the optical reflectivity of neutrons from planar surfaces with a calculation based on Parratt's recursion scheme for stratified media [23, 24]. The film is modelled as consisting of layers of specific thickness, scattering length density and roughness, which are the fitting parameters. The model reflectivity profile is calculated and compared to the measured one. Then the model is adjusted by a change in the fitting parameters to fit best the data.

Results and Discussions

Freshly Prepared PEM

Immediately after the deposition (before acid/base treatment) the samples were studied in different conditions. The neutron reflectivity curves are summarized in Fig. 2. The swelling and water uptake in the samples and their reversibility were followed by exposing the samples subsequently to liquid D_2O, dry N_2, and D_2O saturated vapors and again to liquid D_2O. Detailed analysis of the obtained results was possible after fitting the data and converting

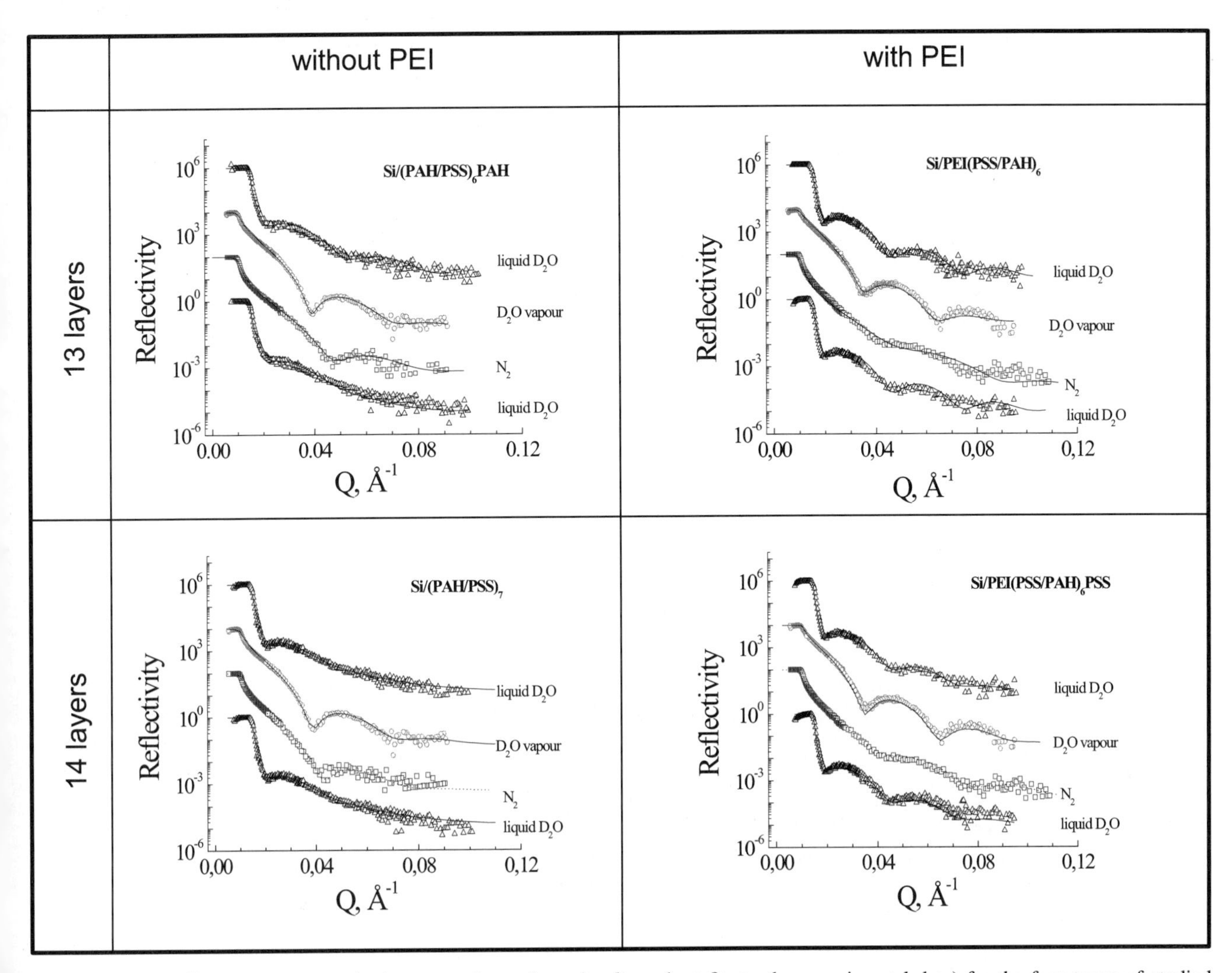

Fig. 2 Neutron reflectometry curves (*points*: experimental results, *lines*: best fits to the experimental data) for the four types of studied samples following the order: freshly prepared sample in liquid D_2O, dry N_2, D_2O saturated vapors and liquid D_2O

them in scattering length density profiles. The best fits to the experimental points are presented as lines in Fig. 2. The fitting parameters are summarized in Table 2 and the respective SLD profiles normal to the film surfaces are presented in Fig. 3.

We chose the samples in dry conditions as a base for further comparison. Samples which have PEI as an anchoring layer are slightly thicker (ca. 2 nm) than those prepared only with PAH/PSS. The SLD is decreased which indicates formation of a less dense structure. Probably the polymer chains can preserve their coiled structure after deposition onto the Si surface modified by PEI in contrast to that in the case without PEI anchoring layer. A difference in the roughness is observed showing that the samples without PEI anchoring layer are rougher than those with a PEI layer (see Fig. 3). A similar influence of PEI anchor layer on the multilayer film structure was already observed in experiments performed in lab atmosphere by ellipsometry and X-ray reflectometry [25]. We interpret this result that branched PEI used as a first layer acts as a uniform anchoring network for the consecutive layers' formation, therefore the uniform layer growth can be observed and the resulting PE film is more homogeneous. A small difference in the film thickness between PSS and PAH terminated layer is observed (ca. 1 nm). The difference is independent of the anchoring layer and coincides with data given in the literature [26]. The film thickness and SLD changed when the films were exposed to liquid D_2O or D_2O vapors. The experimental data and the best fits are also summarized in Figs. 2, 3 and Table 2. All samples swell with ca. 30–40 vol. % and the SLD of the samples increases. The PEM roughness did not change significantly.

When the films were exposed to liquid D_2O or D_2O vapors their thickness became larger than that in dry state. The difference in the film thickness in liquid water and water vapor is negligible for PAH/PSS films. The film thickness of PEM built up with PEI anchoring layer differentiate with ca. 1 nm between the liquid D_2O and D_2O

Table 2 Properties of PEM which were prepared without and with anchoring PEI layer – 13-layer positively charged and 14-layer negatively charged samples. The film thickness (d) and the scattering length density (ρ) were obtained from the best fits to the experimental data shown in Fig. 2. The swelling of the samples is defined as the film thickness in swollen state (d_{swollen}) normalized to the thickness of the dry sample (d_{dry}). The parameter α gives the volume ratio of D_2O in the samples calculated according to Eq. 1

Layers		without PEI thickness, d (nm)	ρ/nm^{-2} $\times 10^8$	swelling $d_{\text{swollen}}/d_{\text{dry}}$	α	with PEI thickness, d (nm)	ρ/nm^{-2} $\times 10^8$	swelling $d_{\text{swollen}}/d_{\text{dry}}$	α
13	D_2O_{liq}	17.4±1.0	5.2±0.2	1.34	0.66	20.7±1.2	5.1±0.2	1.42	0.67
	D_2O_{vap}	17.3±1.0	3.4±0.1	1.33	0.23	19.4±1.4	3.8±0.2	1.33	0.33
	N_2	13.0±0.8	2.9±0.1			14.6±1.7	2.5±0.1		
14	D_2O_{liq}	18.5±1.3	5.0±0.1	1.32	0.59	21.7±1.4	5.0±0.1	1.32	0.63
	D_2O_{vap}	17.9±0.7	4.0±0.1	1.28	0.38	20.2±1.1	4.2±0.1	1.23	0.42
	N_2	14.0±1.0	3.0±0.3			16.4±0.9	2.6±0.1		

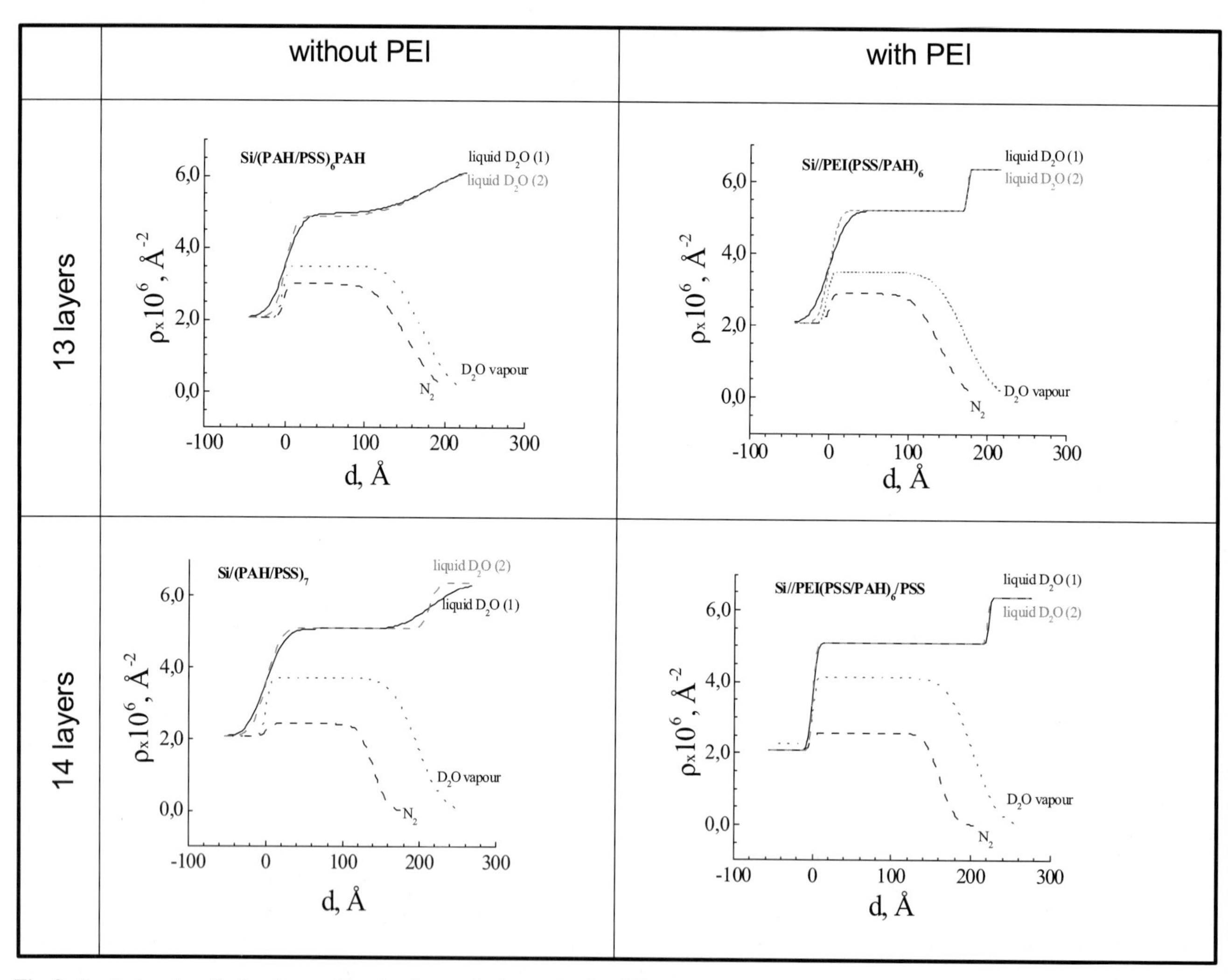

Fig. 3 Scattering length density profiles for the studied samples in different surrounding conditions (liquid D_2O, D_2O vapours and dry N_2). The profiles are obtained after fitting the experimental reflectometry data presented on Fig. 2. The samples were exposed to liquid water immediately after their preparation (liquid D_2O (1)) and after experiments in dry N_2 and D_2O vapors (liquid D_2O (2))

vapor states. This effect can be related to the wetting properties of both types of films (built up with or without PEI). PSS terminated multilayers for PEI/(PSS/PAH) structures are more hydrophobic than those in PAH/PSS films as it was shown in [25]. Therefore, immersion in water is required to fully swell more hydrophobic PEI/(PSS/PAH) multilayer film. On the other hand, dense structure of PAH/PSS measured in dry state leads to lower volume of free space within film where water molecules could be trapped. Hence, regardless if it is liquid D_2O or its vapors, water cannot fill up empty spaces in the film and it has to bind to the polyions causing larger film swelling.

One has to account for the decrease in the PEM density when drawing exact conclusions for the swelling and increase in the SLD because of the uptake of D_2O (either vapors or liquid). More specific conclusions can be drawn from the neutron reflectrometric experiments when scattering densities are accordingly analyzed. In case of samples saturated with water or water vapors, the total experimentally obtained SLD of the sample (ρ^{exp}) is a weighted sum of the SLDs of the dry polymeric film (ρ^{dry}) and that of the D_2O (ρ^{D_2O}) (in our case) according to the equation:

$$\rho^{\text{exp}} = (1-\alpha)\frac{d^{\text{dry}}}{d^{\text{exp}}}\rho^{\text{dry}} + \alpha\rho^{D_2O} , \qquad (1)$$

where α is the water amount taken up by the film. The ratio between the film thickness in dry state d^{dry} and in swollen state d^{exp} accounts for the change in the film SLD due to the decreased polymer density upon the film expansion (swelling).

The data for α (see Table 2) show that the water uptake is equal in the limits of the experimental error independently of the existence of anchoring PEI layer. Careful investigation of the data on Table 2 show different amount of water taken up depending on the outermost layer (PAH or PSS) being more in the case of negatively (PSS) terminated PEM. This is similar to the odd-even effect already reported in the literature [13]. Similar to the literature data, films prepared from PE solutions containing low salt concentration (similar to ours of 0.15 M NaCl) show higher amount of absorbed water when they are terminated with negatively charged PSS compared to those which are positively (PAH) terminated. In contrast, the observed difference in the case of PEM prepared from PE solutions with high salt concentrations is weaker [13].

In contrary to the weak change in the film thickness, a strong difference is observed in the water content of the films when exposed to D_2O vapors or liquid D_2O. The amount of absorbed water increases significantly in case of liquid D_2O compared to that in D_2O vapors. One can speculate that either in water vapors or in liquid water certain active centers in the PE molecules are hydrated which is responsible for the film swelling. The exposure to liquid water only slightly contributes to this hydration (respectively swelling) but it fills up the empty spaces in the PEM structure.

The last experiments of this series were carried out again in liquid D_2O, to check the reversibility of the film structure after drying (swelling). The results of the experiments are presented also in Fig. 3 as SLD profiles. They give evidence of full reversibility of the multilayer structure upon drying/wetting cycle. However, one can observe that multilayers exposed to liquid D_2O after drying were smoother compared to the very same samples measured in liquid D_2O freshly after preparation. The possible explanation of such behavior can be time dependent intrinsic reorganization within multilayers during the wetting/drying process.

Acid/Base Treated PEM

Previous studies show that exposure of the PEM to salt solutions [16, 27] or temperature treatment [28, 29] result in changes in the film structure. The changes were attributed to phase transitions from glassy to liquid state in the PEM. The origin of the effect was found to be a weakening of the electrostatic bonds between the PE chains in the process of ionic treatment. In order to extend the information on behaviour of multilayers in various conditions, the same series of samples were exposed to acid/base solutions with pH of 3/11 and respective ionic strength of 0.001 M. Straight after the treatment the samples were thoroughly washed and exposed to liquid D_2O and NR experiments were performed. The samples consisted of 13- or 14-layer built up without or with PEI anchoring layer. Results of these experiments are presented as reflectometry curves in Fig. 4 and the SLD profiles are summarized in Fig. 5. The parameters for the best fits to the data are collected in Table 3.

No effect related specifically to the acid or base treatment was observed. This is in contrary to our assumptions based on knowledge of the pH dependence of PAH charge density – fully charged in acidic conditions and partially charged in basic conditions. It means that the observed changes in the PEM are not pH related but ionic strength dependent. All samples showed similar response except that prepared with PEI anchoring layer and PSS terminated. We treat these data as a statistical disturbance and because of lack of additional experiments which could confirm or discard the special behavior of that sample we will not speculate on the changes in its structure.

A small change in the film thickness (less than 1 nm) was always registered after the sample treatment. All samples become smoother after the acid/base treatment. A strong decrease is registered in the SLD of the films. Following Eq. 1 this means that the film density increases and/or D_2O was expelled from the film. The small change in the film thickness does not prove a high change in the film density and respectively the PEM SLD. Obviously, the only explanation of the effect is related to the decreased amount of D_2O in the film. The water amount in the film (α) was calculated and the data is presented in Table 3. It decreases from ca. 60 vol. % in the non-treated samples to ca. 30 vol. % in the treated samples. A minor uncertainty

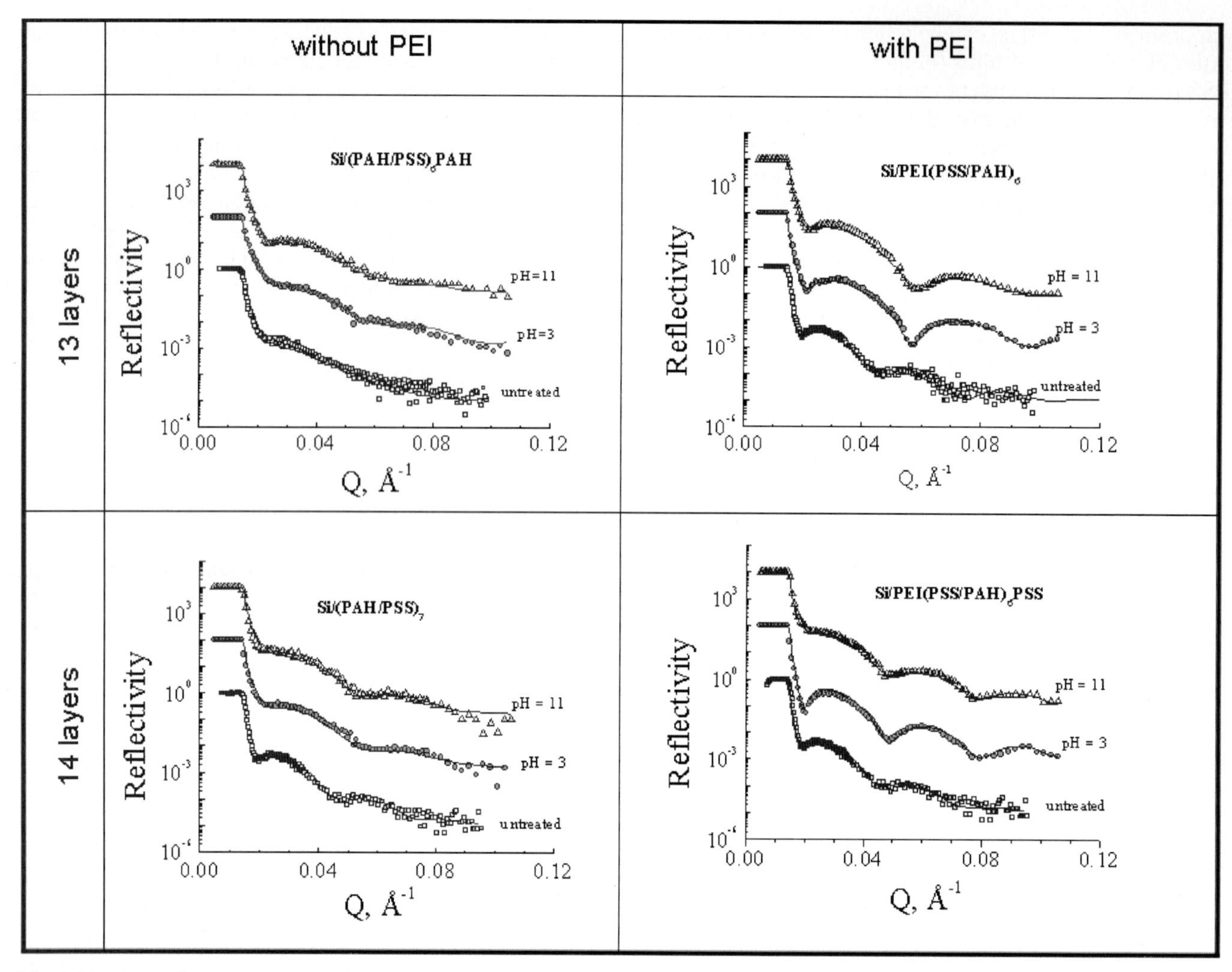

Fig. 4 Neutron reflectometry curves (*points*: experimental results, *lines*: best fits to the experimental points) for the four studied samples exposed to liquid D_2O after treatment with acid/base solutions. The reflectometry curves for the untreated samples also in liquid water are shown for comparison

Table 3 Studied samples (see Table 1) against liquid D_2O after acidic/basic treatment. Parameters which give the best fits to the experimental data in Fig. 4 and presented as SLD profiles in Fig. 5

Layers		without PEI thickness (nm)	$\rho/\text{nm}^{-2} \times 10^8$	α	with PEI thickness (nm)	$\rho/\text{nm}^{-2} \times 10^8$	α
13	untreated	17.0 ± 1.2	5.1 ± 0.1	0.66	21.5 ± 1.5	5.1 ± 0.2	0.67
	pH = 3	16.3 ± 1.7	4.1 ± 0.3	0.35	18.5 ± 1.6	4.1 ± 0.3	0.41
	pH = 11	16.5 ± 2.3	3.9 ± 0.2	0.29	18.5 ± 1.7	4.0 ± 0.5	0.39
14	untreated	18.2 ± 1.1	5.1 ± 0.2	0.59	22.2 ± 1.4	4.9 ± 0.1	0.63
	pH = 3_p	17.8 ± 2.1	3.8 ± 0.3	0.24	20.6 ± 1.3	5.0 ± 0.3	0.64
	pH = 11	18.5 ± 2.6	3.8 ± 0.5	0.24	20.8 ± 0.9	5.3 ± 0.6	0.72

of the result could be expected because for the film thickness and SLD of the dry sample the values obtained for the freshly prepared samples were used. Indeed the samples which were acid/base treated were different than those used to study the structure of freshly prepared PEM. Usually, the preparation of the PEM is very reproducible and such exchange can be used as a good estimation.

The decrease of the amount of D_2O in the films means that PEM are sensitive to any exposure to electrolyte solutions which affects their internal structure. Based on

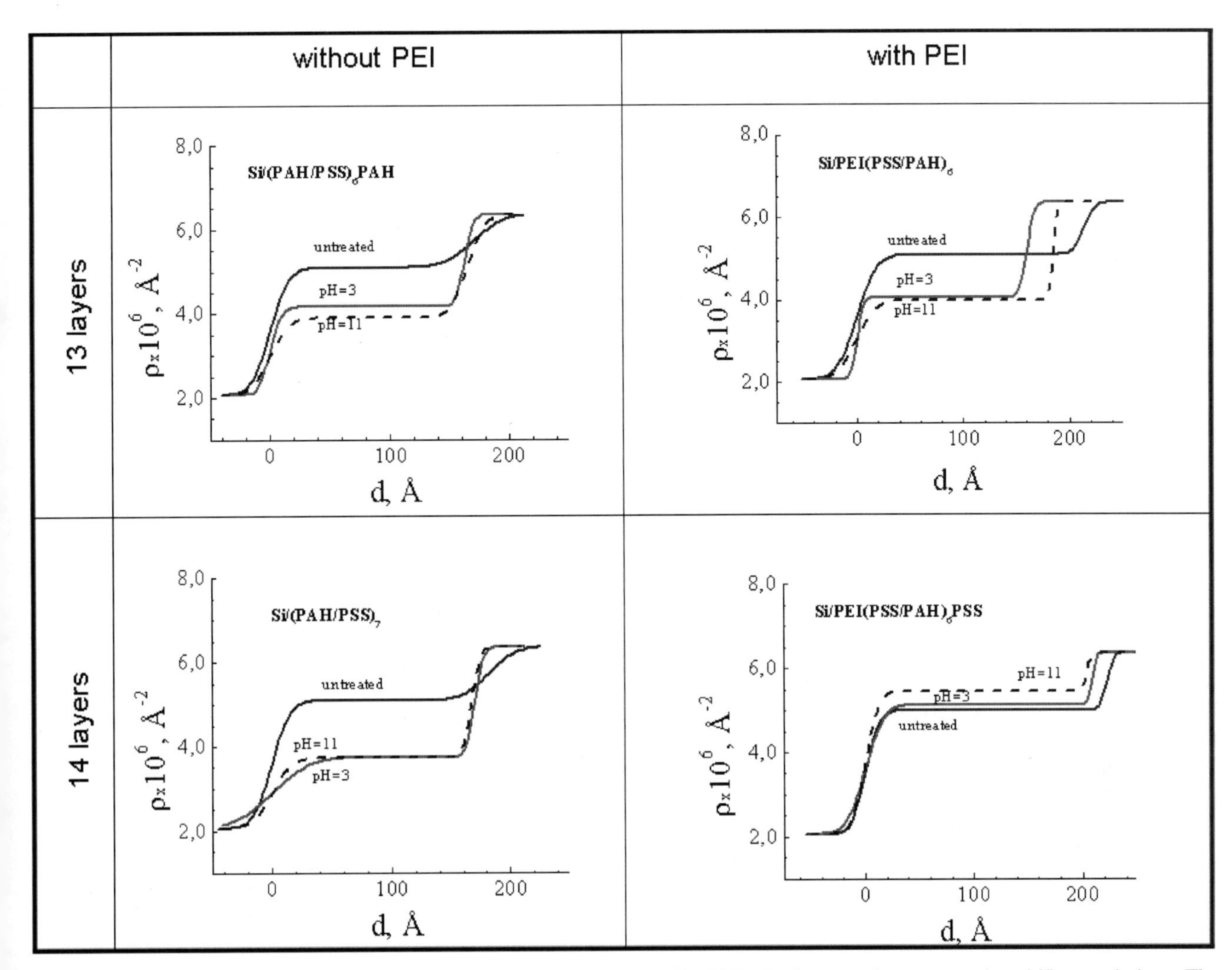

Fig. 5 Scattering length density profiles for the studied samples against liquid D_2O after sample treatment in acid/base solutions. The profiles are obtained after fitting the experimental reflectometry data presented on Fig. 4

our results we can speculate that exposure to electrolyte solution causes exclusion of D_2O from polyelectrolyte films probably due to an increase of polymer chains density within films thus making the films interior less hydrophilic.

Conclusions

Due to the broad application and versatility of polyelectrolyte multilayers in novel multifunctional materials, PEMs have become interesting structures from scientific and practical point of view. However, practical application of PEMs requires full understanding of the role of main physicochemical factors affecting PEMs structure and interfacial behavior. It is worth to stress, that due to a broad range of polyelectrolytes which can be used for multilayer formation in various conditions, one has to restrict studies to characterize certain PEM systems. When the PE multilayer films are formed by LbL technique, the PEI layer is used as a first layer to anchor subsequent layers to the support material. However, there is lack of systematic investigations leading to full understanding of the role of that PEI layer. Therefore, in our studies, selected PAH/PSS multilayers were compared to analogues ones, but formed on the PEI anchor. Additionally the effect of the charge of the outermost layer on film structure was determined. The samples prepared with first anchoring PEI layer were slightly thicker than those without PEI layer (note that the number of PE layers was always the same). Also the PEI supported samples were less rough. The first PEI layer acts as a uniform anchoring network for the consecutive layers formation and the resulting PE film is more homogeneous. A small difference in the film thickness between PSS and PAH terminated layers was observed (ca. 1 nm) and it was independent of the anchoring layer and coincides with data given in the literature [26]. When the samples were exposed to D_2O vapors or liquid D_2O they swelled with ca. 30 vol. %. The swelling was complemented with an increase of the D_2O content in the sample. The striking difference is that if the water content is ca. 30 vol. % when the samples were exposed to D_2O vapors it increases to above 60 vol. % when exposed to the liquid D_2O even though the thickness of the samples did not change. We speculate that the PEM have certain active centers which are hydrated by water vapors or liquid water. These centers are responsible for the film swelling. At high water concentrations (e.g. PEM in contact with liquid D_2O) together with the hydration of these active centers the water can fill up the empty spaces in the PEM structure which will reflect in higher water content at similar or even equal PEM thickness. The model is similar to models which accept existence of different bound water types and introduce the idea of weakly and strongly bound water.

Previous studies show that exposure of the PEM to salt solutions or temperature treatment result in changes in the film structure. In order to extend information on the behaviour of multilayers in various conditions we treated PAH/PSS samples prepared with and without anchoring PEI layer in acid/base solutions with pH of 3/11 and respective ionic strength of 0.001 M. Changes in the PEM structure were obtained but they were not pH specific. This means that the observed changes were related to the ionic strength of the solutions. A small change in the film thickness (less than 1nm) was always registered after the sample treatment. All samples become smoother. A strong decrease is registered in the SLD of the films after the acid/base treatment. This response can be only explained by a decrease of the amount of D_2O in the film. The water amount decreases from ca. 60 vol. % in the non-treated samples to ca. 30 vol. % in the treated samples. The decrease of the amount of D_2O in the films means that PEM are sensitive to any exposure to electrolyte solutions which affects their internal structure. Based on our results we can speculate that exposure to electrolyte solution causes exclusion of D_2O from polyelectrolyte films probably due to increase of polymer chains density within films thus making the films interior less hydrophilic.

Acknowledgement This work was partially supported by EU projects NMP4-CT-2003-001428 "NANOCAPS" and Grant MNII 3 T09A 085 27. Neutron reflectometric studies were performed at the Swiss spallation neutron source SINQ, Paul Scherrer Institute, Villigen, Switzerland and at The Berlin Neutron Scattering Center, Hahn Meitner Institute, Berlin, Germany. The financial support by the European Commission under the 6th Framework Programme through the Key Action: Strengthening the European Research Area, Research Infrastructures. Contract no.: RII3-CT-2003-505925 (NMI3) is highly acknowledged. M.K. acknowledges the partial financial support in preparation of the work as a fellowship in the frame of the Max Planck Society initiative at FRM II, Garching, Germany.

References

1. Decher G, Schlenoff JB (eds) (2003) Multilayer Thin Films. Wiley, VCH
2. Bertrand P, Jonas A, Laschewsky A, Legras R (2000) Macromol Rapid Commun 21:319
3. Hammond PT (2000) Curr Opin Colloid Interf Sci 4:430
4. Schönhoff M (2003) Curr Opin Colloid Interf Sci 8:86
5. Voigt A, Buske N, Sukhorukov GB, Antipov A, Leporatti S, Lichtenfeld H, Baumler H, Donath E, Möhwald H (2001) J Magn Magn Mater 225:59
6. Antipov A, Sukhorukov GB, Donath E, Möhwald H (2001) J Phys Chem B 105:2281
7. Varde NK, Pack DW (2004) Exp Opin Biol Ther 4:35
8. Antipov A, Sukhorukov GB, Fedutik YB, Hartmann J, Giersig M, Möhwald H (2002) Langmuir 18:6687
9. Lvov Y, Antipov A, Mamedov A, Möhwald H, Sukhorukov GB (2001) Nano Lett 1:125
10. Antipov A, Sukhorukov GB (2004) Adv Colloid Interf Sci 111:49
11. Von Klitzing R, Steitz R (2002) Internal Structure of Polyelectrolyte Multilayers. In: Triphaty SK, Nalva HS (eds) Handbook of Polyelectrolytes. American Scientic Publishers, Japan, pp 313–334
12. Kügler R, Schmitt J, Knoll W (2002) Macromol Chem Phys 203:413
13. Carrière D, Krastev D, Schönhoff M (2004) Langmuir 20:11465
14. Von Klitzing R, Leiner V, Steitz R (2001) Neutron reflectivity measurements on polyelectrolyte multilayers at the solid/liquid interface. ILL Millenium Symposium & European User Meeting, Proceedings, pp 73–75
15. Von Klitzing R (2006) PCCP 8:5012
16. Wong JE, Rehfeldt F, Haenni P, Tanaka M, Von Klitzing R (2004) Macromolecules 37:7285
17. Kolasińska M, Warszyński P (2005) Bioelectrochemistry 66:65
18. Delajon C, Gutberlet T, Steitz R, Möhwald H, Krastev R (2005) Langmuir 21:8509
19. http://www.hmi.de/bensc/instrumentation/pdf_files/BENSC_V6.pdf
20. http://kur.web.psi.ch/amor/
21. Krasteva N, Krustev R, Yasuda A, Vossmeyer T (2003) Langmuir 19:7754
22. Braun C (1999) Fitting Routine Paratt 32, version 1.5. HMI, Berlin
23. Parratt LG (1954) Phys Rev 95:359
24. Tolan M (1999) X-Ray Scattering from Soft-Matter thin Films. Springer, Berlin, Heidelberg
25. Kolasińska M, Krastev M, Warszyński P (2007) J Colloid Interf Sci 305:46
26. Steitz R, Leiner V, Siebrecht R, Von Klitzing R (2000) Colloids Surf A 163:63
27. Kovacevic D, Van der Burgh S, De Keizer A, Stuart MAC (2002) Langmuir 18:5607
28. Köhler K, Shchukin D, Möhwald H, Sukhorukov GB (2005) J Phys Chem B 109:18250
29. Nazaran P, Bosio V, Jaeger W, Anghel D, Von Klitzing R (2007) J Phys Chem B 111:8572

Progr Colloid Polym Sci (2008) 134: 39–47
DOI 10.1007/2882_2008_085

Published online: 14 August 2008

Lijuan Zhang
Michael Kappl
Günter K. Auernhammer
Beate Ullrich
Hans-Jürgen Butt
Doris Vollmer

Surface-Induced Ordering of Liquid Crystal on Modified Surfaces

Lijuan Zhang · Michael Kappl · Günter K. Auernhammer · Beate Ullrich · Hans-Jürgen Butt · Doris Vollmer (✉)
Max Planck Institute for Polymer Research, Ackermannweg 10, 55128 Mainz, Germany
e-mail: vollmerd@mpip-mainz.mpg.de

Abstract Surface-induced ordering of 4-*n*-octyl-4′-cyanobiphenyl (8CB) near the isotropic–nematic phase transition was investigated using temperature-controlled atomic force microscopy (AFM). The glass surfaces in contact with liquid crystal were modified by an adsorbed silane surfactant, a deposited 8CB monolayer, or a deposited 8CB trilayer. All three surface modifications induced homeotropic anchoring of the liquid crystal molecules. Even in the isotropic phase, the surface-induced alignment can cause layering of the molecules close to the modified surfaces. The amplitude and range of this surface-induced alignment not only depends on temperature, but also on the modifying agent and method. Both amplitude and range of presmectic order are smallest for surfaces modified by adsorption of a monolayer of 8CB and largest for surfaces modified by the silane surfactant *N*,*N*-dimethyl-*N*-octadecyl-3-amino-propyltrimethoxysilyl chloride (DMOAP). The different extent of presmectic alignment is also reflected in the elastic modulus of the first layer.

Keywords Atomic force microscopy · Liquid crystal · Surface-induced ordering · Presmectic ordering · Landau–de Gennes

Introduction

The molecular alignment of liquid crystals on solid surfaces is not only of fundamental interest in physics [1] but is also relevant for practical applications, for example in optoelectronic devices. In liquid crystal displays the molecules are confined between two surfaces. To minimize the number of defects, surfaces are favored that induce a high degree of orientation of the molecules. Different surface treatments are used to induce and control the orientation of the molecules. A homeotropic (perpendicular) alignment is favored on hydrophobic surfaces that are rough on a molecular level. This is observed in adsorbed monolayers of a surface-active compound such as lipids or surfactant molecules on glass, both for energetic and sterical reasons. Surface modifications can alter the positional order and molecular orientation of liquid crystalline molecules deep into the bulk [2–5]. For this reason the treatment of solid surfaces is the key factor to optimize the alignment of liquid crystal molecules.

Recently it has been shown that surface-induced molecular orientation can be determined by atomic force microscopy (AFM), both in the isotropic and nematic phase of thermotropic liquid crystals [6–10]. At separations of several nanometers between the homeotropically modified glass surface and the AFM tip or homeotropically modified glass microsphere, respectively, a temperature-dependent short-range attractive (prenematic) or on average repulsive, but oscillatory, (presmectic) force was observed [6, 7, 9].

In this paper, we explore the effect of different surface treatments upon surface–induced ordering of the liquid crystal 4-*n*-octyl-4′-cyanobiphenyl (8CB). Cyanobiphenyls have been intensively studied [11–13]

because of their high chemical stability, the fact that mesophases occur at low temperature, and low production costs. As common to thermotropic liquid crystals, 8CB shows a temperature-dependent phase behavior, including an isotropic, a nematic, and a smectic phase. In the nematic and smectic phase the overlapping cyanobiphenyl groups form dimers with a rigid, polar center and hydrocarbon chains protruding to both sides [13].

We modified glass surfaces by either chemisorption of the silane surfactant DMOAP, or physisorption of a monolayer or trilayer of 8CB using Langmuir–Blodgett transfer. In all cases AFM force spectroscopy demonstrated that a modified surface induces alignment of 8CB in the isotropic phase. The amplitude and the range of the alignment, however, depend sensitively on surface treatment. Whereas on DMOAP-modified surfaces the first aligned layer remains oriented up to several degrees above isotropic–nematic phase transition T_{IN}, it rapidly vanishes on the physisorbed layers.

Experiment

Materials

8CB was purchased from Synthon Chemicals GmbH & Co. KG, Germany, and used as received. With differential scanning calorimetry we determined the phase transition temperatures to be: isotropic (40.5 °C), nematic (33.4 °C), smectic (22.5 °C) crystalline. The silica surfaces and some of the glass surfaces were modified by *N*,*N*-dimethyl-*N*-octadecyl-3-amino-propyltrimethoxysilyl chloride (DMOAP), obtained from Sigma-Aldrich, Germany.

Surface Modification

Glass substrates were cleaned in a solution containing 5 : 1 : 1 (volume) $H_2O/H_2O_2/NH_4OH$ at 80–85 °C for 15 min [14]. Then they were rinsed with milli-Q water (resistivity of 18.2 MΩ cm^{-1}), sonicated for 5 min in milli-Q water, and rinsed again with milli-Q water. Finally, the surfaces were cleaned in Argon plasma (Plasma Cleaner, Harrick, PDC 002) at 25 °C with a RF power of 200 W for 10 min to remove remaining organic substances.

In order to study the influence of surface treatment on presmectic layering of liquid crystal, we used two methods to modify the glass surfaces, (i) Langmuir–Blodgett (LB) transfer of 8CB, and (ii) absorption of a monolayer of DMOAP onto the glass surface. For Langmuir–Blodgett transfer we used a Teflon® Langmuir trough system (KSV, Helsinki, Finland) equipped with two moving barriers and a micro-roughened Wilhelmy plate. The trough was filled with water and maintained at 20 ± 0.5 °C using a circulating water bath system. Cleaned glass substrates were immersed into the water subphase prior to monolayer formation. About 80 µl of a 2.42 mol/l solution of 8CB dissolved in HPLC grade chloroform (Sigma-Aldrich, Germany) was spreaded onto the air–water surface. We allowed a time span of 20 min for full evaporation of chloroform before compressing the monolayer at a constant rate of 7.5 cm^2 min^{-1}. Compression was continued until the desired surface pressure was reached. The 8CB layer was allowed to equilibrate for 10 min. Then it was transferred onto the glass substrates at a constant surface pressure by a vertical upstroke through the film at a constant rate of 0.2 mm/min (hydrophilic transfer).

For comparison, some glass surfaces were silanated by chemisorption of DMOAP from a water/methanol solution. The glass plates were left in the freshly prepared water solution containing 1% DMOAP-methanol (50 : 50) for 5–10 min [8]. In order to remove excess DMOAP, the glass plates were thoroughly rinsed with milli-Q water, and dried at 110 °C for about 1 h to evaporate the solvent and anneal the monolayer. Silica spheres were hydrophobized using the same procedure.

Force Measurements

A Multimode Nanoscope IIIa AFM (Digital Instruments, Veeco Metrology Group, NY, USA) was used to record force-versus-distance curves (short: force curves). Two types of cantilevers were used: (1) sharp NP-silicon nitride tip (NP, Veeco Instrument, CA, USA), having a spring constant of 0.58 N/m and a typical tip radius of 20 nm, (2) V-shaped silicon nitride tipless cantilevers (NP-O, Veeco Instrument, CA, USA) with a nominal spring constant of 0.58 N/m. Silica microspheres with a radius of 3.3 µm (Bangs Laboratories, Inc., USA) were glued to the end of tipless cantilevers using a small amount of a resin (UHU plus) by means of a 3D micropositioning micromanipulator (Narishige, MMo 203, Japan) under an optical microscope. Firstly particles were deposited on a glass slide on top of a flat stage, then a small amount of resin was placed near the particles. A tiny amount of glue was taken up by touching the resin with the end of a tipless AFM cantilever. The selected particles could be picked up with the resin-coated end of the cantilever by gently touching it from above.

Silica microspheres used were imaged with a low-voltage scanning electron microscope (SEM) (LEO 1530 Gemini, Oberkochen, Germany) after experiments to determine their shape and size. The radii of the silica microspheres were in range of $R = 3.2$–3.4 µm. The tips turned out not to be completely spherical. Therefore, their radii could not be determined accurately. Values around 60 ± 15 nm for the tips used to measure DMOAP and 50 ± 15 nm for the tips used to determine the elastic modulus of the trilayer were estimated.

In force measurements, the glass surface was periodically moved up and down at constant velocity while the cantilever deflection was measured. The result is a graph of the cantilever deflection versus the height position of the

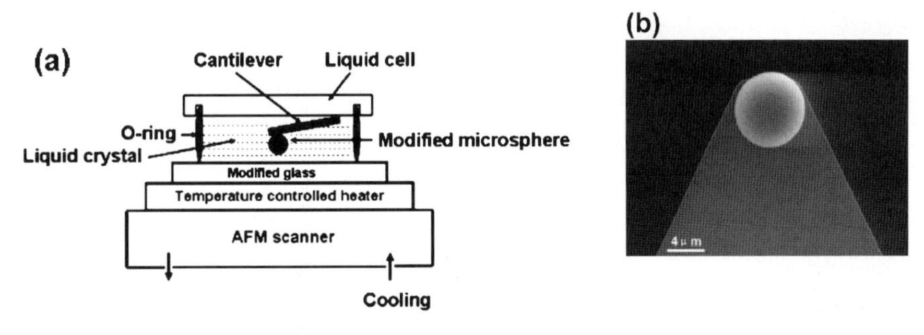

Fig. 1 **a** Sketch of the temperature-controlled AFM setup. The entire AFM was placed in a temperature controlled box to keep the top part of the sample in the isotropic phase. This permits to determine the deflection of the cantilever from the reflection of a laser beam. The temperature of liquid crystal close to the substrate was controlled using a second heater, placed on top of the piezo scanner. To protect the piezo from overheating the AFM scanner was cooled. **b** Scanning electron microscope image of a silica sphere of radius 3.3 µm, glued to the cantilever with a resin. The microsphere is attached to the cantilever using a 3D micropositioning system under an optical microscope

scanner. From this, a force curve was calculated by multiplying the cantilever deflection with the spring constant to obtain the force and subtracting the cantilever deflection from the height position to obtain the distance. Zero distance was derived from the linear contact part of force curves as described previously [15].

The sample was placed on a heating plate attached on top of the piezo scanner, providing temperature stability better than ± 0.05 K (Fig. 1). To avoid capillary forces at the air–liquid interface the cantilevers were totally immersed in the liquid crystal. In order not to disrupt the optical beam deflection system of the AFM, the bulk of the liquid crystal had to be isotropic. This was achieved by placing the entire AFM in a heating chamber whose temperature was kept well above T_{IN} and about 1 K above the temperature of the piezo scanner. The temperature within the box was homogenized using a ventilation system. To avoid vibrations, this heating system was turned off just before the measurements.

In each run, the AFM was equilibrated for 2 h after injecting 8CB. At each temperature, at least three force curves were recorded in three regions of the sample.

Theory

In the case of anisotropic molecules on a homeotropically aligning surface, an orientation of the molecules perpendicular to the surface is sterically favored. 8CB shows a surface-induced orientation perpendicular to the surface without (prenematic alignment) or with layers (presmectic alignment) as sketched in Fig. 2. In the nematic and smectic phase 8CB self-assembles into dimers [13].

The type, range, and strength of this surface-induced order depend on the order parameter and the coupling of 8CB to the surface. The order parameter describes the average orientation of the molecules [1]. It increases when approaching the isotropic–nematic phase transition. Depending on surface treatment and temperature the substrate may induce nematic or smectic alignment. In case of strong anchoring, typically the first layers show smectic ordering, followed by a nematic one (see Fig. 2). This surface-induced ordering can be described in mean-field approximation via a Landau–de Gennes expansion of the scalar nematic, S, or complex smectic, $\Psi = \psi e^{i2\pi d(z)/a_0}$, order parameter [2, 3, 16], where $d(z)$ characterizes the displacement of the layers, a_0 is the equilibrium smectic layer thickness and ψ is the amplitude of the smectic density modulation [1]. If two modified solid surfaces come sufficiently close at temperatures above the bulk nematic–isotropic transition, capillary condensation of the nematic phase causes an effective attraction of the two surfaces [17].

To describe the main contribution to the free energy in the vicinity of a phase transition we consider a Landau–de Gennes expansion to second order [8, 16, 18]. This yields

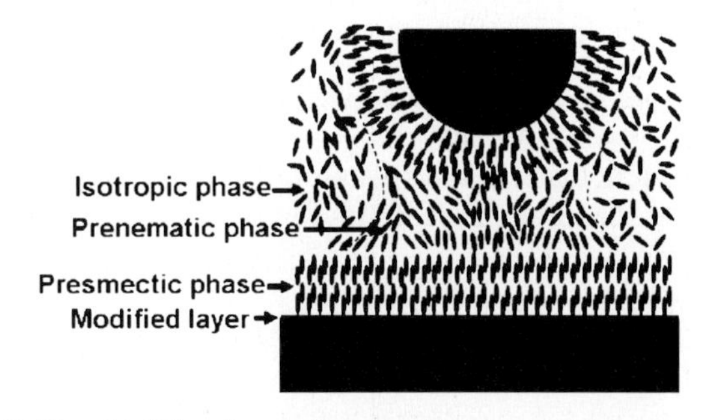

Fig. 2 Sketch of liquid crystal confined in the gap between a modified microsphere and a glass substrate (not to scale). Dependent on the anchoring strength, liquid crystal and temperature the substrate induces prenematic or/and presmectic alignment. If both surfaces approach each other, isotropic liquid crystal condenses into a nematic phase in the gap between both surfaces, causing attraction

for the nematic free energy:

$$F_{\mathrm{nem}} = \int_{-d/2}^{d/2} \left(\frac{1}{2}\alpha(T)S^2 + L_{\mathrm{n}} \left(\frac{\mathrm{d}S}{\mathrm{d}z}\right)^2 + f_{\mathrm{surf}} \right) \mathrm{d}z\,. \tag{1}$$

The temperature difference ΔT between the actual and the phase transition temperature is taken into account via the coefficient of the quadratic term of the order parameter in the Landau expansion $\alpha(T)$, which is proportional to ΔT. L_{n} denotes the elastic coefficient of a nematic liquid crystal. Both coefficients are related via the nematic correlation length, ξ_{n}, which is defined as $\xi_{\mathrm{n}} \equiv \sqrt{L_{\mathrm{n}}/\alpha} \equiv \xi_0/\Delta T$, where ξ_n takes the value $\xi_0$at the surface. Approaching the phase transition causes an increase of the nematic correlation length. The range of the alignment of the molecules furthermore depends on the coupling of the molecules to the surface. The corresponding surface energy density f_{surf} results from the breaking of the translational symmetry of the molecules close to the surface and direct interactions between the substrate and 8CB. Minimization of F_{nem} with respect to S and taking into account the surface coupling for nematic liquid crystal leads to an energy per unit surface area given by:

$$F_{\mathrm{nem}}(d) = w_1^2 \xi_{\mathrm{n}}(T) \frac{1}{w_2 \xi_{\mathrm{n}}(T) + L_{\mathrm{n}} \tanh\left(\frac{d-d_0}{2\xi_{\mathrm{n}}}\right)}\,, \tag{2}$$

where w_1 and w_2 denote the surface coupling constants and $d-d_0$ the separation between the surfaces. Due to the surface modification by adsorption or chemical binding of alkanes, lipids, or other molecules the separation d of the two surfaces is reduced by the effective thickness d_0 of the layer formed by these molecules on the surface. To good approximation this zero-stress separation d_0 is given by the distance where the sphere or AFM tip starts to compress the first adsorbed or chemically bound layer. For our calculations we assume that the AFM tip has a spherical shape. The Derjaguin approximation [19, 20] permits to calculate the force between a sphere of radius R and a flat surface knowing the energy between two plane parallel surfaces:

$$F_{\mathrm{nem}}^{\mathrm{Derj}}(d) = 2\pi R\,[\mathrm{F}(d) - \mathrm{F}(\infty)]\,. \tag{3}$$

For the prenematic liquid crystal confined between a sphere and a flat substrate this leads to [8]

$$F_{\mathrm{nem}}^{\mathrm{Derj}}(d) = 2\pi R w_1^2 \xi_{\mathrm{n}}(T) \left(\frac{1}{L_{\mathrm{n}} + w_2 \xi_{\mathrm{n}}(T)} - \frac{1}{w_2 \xi_{\mathrm{n}}(T) + L_{\mathrm{n}} \tanh\left(\frac{d-d_0}{2\xi_{\mathrm{n}}(T)}\right)} \right)\,, \tag{4}$$

As the force is negative the two surfaces attract each other. The strength as well as the range of the attraction increase with increasing correlation length.

If the molecules have smectic ordering, the free energy depends on how well the smectic layers fit between the walls. If the distance between the walls is not an exact multiple of the smectic period a_0, i.e. $d-d_0 \neq na_0$, strain is applied to the presmectic film, giving rise to a repulsive or attractive force. The mean field energy of presmectic liquid crystal confined between a sphere and a flat substrate can be calculated from a Landau–de Gennes expansion to second order of the order parameter [3, 5]. In the Derjaguin approximation the force is given as [4, 5]

$$F_{\mathrm{smec}}^{\mathrm{Derj}}(d) = 2\pi R \alpha_{\mathrm{s}}(T) \xi_{\mathrm{s}} \psi_{\mathrm{s}}^2 \left(\tanh\left(\frac{d-d_0}{2\xi_{\mathrm{s}}}\right) + \frac{1-\cos\left(\frac{2\pi(d-d_0)}{a_0}\right)}{\sinh\left(\frac{d-d_0}{\xi_{\mathrm{s}}}\right)} - 1 \right)\,, \tag{5}$$

where α_s is the coefficient of the quadratic term of the order parameter Ψ in the Landau expansion. The first term of the free energy is attractive and depends on the smectic correlation length $\xi_{\mathrm{s}} \equiv \sqrt{L_{\mathrm{s}}/\alpha_{\mathrm{s}}}$. The second term is a damped oscillatory function of period a_0, arising from the elastic deformation of the constrained layer. For fixed separation between the surfaces the damping of the oscillations decreases with increasing correlation length. Note that the average of the energy over one oscillation is repulsive, i.e. frustration dominates.

Muševič et al. also observed repulsive electrostatic interactions [21], which hardly depend on temperature and can be quantified from measurements at temperatures several degrees above T_{IN}. We did not observe electrostatic repulsion of the approaching surfaces, probably due to the different plasma treatment. We are working at 200 W compared to 300 W used by Muševič.

If two surfaces approach each other they always feel an attractive van der Waals force:

$$\mathrm{F}_{\mathrm{vdW}} = \frac{R}{6} \frac{A_{\mathrm{H}}}{(d+2d_0)^2}\,.$$

The Van der Waals force between the glass substrate and the silica sphere is given by the superposition of the forces between (i) the glass substrate and the silica sphere acting across an adsorbed layer of 8CB, (ii) the forces between the adsorbed layers and the prenematic 8CB, and (iii) the glass and prenematic 8CB acting across the adsorbed layer [8]. These different interactions are accounted for via the Hamaker constant A_{H}. A precise determination of the Hamaker constant is difficult, because it depends on the optical properties of the medium between the two surfaces, i.e. on the temperature and distance dependent anisotropic dielectric constants. To estimate A_{H} we ignored the anisotropy of the dielectric constant [8]. Dependent on the dielectric constant, the refractive index and the substrate, the Hamaker constant varies between $0.5 \times 10^{-21}\,\mathrm{J} \leq A_{\mathrm{H}} \leq 10 \times 10^{-21}\,\mathrm{J}$.

The compression of the first layer of liquid crystal molecules was analyzed using the solid foundation model [22]. It models the first layer just as a thin elastic layer with an elastic modulus K on top of an incompressible surface. Force and indentation are related by $F = \pi R K \delta^{2/h}$, where δ is the depth of indentation, R the radius of the tip, and h the layer thickness.

Results and Discussion

Surface Modification by Langmuir–Blodgett Transfer of 8CB and Chemisorption of DMOAP

Figure 3 shows a typical Langmuir isotherm of 8CB taken at $20 \pm 0.5\,°C$. The surface pressure gives the reduction of surface tension of the water–air interface due to adsorbed 8CB molecules on the surface. It was recorded while compressing the 8CB film. The isotherm exhibits 5 distinct regions [23, 24] denoted by a–e. At large average area per molecule, the molecules hardly interact with each other (region a). This gas–liquid coexistence undergoes a phase transition to a liquid-like phase when the average area per molecule drops below $0.51\,nm^2$ (region b). Note that this area per molecule corresponds to twice that in bulk 8CB [13]. Further compression causes that the pressure first remains constant (region c) and then rises sharply (region d). The system has passed another first order phase transition from a monolayer to a trilayer. This trilayer is an interdigitated bilayer on top of a monolayer adjacent to the interface [24]. At even higher compression multilayers are formed with stacked interdigitated bilayer domains [24].

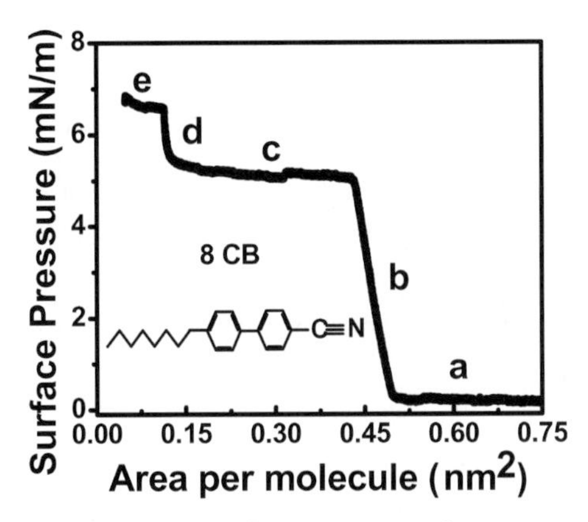

Fig. 3 Dependence of the surface pressure on the average molecular area of an 8CB film on the air–water interface. (a–e) indicate different arrangements of the molecules at the interface. A monolayer is formed in region b, whereas a trilayer is formed in region d. The isotherm is taken at $20 \pm 0.5\,°C$. In the *inset* the structure of 8CB is sketched

To test the influence of the surface treatment on the range and amplitude of presmetic layering, we transferred a monolayer and a trilayer of 8CB on fresh glass substrates at a surface pressure of 4 mN/m (monolayer) and 6.2 mN/m (trilayer), respectively.

An AFM image of a glass surface modified by an 8CB monolayer shows that the surface is homogeneously covered by 8CB, although small islands in the order of a few tens of nanometers are visible (Fig. 4a). The dark areas probably correspond to regions that are not or only partially covered by 8CB. AFM images taken after deposition of a trilayer of 8CB showed similar structures (not shown). A glass surface modified by chemisorption of a monolayer of DMOAP looks smoother, i.e. the regions of equal height are larger (Fig. 4b). This difference is reflected in a roughness of only 0.2 nm, compared to a roughness of 0.7 and 0.5 nm after deposition of an 8CB monolayer and trilayer, respectively.

Monitoring Prenematic and Presmectic Ordering by Force–Distance Spectroscopy

To quantify the influence of surface treatment, we recorded force curves on all three types of modified surfaces as

Fig. 4 AFM images of glass surfaces after modification with a monolayer of 8CB (**a**) and DMOAP (**b**). The images are recorded in tapping mode in air. Glass surfaces modified by monolayer of 8CB are three times as rough compared to surfaces modified by a monolayer of DMOAP

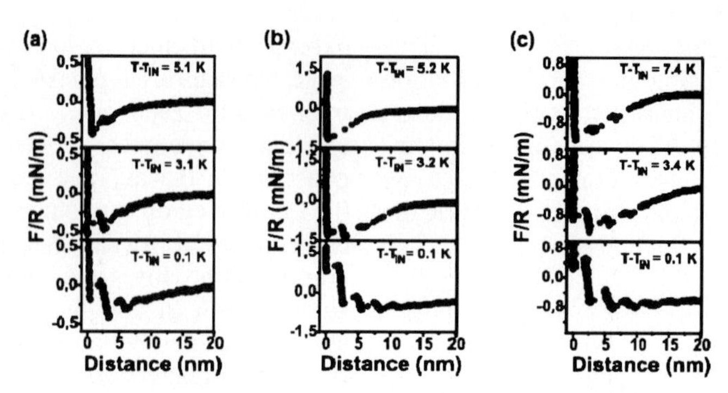

Fig. 5 Force/Radius (F/R) as a function of the distance between a 3.3 μm silica microsphere and glass substrates deposited 8CB monolayer (**a**) and trilayer (**b**) and DMOAP (**c**) in the bulk isotropic phase when temperature approaches to isotropic–nematic phase transition. In all three cases the silica microsphere was modified by chemisorption of a monolayer of DMOAP

well as on untreated surfaces. Figure 5 shows force curves measured in the isotropic phase of 8CB between a microsphere modified by DMOAP and a glass plate modified by a monolayer (Fig. 5a) or trilayer (Fig. 5b) of 8CB or a monolayer of DMOAP (Fig. 5c). From top to bottom, the temperature decreases from several degrees above the isotropic–nematic phase transition temperature T_{IN} to close to T_{IN}. At the highest temperature the force on the AFM cantilever is nearly zero at large separations. However, as the sphere–surface separation decreases below about 20 nm, the sphere is attracted by the surface (Fig. 5a). This attraction is caused by capillary condensation of nematic phase [17, 25]. Approaching T_{IN}, a repulsive force shows up and very close to the surface jumps in the force spectra are visible. They are due to mechanical instabilities of the presmectic layers when the force gradient in the vertical direction gets larger than the spring constant of the presmectic layers. This is the case, if the sphere breaks through a layer formed by 8CB dimers. The number of jumps corresponds to the range of pres-

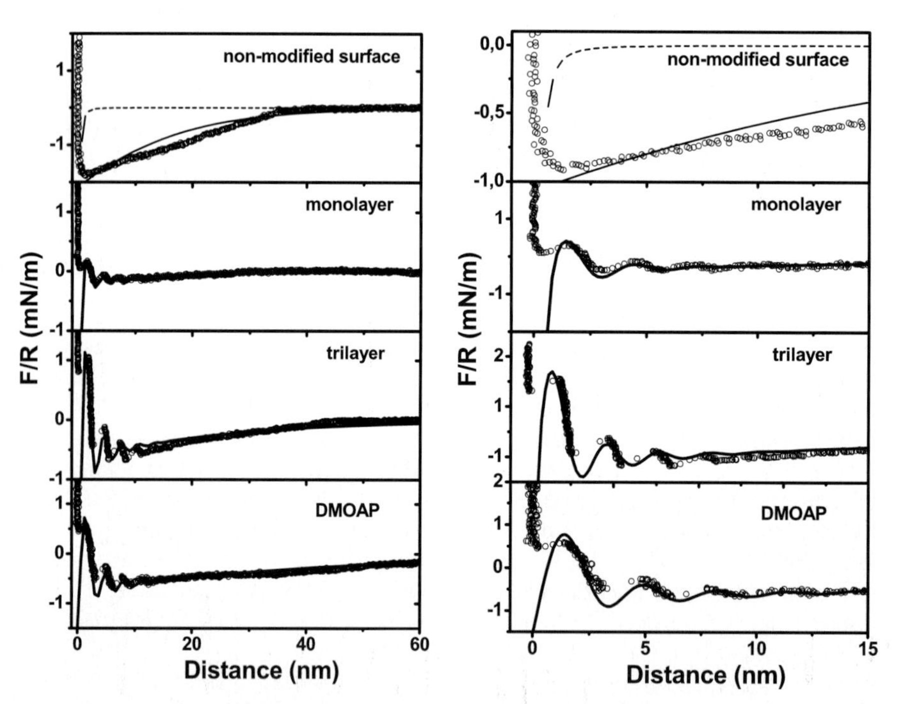

Fig. 6 Force/Radius F/R as a function of the distance between a 3.3 μm silica microsphere and (from *top* to *bottom*) a freshly cleaned silicon surface, 8CB monolayer, 8CB trilayer, and DMOAP modified glass substrates in the bulk isotropic phase, respectively. *Right hand images* focuses on the presmectic ordering, whereas the *left images* show the range of capillary attraction. At large distances the forces are zero. The *dotted line* gives the contribution of the Van der Waals interaction, using a Hamaker constant of $A_{\mathrm{H}} = 1 \times 10^{-21}$ J. The *solid lines* show the results obtained from the Landau–de Gennes free energies. The nematic elasticity is taken to be $L_{\mathrm{n}} = 6$ pN [8]. The spectra are recorded 0.1 ± 0.05 K above the bulk isotropic–nematic phase transition

Table 1 Results of the fit of the force curves with the sum of the prenematic (Eq. 2) and presmectic force (Eq. 4)

	$\xi_n = \xi_0/\sqrt{\Delta T}$ [nm K^{-05}]	w_1 [10^{-4} J/m^2]	w_2 [10^{-4} J/m^2]	d_0 [nm]	ξ_s [nm]	a_0 [nm]	$\alpha\psi^2$ [10^6 J/m^3]
Monolayer	15 ± 5	0.3 ± 0.1	9 ± 3	0.2 ± 0.1	2 ± 0.5	2.9 ± 0.2	0.015 ± 0.005
Trilayer	25 ± 7	0.34 ± 0.1	4.5 ± 2	0.2 ± 0.1	3 ± 0.5	3.0 ± 0.2	0.025 ± 0.02
DMOAP	35 ± 8	0.32 ± 0.1	3.1 ± 1	0.2 ± 0.1	3.3 ± 0.5	3.2 ± 0.1	0.02 ± 0.01

mectic alignment into bulk 8CB and its amplitude to the stability of the layers. Both range and amplitude of the presmectic ordering are enhanced on surfaces modified by trilayers compared to monolayers of 8CB and are largest on surfaces treated by DMOAP. In the latter case, even 7 °C above the phase transition temperature to the nematic phase, presmectic alignment was observed (Fig. 5c). Just above T_{IN}, DMOAP treated substrates show five equidistant jumps and the capillary bridge reaches out towards 70 nm (not visible on this scale).

The dependence of the range and amplitude of presmectic alignment on surface treatment is shown in Fig. 6. The curves are taken 0.1 ± 0.05 K above the isotropic–nematic phase transition temperature. On an untreated surface the force is zero at large distances and attractive at smaller ones. However, no presmectic alignment can be detected. The long-range attraction is caused by capillary condensation of nematic phase. A short-range presmectic alignment is observed on surfaces modified with an 8CB monolayer. Its range increases to almost 12 nm on trilayer-modified substrates and to more than 15 nm on DMOAP-modified ones.

A fit of the force curves with the sum of the prenematic (Eq. 2) and presmectic force (Eq. 4) permits to determine the correlation length of prenematic ξ_n and presmectic ξ_s ordering, the anchoring energies w_1, w_2, as well as the repeat distance a_0 (see Table 1).

The prefactor $\alpha\psi^2$ depends on the amplitude of the smectic density modulations ψ. As we can determine the absolute temperature only up to ±0.05 K, we have given the nematic correlation length instead of its value at the surface ξ_0. Furthermore, certain parameters can not be determined unambiguously, as different combinations of them result in similar curves. For example an increase of the smectic correlation length can be compensated by a decrease of the smectic amplitude. Similarly, an increase of the ordering parameter w_1 can be counterbalanced by an increase of the disordering parameter w_2, or an increase of the nematic correlation length by a decrease of the ordering parameter. The repeat distance of the presmectic layers is close to those expected for bulk 8CB, i.e. 3.2 nm. DMOAP modified glass substrates show the largest prenematic and presmectic correlation length and the lowest disordering energies. On weakly orienting surfaces the disordering energy is largest and the smectic correlation length, which drives the nematic–smectic phase transition, is smallest. The Landau–de Gennes model assumes a homeotropic anchoring. As this assumption isn't fulfilled on non-modified surfaces no satisfying description could be obtained.

Elasticity of the First Layer of 8CB on Modified Surfaces

To obtain detailed information on the first layer, we used a sharp bare AFM tip and recorded force curves of 8CB confined between the tip and modified glass substrates. As a layer of 8CB dimers can intercalate between the alkyl chains of the DMOAP molecules the first layer is formed by these intercalated dimers [26, 27]. Similarly, in case of surfaces modified by a trilayer the first layer is formed by the two topmost layers of the 8CB trilayer. The elastic modulus of the first layer might be slightly influenced by the adsorbed 8CB or DMOAP monolayer underneath.

A modification by DMOAP gives rise to a short-range repulsion already 10 K above T_{IN} (Fig. 7a). The repul-

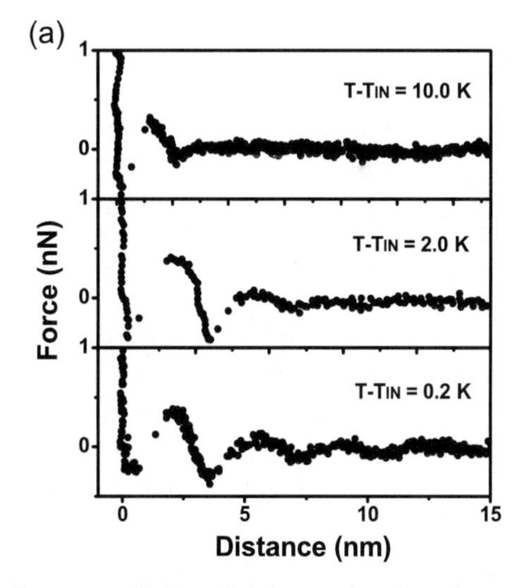

Fig. 7 **a** Force on a Si_3N_4 AFM tip as a function of separation between the tip and the DMOAP silanated glass surface in isotropic phase of 8CB. A first layer is stable up to more than 10 K above T_{IN}. This indicates that the first layer of liquid crystal molecules is strongly coupled to the substrate. Sometimes, even a second and third layer is observed just above the phase transition. As the tip was not completely spherical the tip radius could not be determined accurately

sive force increases until the surfaces are separated by about 1.4 nm. Further approach of the substrates causes that the 8CB layer becomes unstable and the tip jumps into contact with the DMOAP modified glass substrate. Decreasing temperature induces formation of a second and even a third layer. In difference to the findings by Muševič et al. the breakthrough forces for the first and second layer differ only by a factor of two instead of an order of magnitude although the absolute value for the first layer is similar [7].

A weaker adsorption of the first layer is observed on substrates modified by Langmuir–Blodgett transfer of a monolayer and trilayer of 8CB, respectively (Fig. 7b).

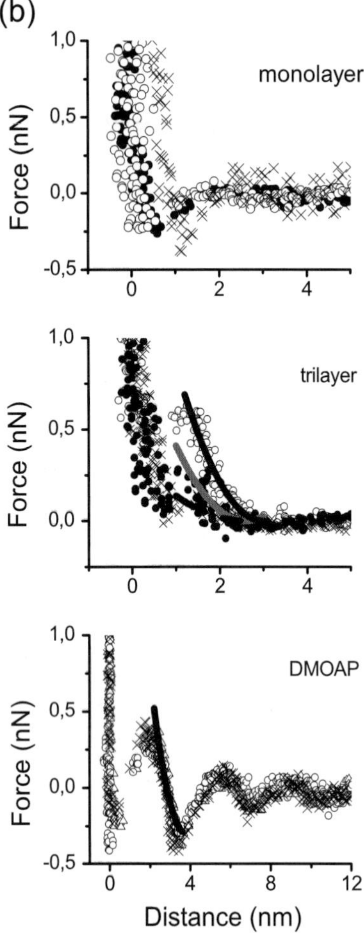

Fig. 7 b Force on a Si_3N_4 AFM tip as a function of separation between the tip and a glass substrate modified by Langmuir–Blodgett transfer of a monolayer of 8CB (*top*), by Langmuir–Blodgett transfer of a trilayer of 8CB (*middle*), and chemisorption of a monolayer of DMOAP (*bottom*). The curves are taken 0.1 ± 0.05 K above the isotropic–nematic phase transition. No first layer could be observed on substrates modified by a monolayer, a weak first layer on substrates modified by a trilayer and a strongly coupled first layer on substrates modified by DMOAP. The *solid lines* show a fit to the curves by the solid foundation model

For monolayers the force curves are zero until the tip touches the substrate. In trilayers a weak repulsion appears at distances of 3 nm. However, the strength of repulsion differs significantly from approach to approach and from position to position on the sample. The elastic modulus K of the first layer can be estimated from the force curve. However, because the tips are not spherical the tip radii can be determined only with an accuracy of about 30% at best. Taking these uncertainties into account, the elastic modulus for trilayer modified substrates is between 0.5×10^6 N/m^2 and 4×10^6 N/m^2 and for DMOAP modified substrates it is between 4×10^6 N/m^2 and 8×10^6 N/m^2. The elastic modulus of the first layer for substrates modified by a monolayer could not be determined. Remarkably, the adsorbed trilayers are still so strongly bound to the glass surface that they do not seem to rearrange considerably to optimize their packing on the time scale of our experiment. The lower elastic modulus for trilayers compared to DMOAP is reasonable, as the trilayers are only physically adsorbed, whereas DMOAP is chemically bound.

Summary

We have investigated prenematic and presmectic ordering of the liquid crystal 8CB on surfaces homeotropically modified by adsorption and chemical binding of a monolayer or a trilayer of 8CB or of DMOAP, respectively. Close to the isotropic–nematic phase transition all three substrates modifications lead to presmectic ordering of isotropic bulk 8CB. The range and amplitude of the presmectic ordering sensitively depends on temperature and surface modification. Adsorbed 8CB monolayers induce weak presmectic layering, as the disordering surface energy is large. A few degrees above T_{IN} the presmectic ordering vanishes but prenematic ordering still remains. This leads to short range attraction due to the formation of capillary bridges of nematic 8CB between the silica sphere and the substrate. The difference in range and strength of substrate-induced alignment for mono- and trilayer modified surfaces can be understood taking into account the average area per adsorbed 8CB molecule. For trilayers the average area per molecule corresponds to that of bulk 8CB, whereas in case of a monolayer it exceeds the bulk packing by almost a factor of two, i.e. the molecules are poorly ordered on the substrate. Therefore, they can hardly induce any prenematic orientation

The elastic modulus of the first layer has been determined using sharp Si_3N_4 tips. On substrates modified by an 8CB monolayer the elastic modulus of the first layer was too small to be determined by AFM. Even the elastic modulus of the first layer in substrates modified by an 8CB trilayer is subject to large errors. This poor reproducibility is explainable for physically modified substrates, as the tip may push the molecules aside if the anchoring strength is not sufficient. Substrates modified by chemical binding of DMOAP show a well-aligned first layer, even more than 7 °C above T_{IN}. Close to T_{IN}, formation of a second and third layer was observed.

Acknowledgement The authors like to thank Karlheiz Graf for help with the Langmuir–Blodgett transfer and Denis Andrienko for stimulating discussions. D.V. and G.K.A. acknowledge funding by the SFB-TR6 and LEA and B.U. by the German Science Foundation (DFG Vo 639-10).

References

1. De Gennes PG (1974) The physics of liquid crystals. Clarendon Press, Oxford, UK
2. De Gennes PG (1971) Short range order effects in the isotropic phase of nematics and cholesterics. Mol Cryst Liq Cryst 12:193
3. De Gennes PG (1990) Interactions between solid surfaces in a presmectic fluid. Langmuir 6:1448
4. Moreau L, Richetti P, Barois P (1994) Direct measurement of the interaction between two ordering surfaces confining a presmectic film. Phys Rev Lett 73:3556
5. Richetti P, Moreau L, Barois P, Kékicheff P (1996) Measurement of the interactions between two ordering surfaces under symmetric and asymmetric boundary conditions. Phys Rev E 54:1749
6. Kočevar K, Blinc R, Muševič I (2000) Atomic force microscope evidence for the existence of smecticlike surface layers in the isotropic phase of a nematic liquid crystal. Phys Rev E 62:R3055
7. Kočevar K, Muševič I (2002) Surface-induced nematic and smectic order at a liquid-crystal–silanated-glass interface observed by atomic force spectroscopy and Brewster angle ellipsometry. Phys Rev E 65:021703
8. Kočevar K, Muševič I (2001) Forces in the isotropic phase of a confined nematic liquid crystal 5CB. Phys Rev E 64:051711
9. Carbone G, Barberi R, Muševič I, Kržič U (2005) Atomic force microscopy of presmectic modulation in the nematic and isotropic phases of the liquid crystal octylcyanobiphenyl using piezoresistive force detection. Phys Rev E 71:051704
10. Shinto H, Kobayashi K, Hyodo T, Ishida N, Higashitani K (2005) Capillary forces between planar anchoring surfaces in the isotropic phase of a nematic liquid crystal. Chem Lett 34:1318
11. Demus D, Goodby JW, Gray GW, Spiess HW, Vill V (eds) (1998) Handbook of Liquid Crystals. Wiley-VCH, Weinheim
12. Ruths M, Steinberg S, Israelachvili JN (1996) Effects of confinement and shear on the properties of thin films of thermotropic liquid crystal. Langmuir 12:6637
13. Leadbetter AJ, Durrant JLA, Rugman M (1977) The density of 4-*n*-octyl-4-cyano-biphenyl (8CB). Mol Cryst Liq Cryst 34:231
14. Chunhung W, Tianbo L, Henry W, Benjamin C (2000) Atomic force microscopy study of $E_{99}P_{69}E_{99}$ triblock copolymer chains on silicon surface. Langmuir 16:656
15. Butt H-J, Cappella B, Kappl M (2005) Force measurements with the atomic force microscope: Technique, interpretation and applications. Surf Sci Rep 59:1
16. Sheng P (1976) Phase transition in surface-aligned nematic films. Phys Rev Lett 37:1059
17. Stark H, Fukuda J-I, Yokoyama H (2004) Capillary condensation in liquid-crystal colloids. Phys Rev Lett 92:205502
18. Sheng P (1982) Boundary-layer phase transition in nematic liquid crystals. Phys Rev A 26:1610
19. Derjaguin BV (1934) Untersuchungen über die Reibung und Adhäsion, IV. Kolloid Z 69:155
20. Israelachvili J (1992) Intermolecular Surface Forces. Academic Press, San Diego
21. Kočevar K, Muševič I (2002) Observation of an electrostatic force between charged surfaces in liquid crystals. Phys Rev E 65:030703
22. Johnson KL (1985) Contact Mechanics. Cambridge University Press, New York
23. Xue J, Jung CS, Kim MW (1992) Phase transitions of liquid-crystal films on an air–water interface. Phys Rev Lett 69:474
24. De Mul MNG, Mann JA (1994) Multilayer formation in thin films of thermotropic liquid crystals at the air–water interface. Langmuir 10:2311
25. Bračič AB, Kočevar K, Muševič I, Žumer S (2003) Capillary condensation in a confined isotropic–nematic liquid crystal. Phys Rev E 68.011708
26. Huang JY, Superfine R, Shen YR (1990) Nonlinear spectroscopic study of coadsorbed liquid-crystal and surfactant monolayers: Conformation and interaction. Phys Rev A 42:3660
27. Ruths M, Steinberg S, Israelachvili JN (1996) Effects of confinement and shear on the properties of thin films of thermotropic liquid crystal. Langmuir 12:6637

Progr Colloid Polym Sci (2008) 134: 48–56
DOI 10.1007/2882_2008_087

Published online: 19 July 2008

SURFACES AND INTERFACES

José Marqués-Hueso
Hans Joachim Schöpe

Regular Horizontal Patterning on Colloidal Crystals Produced by Vertical Deposition

José Marqués-Hueso ·
Hans Joachim Schöpe (✉)
Institut für Physik, Johannes Gutenberg-Universität Mainz,
Staudingerweg 7, 55099 Mainz, Germany
e-mail: jschoepe@uni-mainz.de

Abstract Colloidal particles have proved to be a suitable precursor to the formation of nanoscaled materials. More explicitly, they are a suitable way to create photonic band gap materials in 3D. Several methods have been developed to assemble colloidal multilayer systems, and have yielded various levels of success. The vertical deposition method has shown itself to be one of the best in terms of time, control of the final product, crystal size and homogeneity. Despite this, the resulting crystals often present point defects, dislocations, cracks and polycrystallinity, as well as a horizontal modulation of film thickness. These defects compromise the possible utilities of the crystals. The study of this last kind of defect and the influential factors are the gravity point of this work. For first time in our knowledge, several of these factors where quantitatively studied and new ones were found. A new growth method was developed for a better analysis of the formation process.

Keywords Colloids · Film formation from liquid phases · Photonic crystals

Introduction

The formation of periodical structures in the nanoscale is a busy field in the physics of materials. Submicrometer structured materials have, and are expected to have, various applications [1–4], like optical filters and gratings, antireflective surface coatings, high density data storage, selective solar absorbers, microelectronics, optical switches, waveguides with low lost, chemical and biochemical sensors and resonant cavities for small lasers.

In 1987, Yablonovitch [5] and John [6] proposed independently that one of the interesting applications of such structures is that they could behave like complete photonic band gap crystals, where the photons are located at special places inside the photonic crystal [7]. This is due to the variation of the dielectric constant which changes spatially with submicrometer periodicity which confers them their special diffractive optical properties. Therefore it is possible to obtain materials, through which the light can only propagate in a determinate direction.

A lot of different experiments were carried out to obtain such materials. On the one hand, we have the lithographic approach. Photolithography [8], e-beam writing [9] and microcontact printing [10] are precise methods, but they have the problem that they are expensive, complicated and produce hardly 3D structures. On the other hand, we have the colloidal approach, which is easier, faster and cheaper.

The crystallization of colloidal latex particles on a substrate in thin crystals provides us with the first step in the production of photonic band gap materials. But at the moment it presents the problem of precision: point defects, dislocations, cracks and polycrystallinity are always present, and they negatively affect the properties of the crystals.

The quality of such arrays must be extremely high in order to ensure that these optical properties [11] are kept. Consequently, in the last years, many groups have invented several fabrication processes to improve the quality of the colloidal crystallization, induced by gravitational [12], capillary [13] or convective [14] forces, or even by geometrical [15], chemical [16] or optical [17] patterning.

In 1996, Nagayama and co-workers proposed a method called vertical deposition [18], where a substrate is dipped into the suspension and pulled out slowly or even not at all. The wettability of the substrates thus assures that the suspension will form a meniscus on the substrate. As the evaporation in a meniscus is higher than in the surface of the suspension, a flow of particles is created towards the meniscus, similar to the mechanism of the horizontal deposition. Vertical deposition has the advantages that it produces random hexagonal packed crystalline layers with a grain size of several mm^2 with the (111) direction perpendicular to the substrate. Large areas of homogeneous colloidal crystals can be obtained and the thickness of the crystal can be nearly controlled by adjusting the volume fraction φ and the evaporation rate of the suspension, the pulling velocity v of the substrate, the wettability of the substrate and the interaction between the particles, and by keeping all these parameters constant during the experiment (the small change of φ is particularly remarkable here, since this is not usually the case with other methods). But also by using this technique, crystalline defects like cracks, point defects, stacking faults, dislocations and a non constant thickness are still present.

One of the less studied defects in the case of colloidal crystals produced by vertical deposition is the modulation of the film thickness. Often a near regular periodic horizontal modulation in the thickness of the crystalline film can be observed. A thickness modulation with a horizontal frequency on a mm scale of colloidal crystals prepared by vertical deposition was reported [19, 20] but never analyzed in a systematic way. Im and coworkers [20] reported experiments on vertical deposition at a tilted angle using 265 nm PS-particles in water at $\varphi = 0.2\%$ at a temperature of 60 °C and an evaporation rate of 0.155 μm/s. They obtained a thickness modulation in all films which could be reduced at much lower evaporation rates (4.5 nm/s) and by tilting the substrate at an angle greater than 10°.

Koumoto and coworkers used the effect of thickness modulation to create a pattern of particle wires by vertical deposition [21]. For their experiments they used 1000 nm silica particles dispersed in ethanol at low volume fractions. Particles were deposit on a glass slide by vertical deposition at 70 °C obtaining particle wires of 150 μm width and a regular interval of the particle wires of 200 μm.

The formation of stripe patterns after evaporation of a solution or suspension ("coffee spot problem") was mainly investigated using simple droplet evaporation. In 1995, Adachi et al. [22] analyzed the stripe patterns of monolayers formed by 144 nm particles on glass and suggested a model to obtain sinusoidal patterns from drying drops. The particle flow to the contact line affects the motion of the contact line as a source of friction. The shape of the droplet edge depends on the position of the contact line, the evaporation rate and the surface tension and affects the self assembly of the particles. This coupled system originates a sinusoidal like periodic slip–stick motion of the perimeter of the drop and is mainly based on the concurrence between the droplet's surface tension, the wetting-film's surface tension and the friction force at the contact line, which is influenced by the evaporation rate, the film height the viscosity and other parameters. Using their model they were able to describe the average stripe density of the observed monolayer stripes.

In 1997, Deegan and coworkers [23, 24] proposed a pinning–depinning model to calculate the mass deposited in each ring formed by the evaporation of the solvent of a suspension's drop. The origin of the stripe pattern is the contact line pinning and the evaporation rate from the edge of the drop. The evaporation rate depends on the curvature of the surface and on the velocity of the fluid towards the drying area on the droplet's radius. In this pinning–depinning model, the pinning points are supposed to be imperfections of the substrates, and so a non-periodical pattern is obtained.

In 2002, Shmuylovich et al. [25] carried out experiments concerning the drying process of drops of 880–3150 nm PS-particles with $\varphi = 0.008$–0.01% on precleaned glass substrates under normal laboratory conditions. It was found that the limits of the air–water-substrate did not move with constant velocity. The perimeter of the drop of the suspension of 880 nm PS-particles moved for approximately 17 s and stayed stationary for 50 s. At higher volume fraction, this difference in time was lower. These stick–slip motions were observed to be stochastic, which could be due to the non-treated substrates used, or to the lack of control over the evaporation process.

The results reported using simple droplet evaporation are inconsistent concerning the periodicity of the obtained pattern and of the origin of the stick slip motion. No systematic investigations have been done so far analyzing the stripe pattern formation obtained by vertical deposition; using drop drying only the volume fraction was varied in a systematic way.

Therefore in the present work several series of experiments with vertical deposition were carried out to discover new clues about the origin of the crystal thickness modulation. The volume fraction and the salt concentration of the suspension were varied, as well as the tilting angle of the substrate, the substrate material and the pulling speed. We present a quantitative analysis of the lateral periodicity of the obtained line pattern of the dried colloidal crystals. In the experiments it was found that the created colloidal crystals showed a near regular periodic horizontal modulation in thickness with a periodicity between 0.3 and 1.8 mm and this thickness modulation is influenced by all varied parameter.

Experimental

For our experiments we used a charge stabilized suspension of polystyrene spheres dispersed in ultrapure water (Batch No. PS-F-3390, Berlin Microparticles GmbH Germany). The diameter was determined by electron microscopy to be 590 nm. The size polydispersity was determined to be 5.8%. The particles are stabilized with COOH- and HSO_4-groups and the effective charge was measured by conductivity to be $Z^* = 3000 \pm 100$. For diluting of the stock solution to a definite volume fraction deionized water of a MilliQ water system was used. To adjust the salt concentration of the suspension NaCl was added to screen the interaction of the particles (typically 1 mM).

Glass microscope slides ($75 \times 25 \times 1\,\mathrm{mm}^3$, Superior Marienfeld GmbH) as well as Quartz substrates ($45 \times 12.5 \times 1.25\,\mathrm{mm}^3$, Starna GmbH) were used as substrates. The substrates are first cleaned with hot sulfuric acid for 12 h, and then they are rinsed with ultra pure water and dried under a stream of nitrogen. The wettability of the substrates is kept constant by this cleaning procedure and was checked measuring the meniscus height when the substrate was dipped vertically in ultrapure water.

Two different setups were used to carry out the experiments. The first setup was a conventional one for vertical deposition. Here an ac engine connected to a high precision gearbox system (both RS-components) pulls up the substrates, which are immersed into the suspension filled inside a glass vessel. We carefully checked that the adjusted pulling speed was constant during the experiment, because fluctuations could produce patterns on the substrates. Therefore, a home-made motor controller was used to monitor the speed of the motor versus time, and the movement of the translation stage was measured as a function of time using a micrometer dial gage. Up to three substrates are mounted on the motorized translation stage. The translation stage is fixed on a home made goniometer which is mounted vertically on massive table, so that the inclination angle of the substrates can be varied.

To avoid a possible local difference of particle number density in the suspension, the colloidal suspension was homogenized by the use of a magnet stirrer at a low spin velocity. We always used the same stirring speed and the same filling height of the reservoir. The stirring speed was set to value that the surface of the suspension was still flat, but fast enough that particle sedimentation could be neglected. The whole setup was covered by a box made of PMMA ($1 \times 0.8 \times 0.7\,\mathrm{m}^3$) to get a constant humidity and to avoid any unknown air flow at the sample. The humidity in the box was adjusted at $81 \pm 2\%$ by placing a large vessel of a salt saturated KCl solution inside the box. The atmosphere inside the box, and so the evaporation rate, was also homogenized using a small ventilator and was 5.62 nm/s. This setup was placed inside a thermostated lab so that the temperature during the experiment was $20 \pm 1\,°\mathrm{C}$. Humidity and temperature were monitored using an electronic hygrometer and thermometer.

Four series of experiments with volume fractions between 1.00 and 8.32% and pulling velocities between 0.008 and 2 μm/s were done to study the influence of the pulling velocity and the particle number density. The concentration of NaCl in all the suspensions was 1 mM. To investigate the influence of the interaction between the particles the screening was varied. Here a new series of experiments at different pulling velocities with a suspension of volume fraction $\varphi = 1.0\%$ and 10 mM NaCl was carried out and the results compared with the correspondents to 1 mM NaCl. The influence of the inclinations angle was checked with a new series with 0, 10, 20, 30 and 40° of inclination in respect to the perpendicular. The pulling velocity was $0.0505 \pm 0.0004\,\mu\mathrm{m/s}$, the salt concentration 1 mM and the volume fraction $\varphi = 1.00\%$ in all the experiments. The dependence on the material substrate was also checked. Five experiments were carried out, each of them with two glass substrates and one quartz slide. The pulling velocity was $0.099 \pm 0.004\,\mu\mathrm{m/s}$, the salt concentration 1 mM and the volume fraction $\varphi = 1.00\%$ for the five experiments. The resultant samples were examined with SEM, optical microscopy, AFM and low resolution transillumination microscopy [26].

In order to get more information about the crystal formation, a new experiment, which for clarity in this text we will call inverse vertical deposition, was developed to record the film growth (see Fig. 1). This device uses the walls of the suspension container as substrate for the deposition. This is reached because in this device the substrates do not move, but the level of the suspension decreases due to a hydrodynamic system. This allows us to record the crystal growth on the recipient walls. To do so we constructed a special sample cell: here two glass slides are mounted onto a frame made of stainless steel acting as container walls. Viton sheets are used to seal the connection. A teflon tube (connected at the bottom of the sample cell) connects the sample cell with a reservoir. The sample cell is mounted on a motor driven translation stage moving in vertical direction. Increasing the height of the sample cells means that there is a flow of the suspension out of the sample cell into the reservoir because the height of the reservoir is fixed. The level of the suspension inside the sample cell is lowered and so crystal formation at the container walls can be observed. Therefore a microscope using a 100× objective (Leica PL FLUOTAR L $100 \times /0.75\ \infty/-$) with a working distance of 4 mm mounted horizontally was used. The whole setup was placed inside the PMMA box to control humidity and air flow around the suspension.

The study of the pictures obtained with this system allowed us to determine the growth velocity of the film.

An experiment with resultant pulling velocity $0.049 \pm 0.003\,\mu\mathrm{m/s}$, volume fraction $\varphi = 1.00\%$ and 1 mM NaCl was carried out in order to obtain a thin colloidal film. The

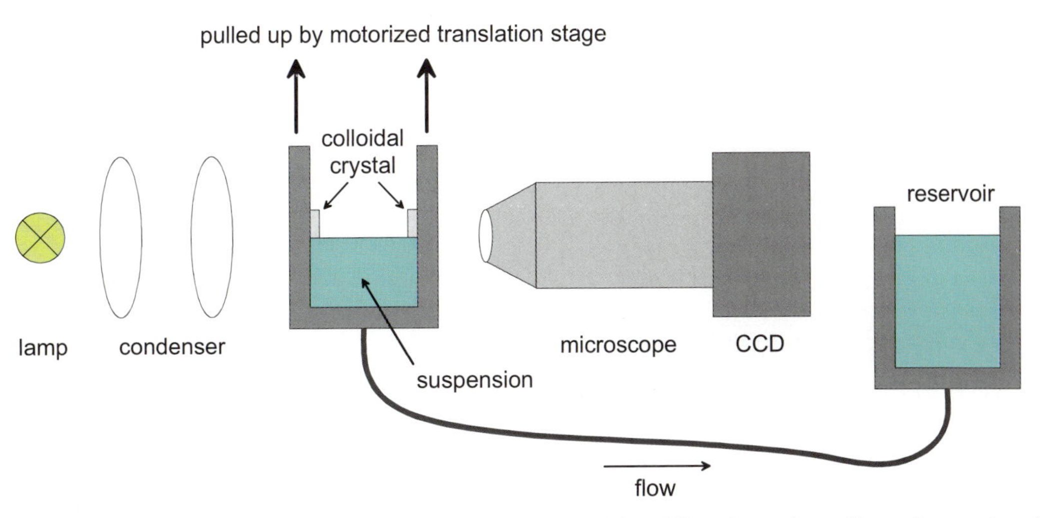

Fig. 1 Setup for recording the crystal growth during the inverse vertical deposition. When the engine pulls up the sample cell, the suspension flows towards the reservoir. The resultant colloidal crystal at the walls of the cell is illuminated by the condenser and recorded by the CCD camera

growth of the crystal was analyzed during three intervals of 8–15 min. The measurements started 150 min after the commencement of the experiments in order to ensure that the system was stable. The second measurement was made 180 min after the beginning, and the third ones, 270 min after the launch.

Results and Discussion

In all experiments performed in this investigation, large scale dried colloidal crystals of different quality were not only obtained, but also a thickness modulation which originated horizontal stripes showing a regular pattern has been observed throughout investigations (see Fig. 2). These stripes are oriented perpendicular to the growth direction of the crystalline layer and show a typical periodicity which can be determined measuring the average stripe size or stripe width.

But first we will show the influence of homogenizing the suspension as well as the surrounding atmosphere in a qualitative way. As can be seen in Fig. 2 the sample prepared by vertical deposition by continuous mixing of the suspension (right side) shows a better periodicity of stripes as without homogenizing the suspension (left side).

The averaged stripe width was determined using image analysis. Figure 3 presents on the left a magnified region of dried colloidal crystal. The stripe size L is measured be-

Fig. 2 Samples prepared with 590 nm PS-particles, 1.0% volume fraction, 1 mM NaCl, pulling speed 0.05 μm/s. The samples obtained with magnet stirrer and ventilator (*right sample*) showed a better periodicity of the stripes than those ones without these improvements (*left sample*). The samples are illuminated with white light from side. The gradual change of color in the reflected light indicates that the structure of the crystal is preserved in the whole film. Width of the substrate: 25 mm

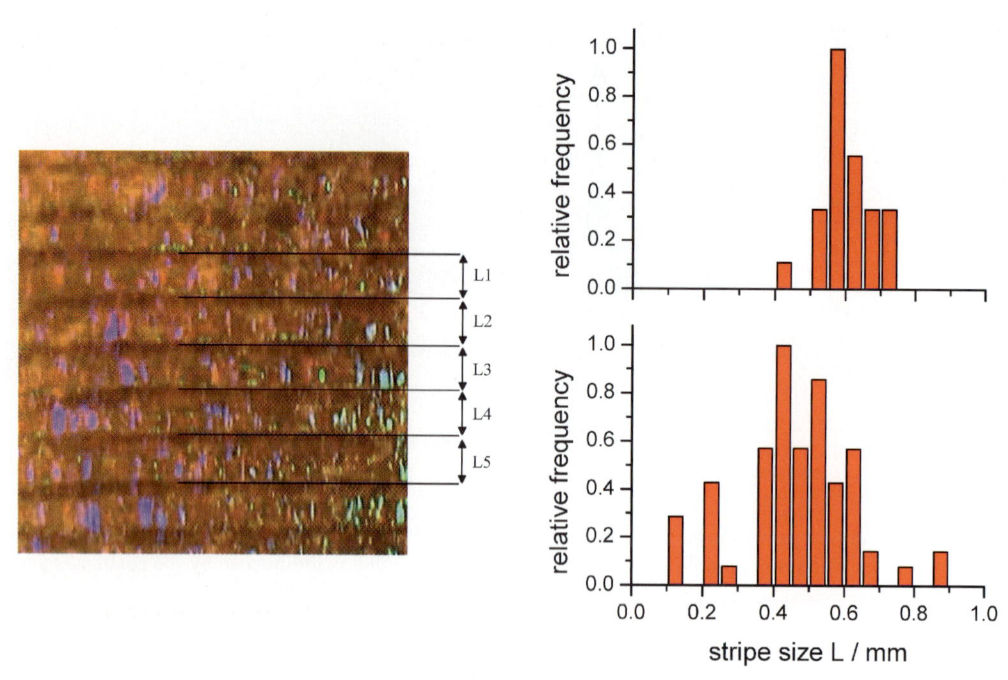

Fig. 3 *Left*: Definition of the stripe size L (picture size 8×8 mm^2). *Right top*: Stripe size distribution of the sample obtained with magnet stirrer and ventilator, *right bottom*: stripe size distribution of the sample obtained without magnet stirrer and ventilator

tween the beginnings of two nearest stripes. On the right the stripe size distribution for the two samples of Fig. 2 are shown. The sample prepared homogenizing the suspension as well as the surrounding atmosphere (on top) has a much narrower distribution than the sample without these improvements. This simple experiment shows that the substrate properties have much less influence on the periodicity of the formed pattern than the suspension properties or the evaporation rate.

Figure 4 shows more detailed information about the stripe morphology and its structure using low resolution transillumination microscopy, high resolution microscopy

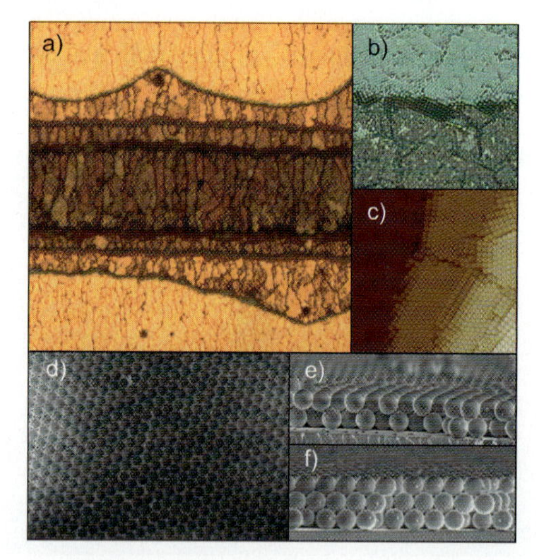

Fig. 4 Details of the stripe morphology and crystalline structure using 590 nm particles, 8.32% volume fraction, pulling speed 2 μm/s, 1 mM NaCl. **a** low resolution microscopy picture showing the morphology of the stripes (picture size 577×577 μm^2); **b** high resolution microscopy picture showing the crystal structure of the multilayer system from two to three layers (picture size 30×30 μm^2); **c** AFM picture at the transition from two to four layers (picture size 17.7×17.7 μm^2); **d** SEM picture showing the quality of the prepared photonic crystal in a region of hexagonal layers; **e** and **f** SEM picture showing the quality of thin hexagonal layers on an edge. Particle diameter: 590 nm

and AFM. Starting from the minimum layer number, the thickness of the film increases progressively to its maximum before decreasing again to its minimum number. In the shown low resolution picture the minimum number is a bilayer and the maximum thickness is five layers. The thickness modulation is observed to be up to 5 layers for thin films (minimum thickness: monolayer) and up to the double of the minimum layer thickness for thick films (around 50 layers). By changing the number of layers a change in the crystalline structure is observed. The system changes from a hexagonal structure to a square one and back to a hexagonal symmetry $n\Delta \rightarrow (n+1)\Box \rightarrow (n+1)\Delta$, were n denotes the number of layers and the symbol giving the symmetry (Fig. 4b,c). This sequence was first noted by Pieranski [27] observing the increasing layer number of a thin colloidal crystal in a colloidal suspension in confinement and was expanded by Neser [28] and by Schöpe [29]. As the number of layers increases, the number of possible stacking sequences for the hexagonal layers increases, as can be seen in the color code in Fig. 4a. Details of the color coding are explained elsewhere [26]. Figure 4d–f show the quality of the prepaired hexagonal crystals from top and from the side. In the following we will not go in details of the thickness modulation but we will deal with the analysis of the lateral periodicity of the formed pattern.

Figure 5a shows the stripe distance as a function of the applied pulling speed for four different volume fractions. At higher pulling velocity, the distance of the stripes becomes smaller. This is also observed decreasing the volume fraction. Both factors are related to the film thickness which can easily be seen from the following formula proposed by Nagayama [18]:

$$N = \frac{\beta l j_e \varphi}{0.605 v D (1-\varphi)} \,. \tag{1}$$

The number of obtained dried layers (N) is related to the pulling velocity (v), the evaporation flow (j_e) at the meniscus, the diameter of the particles (D), the meniscus height (l), the volume fraction (φ) and a factor β which represents the ratio of the velocity of the particles towards the meniscus and the velocity of the solvent. The interaction between the particles influences the velocity towards the crystal interface and so the attachment rate. The stronger the interaction between particles, the slower their motion, and the smaller the value of β. Figure 5b shows the averaged layer thickness for four different volume fractions as a function of the pulling speed. The layer number was determined in a combination of low resolution microscopy [26], scanning electron microscopy and transmission spectroscopy. The latter method allows measuring the layer thickness quite accurate determining the oscillations in the transmission spectrum close to the *fcc* (111) dip stemming from thin film interference. The thin line is just a guide to the eye representing the expected $1/v$-dependence. The datapoint measured at the lowest concentration are following the theoretical prediction inside the experimental uncertainties, whereas there are deviations observable at higher volume fractions and low pulling speeds. A more detailed look at Fig. 5a,b reveals that there is no clear L- and thickness dependence at low pulling speeds in the case of $\varphi = 2.00$ and 3.61%. Here the stripe distance and the layer thickness decreases again. The origin of this observation is unclear. But we can summarize that for thick films, the size of the stripes is larger than for thin colloidal multilayer crystals.

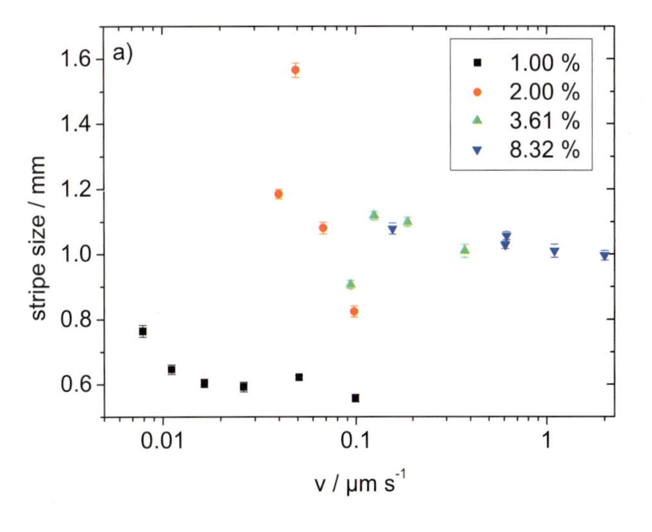

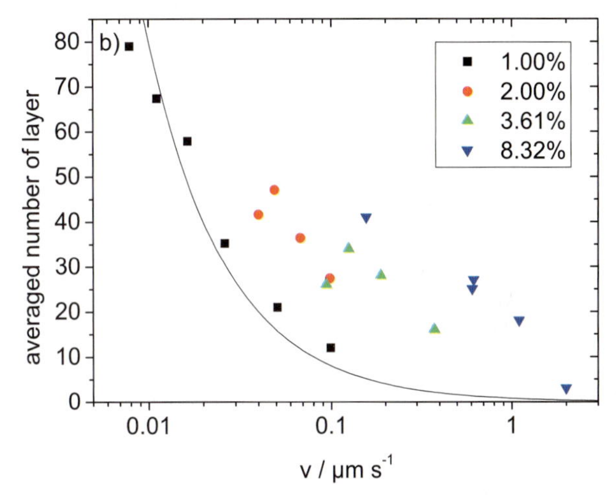

Fig. 5 **a** Stripe size as a function of the pulling velocity at several volume fractions for 590 nm PS suspensions with a salt concentration of 1 mM NaCl. **b** Sample thickness as a function of the pulling velocity at several volume fractions for 590 nm PS suspensions with a salt concentration of 1 mM NaCl. The *thin line* is a guide to the eye showing the expected $1/v$-dependence of the sample thickness

The influence of different screening parameters on the stripe formation can be seen in Fig. 6. The comparison between the 1mM NaCl series and the series with 10mM NaCl shows that the stripe distance increases with decreasing particle interaction. As can be seen from Figs. 5 and 6 there is a relationship between pulling speed and stripe for-

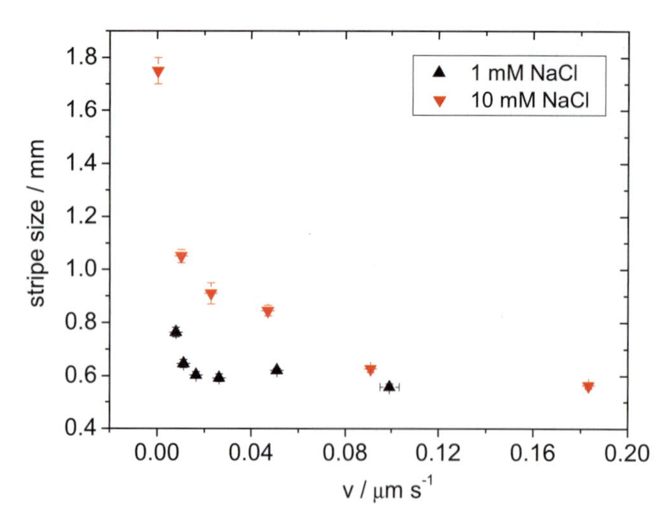

Fig. 6 Stripe size as a function of the pulling velocity at several salt concentrations of 590 nm PS suspensions with 1% volume fraction

mation. The size of the stripes increases by decreasing the crystal growth velocity. The stripe width seems to reach an asymptotic minimum value at high pulling speeds. In both figures, a higher availability of particles at the meniscus (due either to a high particle concentration or to a strength screening) originates a larger stripe distance.

The stripe size is strongly influenced by the inclination angle of the substrate in respect to the perpendicular (see Fig. 7). By increasing the tilt angle the apparent wetting angle is reduced. When the substrate is perpendicular to the suspension surface, the samples present more stripes per unit length than if the substrates are tilted at an angle α.

This could be explained as a geometrical question. If the mechanism which originates the stripes is periodic and it only depends on the meniscus position, a tilting of the substrates would increase the distance between two consecutive stripes.

If L is the period of the stripes in the case of substrates perpendicular to the suspension surface, the period L' of the stripes in the tilted case should be $L' = L/\cos(\alpha)$, been α the angle between the substrate and the perpendicular. The linear fit of L' in front of $1/\cos(\alpha)$ was made in order to test this hypothesis. The resulting good correlation of the linear fit ($R^2 = 0.992$) shows that our hypothesis is reasonable (see insert in Fig. 7). This experiment demonstrates that the stripe size is increased by increasing the length of the meniscus by reducing the meniscus curvature.

In Fig. 8 we show the influence of the used substrate material. Here a glass and a quartz substrate are mounted on the sample holder and are pulled out of the suspension. This experiment was repeated 5 times. The scatter of the measurements is quite small for such kind of experiments (in the case of glass less than 10%) demonstrating the performance of our set up. In the case with quartz, one finds more stripes per unit length than in the case with glass. Both materials produce different meniscus shape. In the case with glass, the meniscus is about 15% higher. The distance between stripes depends on the wetting angle of the substrates material. This experiment is further proof that a longer meniscus leads to longer stripes and that the wetting properties do influence the periodicity of the pattern.

The experiments performed using the new experimental setup allows us to measure the crystal growth velocity as a function of time. An important observation is that the

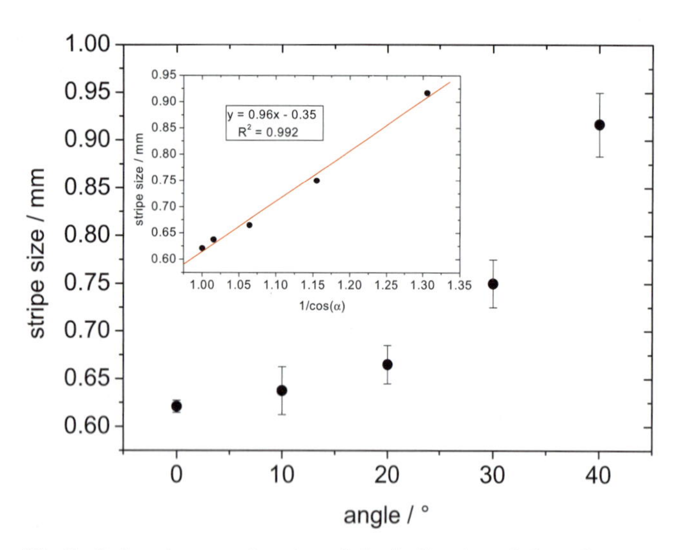

Fig. 7 Stripe size as a function of the inclination of the substrates in respect to the vertical axis. The *insert* shows the stripe thickness versus $\cos(\alpha)^{-1}$ to check the geometrical influence of the meniscus. The pulling velocity was 0.0505 ± 0.0004 µm/s, the volume fraction $\varphi = 1.00\%$ and the NaCl concentration 1 mM

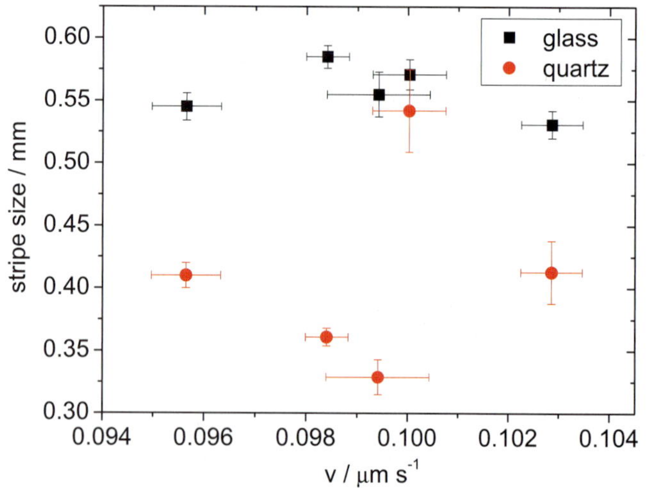

Fig. 8 Reproducibility of the results and difference of values for glass and quartz substrates. For the five experiments the pulling velocity was 0.099 ± 0.004 µm/s, the volume fraction $\varphi = 1.00\%$ and the salt concentration 1 mM

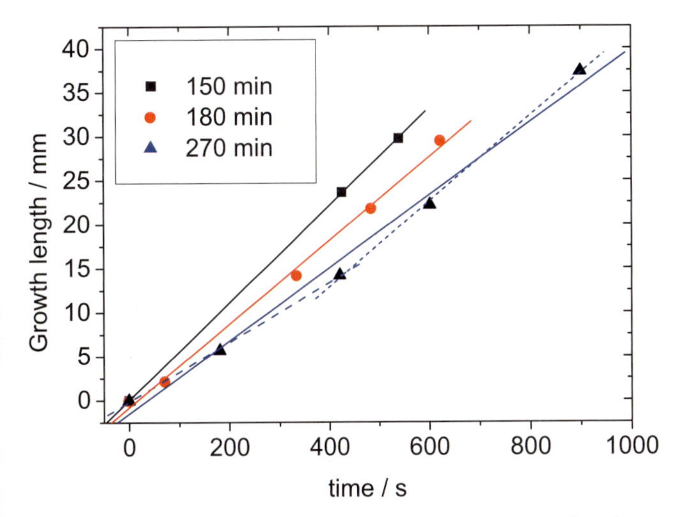

Fig. 9 Increase of length of the crystal at three different time intervals (540–900 s each one), recorded 150, 180 and 270 min after the beginning of the crystallization process. The first point of each series is displaced to (0, 0) for clarity. The *dashed lines* represent the velocity at the beginning and end of the third series of data, which are clearly different

growth velocity of the crystal varies with time, as shown in Fig. 9. Here the length of the growing crystal is plotted as a function of the elapsed time. The three data series correspond to different time intervals of the crystal formation. The obtained slopes represent the short-term growth velocity in these time intervals, velocities which are 0.056, 0.047 and 0.041 μm/s respectively. In the third measurement, it is even possible to recognize by eye that the velocity increases during the small interval of 15 min. From the first to the third point of this third series, the averaged velocity calculated is 0.033 μm/s, while for the last three points, 0.049 μm/s. That means that the velocity increases 48% in only 12 min. The slower growth velocity calculated from the data obtained from two consecutive pictures was 0.031 μm/s, while the faster was 0.056 μm/s. The velocity of the particles towards the interface was approximately 4 μm/min in the immediate surrounding of the assembly area. The particles approached this area with Brownian motion. Temporal and permanent aggregates were also observed in the suspension, causing sometimes dislocations or grain boundaries but having no influence on the stripe formation. It was observed that the thickness of the colloidal film was constant during the whole reported part of the experiment in which the crystal was growing about 0.3 mm. This is consistent with the fact that the expected distance between stripes was 0.6 mm (see Fig. 6).

Conclusions

The observed thickness modulation of the prepared photonic crystals was observed to be reproducible for the same set of experimental parameters (volume fraction, salt concentration etc.). It was made sure that the relevant experimental parameters (pulling velocity, humidity etc.) were kept constant during the performed experiments. Patterns of constant periodicity have been observed if the surrounding atmosphere and the suspension were homogenized: Periodic patterns are formed when the evaporation rate and the particle concentration at the meniscus are almost constant. The substrate properties do influence the wetting properties and so the stripe distance but does not influence the quality of the periodicity. This means that the stripe formation is not caused by particular defects of the single glass substrates, but something which affects all the samples simultaneously in a systematic kind of manner.

The thickness of the prepared colloidal crystal is influenced by the volume fraction, the interaction between the particles, the substrate material and, due to geometrical reasons, by the tilted angle. It was observed that the stripe distance is increasing with increasing layer thickness. Thicker films are obtained by increasing the particle concentration, by decreasing the particle interaction, by slowing down the growth velocity and by increasing the meniscus height.

The variation of the growth velocity observed in the last experiment suggests an inconstant behavior of the meniscus with a stick–slip motion like observed in the droplet drying experiments [22–25, 30]. We suggest that a stick–slip motion of the meniscus during the evaporation process could explain the stripes morphology in the case of colloidal crystals. This slip–stick motion is caused by a competition between a friction force pinning the contact line F_{pin} and a force caused by the surface tension moving the contact line F_{move} when the water evaporates. Due to the pinning process and the decreasing level of the suspension, there is progressive deformation of the meniscus which makes it thinner. A thin (but high) meniscus is supposed to create a thick colloidal crystal, especially when there is enough time to assemble particles in one area. When the water evaporates the meniscus gets flatter and the contact line moving force parallel to substrate increases. Finally when $F_{move} > F_{pin}$, a slip motion moves the meniscus before it is pinned again. A short meniscus creating again a thin colloidal crystal will be the result. This mechanism could modulate the thickness of the growing crystal. An additional factor could act in thick films, due to the underlying geometry, which could act like a chimney, which would increases the evaporation and the thickness of the crystal, as a consequence.

The results obtained in our experiments are in agreement with the pinning depinning model published by Adachi and coworkers [22] where the meniscus is pinned by the assembling colloidal crystal. This model was designed to describe the stripe pattern observed in monolayer formation prepared by horizontal deposition. Also here the stick–slip motion is explained by a competition between a friction force pinning the contact line and a force caused by the surface tension moving the contact line when the water evaporates. As a result they present the following formula describing the number of stripes per unit length N_S:

$$N_S \approx -\frac{\eta}{\pi(\gamma_f - \gamma_L \cos\theta)h} \frac{\sqrt{c_0 + c_1\varphi + c2\varphi^2}}{\varphi}, \quad (2)$$

where η is the viscosity of the liquid, γ_f the surface tension of the film, γ_L the surface tension of the drop, θ the wetting angle, h the film thickness and φ the volume fraction. The expressions c_0, c_1 and c_2 are given as follows:

$$c_0 = \frac{4P_1P_2 - P_2^2}{\beta^2},$$

$$c_1 = \frac{2P_1P_2}{\beta} + 2(\beta - 1)c_0,$$

$$c_2 = -P_1^2 + (\beta - 1)(c_0 - c_1).$$

The parameters P_1 and P_2 are written as:

$$P_1 = j_e V_w / h \quad P_2 = 2\eta/\rho^* h.$$

Here β is the ratio of the velocity of the particles the velocity of the solvent as already mentioned above, j_e is the evaporation flow, V_w is the volume of a water molecule and ρ^* is the effective mass density of the contact line.

The results obtained from our experiments do support Adachi's model in a qualitative way as follows:

(i) The stripe size is increasing by increasing the film height.
(ii) The stripe size is increasing by decreasing the apparent wetting angle.
(iii) The stripe size is increasing by increasing the meniscus height which also means by increasing the surface tension.
(iv) The stripe size is increasing by increasing the volume fraction.
(v) The stripe size in increased by increasing beta (decreasing the interaction).

In summary, we can state that a periodic pattern can be obtained by homogenizing the suspension and keeping the evaporation rate constant. We propose that the stripe pattern will be removed by decreasing the evaporation rate (growth velocity) and by creating a long meniscus with low curvature.

Acknowledgement The authors thank Thomas Palberg for helpful discussions, and the MWFZ Mainz as well as the DFG SFB TR6 D1 for financial support.

References

1. Imhof A (2003) Three-dimensional photonic crystals made from colloids. In: Liz-Marzan LM, Kamat PV (eds) Nanoscale Materials. Kluwer Academic, Boston, MA, pp 423–454
2. Lin SY et al. (1998) Nature 394:251
3. Hayashi S, Kumamoto Y, Suzuki T, Hirai T (1991) J Colloid Interf Sci 144:538
4. Petrov EP, Bogomolov VN, Kalosha II, Gaponenko SV (1998) Phys Rev Lett 81:77
5. Yablonovitch E (1987) Phys Rev Lett 58:2059
6. John S (1987) Phys Rev Lett 58:2486
7. Joannopoulos JD, Meade RD, Winn JN (1995) Photonic crystals: Molding the flow of light. Princeton University Press, New Jersey
8. Renak ML, Bazan GC, Roitman D (1997) Adv Mater 9:392
9. Persson SHM, Dyreklev P, Inganäs O (1996) Adv Mater 8:405
10. Kumar A, Abbott NL, Kim E, Biebuyck HA, Whitesides GM (1995) Acc Chem Res 28:219
11. Vlasov YA, Astratov VN, Baryshev AV, Kaplyanskii AA, Karimov OZ, Limonov MF (2000) Phys Rev E 61:5784
12. Van Blaaderen A, Ruel R, Wiltzius P (1997) Nature (London) 385:321
13. Bunzendahl M, Lee Van Schaick P, Conroy JFT, Daitch CE, Norris PM (2001) Colloids Surf A 182:275
14. Micheletto R, Fukuda H, Ohtsu M (1995) Langmuir 11:3333
15. Yin Y, Xia Y (2002) Adv Mater 14:605
16. Masuda Y, Itoh M, Koumoto K (2005) Langmuir 21:4478
17. Hayward RC, Saville DA, Aksay IA (2000) Nature 404:56
18. Dimitrov AS, Nagayama K (1996) Langmuir 12:1303
19. Goldenberg LM, Wagner J, Stumpe J, Paulke BR, Görnitz E (2002) Langmuir 18:3319
20. Im SH, Kim MH, Park OOK (2003) Chem Mater 15:1797
21. Masuda Y, Itoh M, Koumoto K (2003) Chem Lett 32:1016
22. Adachi E, Dimitrov AS, Nagayama K (1995) Langmuir 11:1057
23. Deegan RD, Bakajin O, Dupont TF, Huber G, Nagel SR, Witten TA (1997) Nature 389:827
24. Deegan RD, Bakajin O, Dupont TF, Huber G, Nagel SR, Witten TA (2000) Phys Rev E 62:756
25. Shmuylovich L, Shen AQ, Stone HA (2002) Langmuir 18:3441
26. Schöpe HJ, Barreira Fontecha A, König H, Marqués-Hueso J, Biehl R (2006) Langmuir 22:1828
27. Pieranski P, Strzelecki L, Pansu B (1983) J Phys (Paris) 44:531
28. Neser S et al. (1997) Progr Colloid Polym Sci 104:194
29. Barreira Fontecha A, Schöpe HJ, König H, Palberg T, Messina R, Löwen H (2005) J Phys Condens Matter 17:S2779
30. Decker EL, Garoff S (1997) Langmuir 13:6321

Progr Colloid Polym Sci (2008) 134: 57–65
DOI 10.1007/2882_2008_084

Published online: 16 July 2008

Elmar Bonaccurso

Microdrops Evaporating on AFM Cantilevers

Elmar Bonaccurso (✉)
Max-Planck Institute for Polymer Research, Ackermannweg 10, 55128 Mainz, Germany
e-mail: bonaccur@mpip-mainz.mpg.de

Abstract The kinetics of evaporation or drying of microscopic, sessile drops from solid surfaces is relevant in a variety of technological processes, such as printing, painting, and heat-transfer applications, besides being of fundamental interest. Drop evaporation has been commonly observed by means of video-microscope imaging, by ultra-precision weighing with electronic microbalances or with quartz crystal microbalances (QCM). Abundant information was gained over the years with these techniques, so that the evaporation of macroscopic drops of simple liquids from inert surfaces is nowadays well understood. The same techniques are, however, not applicable to microscopic drops. Furthermore they do not directly provide a measure of the interfacial stresses arising at the contact area between liquid and solid, which are known to play a key role in the evaporation kinetics of small drops.

Here I show how the use of atomic force microscope (AFM) cantilevers as sensitive stress, mass, and temperature sensors can be employed to monitor the evaporation of microdrops of water. Starting drop diameters are always below 100 μm. The foremost interest lies in exploring the last stages of the evaporation process.

Keywords Atomic force microscopy · Evaporation law · Microdrop evaporation · Micromechanical cantilevers · Surface tension · Vaporization heat · Young's equation

Introduction

Understanding the kinetics of evaporation or drying of microscopic, sessile drops from solid surfaces is a key factor in a variety of technological processes, such as: (i) printing [1–3] and painting [4]; (ii) heat-transfer applications, for example in the electronic industry to cool integrated circuits (ICs) and electronics chips [5–8], or even for fire fighting [9]; (iii) micro lithography, for example on polymer [10–13] or on biomaterial [14] surfaces. Such microscopic drops are primarily generated by spray nozzles and atomizers [15], or by inkjet devices and drop on demand generators [16]. The first type of apparatus are capable of simultaneously producing a big number of drops by one nozzle, but with a large size distribution and no control over the size (from below 1 to above 100 μm). The second type of apparatus are only capable of generating single consecutive drops by one nozzle, but monodisperse and with a good control over the size. In fact, the resolution of inkjet printers is steadily increasing as the size of the drops decreases. A commercial standard inkjet printer has nowadays a resolution of around 1200 dpi, which means that the drops have a diameter of around 20 μm and a volume of around 4 pl. If such a drop would be pure water, instead of a mixture of water and dye, and be deposited on a flat surface it would evaporate in less than 150 ms at NPT and at a RH of 50%. In comparison to inkjetted drops, rain drops have diameters between 1 and 2 mm, while drops in a fog have diameters below 10 μm [17], and correspondingly the evaporation times are longer or shorter.

Drop evaporation has been classically monitored by means of video-microscope imaging [18, 19], by ultra-precision weighing with electronic microbalances [20, 21] and with quartz crystal microbalances (QCM) [22]. Recently also atomic force microscope (AFM) cantilevers have been successfully employed for this purpose [23–25]. Using the first two, since long established techniques a wealth of information was gained and the evaporation of macroscopic drops of simple liquids from inert surfaces is now well understood. These techniques are, however, not sensitive enough to characterize microscopic drops. Furthermore they can not directly measure the interfacial stresses arising at the contact area between liquid and solid, which are known to play a key role in the evaporation kinetics of small drops [18, 20, 21, 26–28], nor are they capable of sensing the heat absorbed by the liquid during evaporation.

Here, I present evaporation studies of microscopic water drops on solid surfaces, and I will show how it is possible to simultaneously measure surface forces, the mass, and the vaporization heat of a drop.

Materials and Methods

Drop in Equilibrium

For the sake of simplicity, let us first consider a drop deposited onto a non-deformable and non-soluble substrate (Fig. 1). In equilibrium, i.e. when the drop is not evaporating, Young's equation must hold. It establishes the relation among the three surface tensions acting at the rim of the drop [29] (three phase contact line, TPCL):

$$\gamma_S - \gamma_{SL} = \gamma_L \cos\Theta \, . \tag{1}$$

γ_L is the surface tension at the interface liquid/gas, γ_S is the surface tension at the interface solid/gas, and γ_{SL} is the surface tension at the interface solid/liquid. Θ is called contact angle, or wetting angle, of the liquid on the solid. Equation 1 is valid in a strict sense only if the drop is not evaporating and if we neglect gravity. We can neglect gravity when the surface tension of the drop is high enough to counterbalance the effect of the hydrostatic pressure, which would flatten the drop. If the drop is spherical or not is thus a function of its size, and the balance of the two counteracting forces is expressed by the capillary length [29]

$$K = \sqrt{\frac{\gamma_L}{\rho g}} \, , \tag{2}$$

where ρ is the density of the liquid and g the gravitational acceleration. For water, $\gamma_L = 0.072\,\mathrm{N/m}$, $\rho = 1\,\mathrm{g/cm^3}$ and $g = 9.8\,\mathrm{m/s^2}$, thus the shape of the drop is not influenced by gravity if the radius of curvature is smaller than $\sim 2\,\mathrm{mm}$. This is always true for the experiments presented in this work, since drops smaller than 100 μm were always used. For such sizes the shape of the drop is thus determined solely by surface forces and it has the form of a spherical cap.

In addition to the surface tensions acting at the TPCL, another force plays a major role. Due to its curvature, the pressure inside the drop is higher than outside. The difference between in and out is called Laplace or capillary pressure [29]

$$\Delta P = \frac{2\gamma_L}{R} = \frac{2\gamma_L \sin\Theta}{a} \, , \tag{3}$$

where R is the radius of curvature of the drop, which is related to the contact radius a of the drop with the surface and to the contact angle Θ (Fig. 1).

As an example, for a drop of water forming a contact angle of 60° with a surface, the pressure difference is $\Delta P = 1.2\,\mathrm{mbar}$ when $a = 1\,\mathrm{mm}$, and $\Delta P = 1200\,\mathrm{mbar}$ when $a = 1\,\mu\mathrm{m}$. It is clear that huge forces are acting here.

Summarizing, one can say that when a small, non evaporating microdrop is sitting on a surface and forms a finite contact angle with it, the macroscopic picture is:

(i) The drop has a spherical shape;
(ii) Young's equation accounts for the in plane (horizontal) balance of forces at the TPCL;
(iii) The vertical component of the surface tension $\gamma_L \sin\Theta$ is pulling upwards at the TPCL and is counterbalanced by the Laplace pressure ΔP, which is pushing uniformly downwards over the whole contact area πa^2.

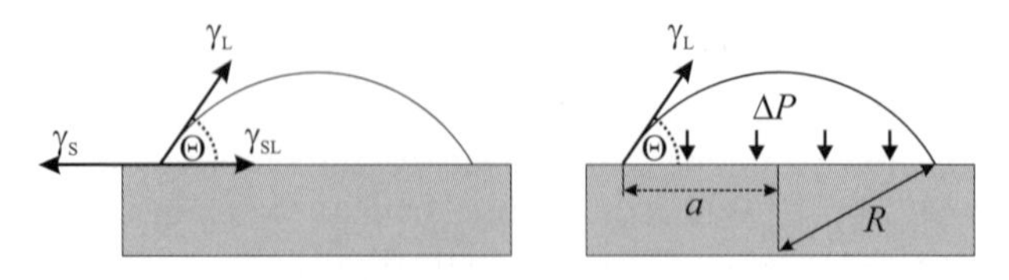

Fig. 1 *Left*: Sessile drop in equilibrium on a solid surface, with contact angle Θ and surface tensions γ_L, γ_S, and γ_{SL}. *Right*: Action of liquid surface tension γ_L and Laplace pressure ΔP

Evaporating Drop

Why do drops evaporate at all? A liquid (condensed phase) with a planar surface evaporates only when its vapor pressure P_0 is higher than the pressure of its vapor (gas phase) in its surroundings. Thus, if the surroundings are saturated with its vapor the liquid does not evaporate. It is in equilibrium, because at any time the number of molecules evaporating from and condensing to the surface is similar. However, drops have a slightly higher vapor pressure in comparison to a planar surface due to their curvature. For this reason, they evaporate also in a saturated atmosphere.

This is stated by the Kelvin equation [29]

$$P_V = P_0 e^{\lambda/R} , \tag{4}$$

where the vapor pressure of the liquid in the drop is P_V, and the parameter λ is a function of the temperature and the nature of the liquid. Thus the vapor pressure increases with decreasing drop size. As an example, a planar water surface has a vapor pressure $P_0 = 31.69$ mbar at NPT. If the surface is curved and the radius of curvature is $R = 1$ µm, the vapor pressure is $P_V = 31.72$ mbar, and if $R = 100$ nm, $P_V = 32.02$ mbar. The difference between the planar and the curved surfaces is small, but it is high enough for the drop to evaporate.

The evaporation law for microscopic drops was derived for drops of pure liquids, assuming $\Theta =$ const. and neglecting the cooling resulting from the vaporization, for two cases [30]:

(i) Drop in its saturated vapor:

$$V_L = V_{L0} - \alpha D P_0 t , \tag{5}$$

where V_L is the drop volume, V_{L0} is the initial drop volume, D is the diffusion coefficient of the molecules in the vapor, P_0 is the vapor pressure, and α is a known parameter which depends on the drop properties, on the contact angle, and on the temperature (for details see [30]). The equation contains no free parameters, and states that the volume of the drop decreases linearly with time.

(ii) Drop in non-saturated vapor:

$$V_L^{2/3} = V_{L0}^{2/3} - \beta D \varphi P_0 t , \tag{6}$$

where φP_0 , with $\varphi < 1$, represents the reduced vapor pressure, and β is a known parameter which depends on the drop properties, on the contact angle, and on the temperature. The equation contains no free parameters, and states that the volume of the drop to the power of 2/3 decreases linearly with time.

Both evaporation processes, in saturated and in non-saturated atmosphere, are "diffusion limited", i.e. the evaporation is limited by the diffusion of the liquid molecules through a saturated vapor layer around the drop.

We examined the evaporation of microdrops of water with different initial volumina on a silicon surface coated with a 30 nm thin fluoropolymer film (perfluoro-1,3-dimethylcyclohexane). The initial contact angle was $\Theta = 90°$ and remained constant for more than half of the evaporation time. During the experiments, the temperature ($T = 25$ °C) and the relative humidity (RH ~ 99%) were constant. We used a video microscope to track the dimensions of the evaporating drop from the side [23, 30].

Figure 2A shows the calculated evaporation times τ versus the initial drop volume V_0 (in double logarithmic scale), as calculated by Eq. 5 for RH = 99%. The similarity between the two upper lines, calculated for contact angles of $\Theta = 90$ and 30°, emphasizes that τ depends stronger on V_0 than on Θ. The slope of the curves is exactly 2/3. The three upper symbols represent evaporation times of microdrops with different initial volumina evaporating with a constant contact angle of $\Theta = 90°$ on the fluoropolymer surface. All other parameters are unchanged. The agreement with the calculated times is very satisfying, especially since we used no free parameters. The model is also applicable for smaller RHs, as shown for two microdrops (drops 4 and 5) evaporating at RH = 30%: the calculations yield respectively $\tau = 4.3$ s and 1.6 s, the measurements yield $\tau = 2.2$ s and 0.6 s. The evaporation time is strongly dependent on the vapor saturation: at RH = 30% a microdrop with a mass of 150 ng evaporates in $\tau = 4.3$ s, at RH = 99% in $\tau = 290$ s, and at RH = 100% in $\tau = 7000$ s. Experimentally, it is very difficult to set a constant RH = 100% for a prolonged time, that

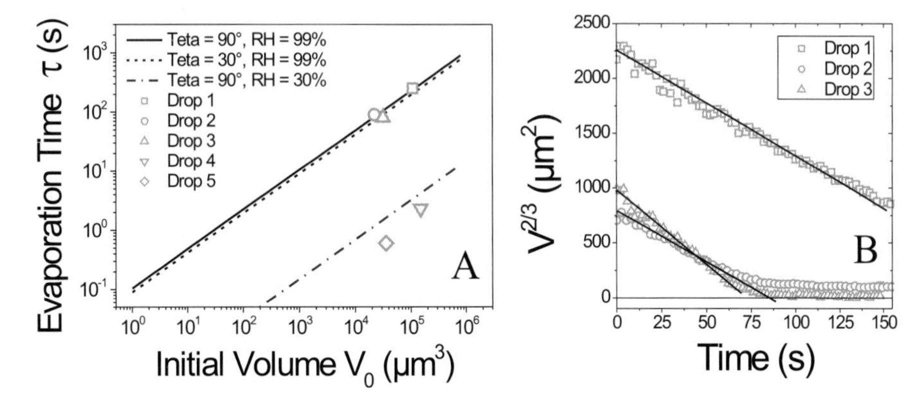

Fig. 2 **A** Calculated evaporation times τ versus initial drop volume V_0 for three cases: $\Theta = 90°$ and RH = 99% (*solid line*), $\Theta = 30°$ and RH = 99% (*dashed line*), and $\Theta = 90°$ and RH = 30% (*dashed-dotted line*). Corresponding experimental evaporation times (*open symbols*) of five drops on the fluoropolymer film, all with an initial contact angle $\Theta = 90°$. For the calculations $\alpha = 2.5 \times 10^{-17}$ m^2 s^2 kg^{-1} and $\beta = 1.2 \times 10^{-8}$ m s^2 kg^{-1} were used (for details see [30]). **B** $V^{2/3}$ versus time for drops 1, 2, and 3 of (**A**); *solid lines* serve only as guides to the eye

is why we were not yet able to conduct such experiments adequately.

Representing $V^{2/3}$ versus time (Fig. 2B), the experiments reveal that at the beginning of the process, when the three evaporating drops are still large, the dependence is linear with time (solid lines). At the end of the process there are deviations and the evaporation appears to slow down. This can be due, e.g. to the presence of solid impurities in the water, which get enriched as the drop evaporates so that the vapor pressure decreases. Or we simply meet the resolution limit of the optical technique. In order to track the evaporation until the end we introduced a new technique which allows to test the evaporation law also for extremely small drops.

Experimental Setup

When a drop is sitting on a surface, its surface tension pulls upwards at the TPCL, while the Laplace pressure pushes uniformly downwards. Can we use this as a sensor principle? If the solid substrate is rigid, after the drop is deposited not much happens after its equilibrium configuration is attained. However, if the substrate is thin enough the surface forces can cause its bending: the thinner the plate, the stronger the bending. This can be used as a sensor principle. A suited thin plate was found and a technique developed for measuring the degree of bending. We used silicon cantilevers, which look like microscopic diving boards (Micromotive GmbH, Mainz, Germany). Eigth of them are attached to one side of a silicon chip (Fig. 3A). Standard cantilever dimensions are: Length $l_0 = 750\ \mu m$, width $w = 90\ \mu m$, and thickness $d = 1\ \mu m$. Using an inkjet capillary, we deposit water microdrops onto the upper side of a cantilever, close to its base (Fig. 3B). The working principle is as follows (Fig. 3C): (i) before drop deposition no forces act on the cantilever, which is thus straight; (ii) upon drop deposition the cantilever bends upwards. This bending can be measured with the light lever technique, as done in AFM, where a laser beam is pointed at the free end of the cantilever; (iii) the measured signal is not the "actual bending", but the "inclination" dz/dx at the free end of the cantilever.

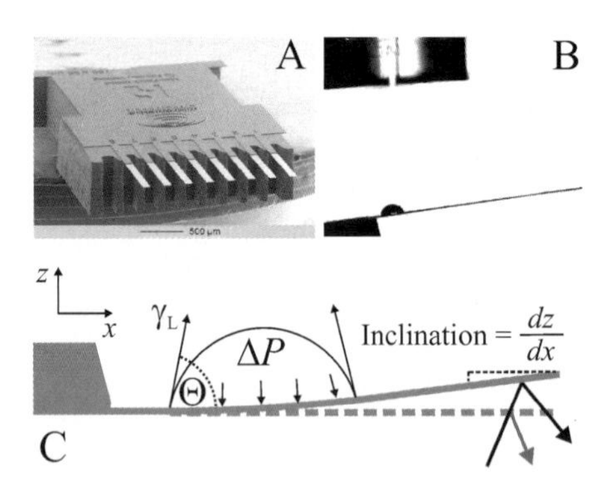

Fig. 3 **A** SEM image of a silicon chip with eight attached cantilevers. **B** Side view of a cantilever with drop deposited at its base and inkjet capillary used for drop generation. **C** Scheme of the equilibrium configuration of a drop on a cantilever: Θ is the equilibrium contact angle, γ_L is the liquid surface tension, ΔP is the Laplace pressure

Results and Discussion

Evaporation Curve

A typical evaporation curve of a water microdrop on a silicon cantilever, acquired at NPT and RH $\sim 30\%$, is a plot of the inclination of the cantilever versus time (Fig. 4A). At the same time, the contact angle $\Theta$and the contact radius a are recorded with a video microscope from the side (Fig. 4B and C). The water microdrop is deposited onto the cantilever at $t = 0$ with the inkjet device, and immediately starts evaporating. The evaporation is over after ~ 0.6 s, as the cantilever's inclination returns to its initial value. In the contact angle and contact radius curves, the black lines are simply guides to the eye. They show that two evaporation modi take place: at the beginning, the drop evaporates in the constant-contact-radius (CCR) mode, and after ~ 0.3 s both, Θ and a, decrease linearly with time. Plots of V and of $V^{2/3}$ versus time demonstrate the agreement with the evaporation law derived in Eq. 6 for a drop evaporating in non-saturated vapor (Fig. 4D).

At this point a model is needed to analyze the acquired inclination data and relate it to surface forces, drop shape, and cantilever properties.

Force Model

As first, for a simplified treatment, we make some assumptions on the drop and on the cantilever:

(i) The drop is in thermodynamic equilibrium, i.e. it does not evaporate, during the acquisition of a single data point. This is verified, since the acquisition time is typically < 1 ms.

(ii) Beam theory (also called elastic theory) is used for modeling the cantilever [31, 32]. For it to apply, $l_0 \gg w, d$, which means that the bending of the cantilever is considered to be one dimensional ($z = f(x)$) and that the transversal cross sections are flat. The Poisson's ratio is zero ($\upsilon = 0$) in beam theory, which means that during its deformation the volume of the cantilever is not conserved.

The inclination at the end of the cantilever, which is given by the overall balance of forces acting on it, is then [23].

$$\frac{dz}{dx} = \frac{3\pi a^3}{Ewd^3}\left[\gamma_L \sin\Theta + \frac{2d}{a}\left(\gamma_L \cos\Theta - \gamma_S + \gamma_{SL}\right)\right], \tag{7}$$

where E is Young's modulus of the cantilever material, while all other parameters have been introduced before.

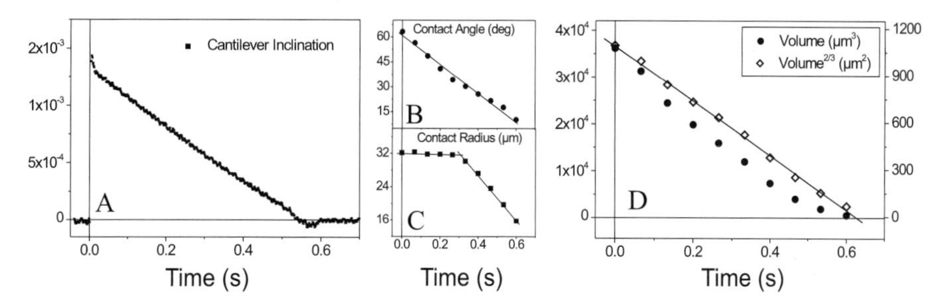

Fig. 4 $T \sim 25\,°\text{C}$, RH $\sim 30\%$; drop data: water, $a = 32\,\mu\text{m}$, $\Theta = 63°$, $\gamma_L = 0.072\,\text{N/m}$, $m_0 = 36\,\text{ng}$. Cantilever inclination (**A**), contact angle (**B**), contact radius (**C**), and volume and volume$^{2/3}$ versus time (**D**). *Solid lines* are guides to the eye

The first term contains the vertical contribution of the surface tension and the Laplace pressure, while the second term is basically Young's equation. It is significant to note here that all parameters in this equation are known, i.e. there are no fit parameters.

The agreement between the evaporation curve (Fig. 4A) and the curve calculated according to Eq. 7 would be satisfying, especially considering that no fitting parameters were used and that the cantilever model is a simplification (Fig. 5). It appeared, however, that all curves acquired in similar experimental conditions lie systematically below the calculated curves. Thus the model needs a refinement. According to the simplifications made, mainly three issues can cause this deviation:

(i) We used beam theory and Poisson's ratio was zero ($v = 0$).
(ii) The cantilever also has a lateral extension. Therefore a drop causes also a transversal bending of the cantilever, which increases its effective stiffness.
(iii) The cantilever is clamped at the base. That edge can thus not be deformed, which increases its effective stiffness.

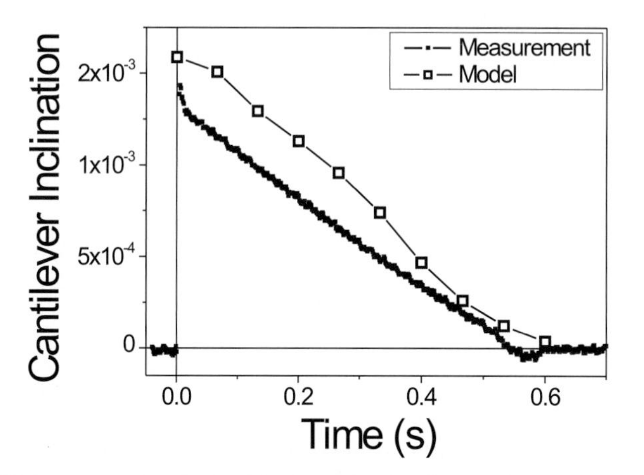

Fig. 5 $T \sim 25\,°\text{C}$, RH $\sim 30\%$; drop data: water, $a = 32\,\mu\text{m}$, $\Theta = 63°$, $\gamma_L = 0.072\,\text{N/m}$, $m_0 = 36\,\text{ng}$; cantilever data: silicon, $E = 180\,\text{GPa}$, $l_0 = 500\,\mu\text{m}$, $w = 100\,\mu\text{m}$, $d = 0.9\,\mu\text{m}$. Experimental and calculated curves in comparison

We implemented a more sophisticated 3D finite element (FE) model of the cantilever taking into account these three issues [24]. We used a Young's modulus $E = 180\,\text{GPa}$ and a Poisson's ratio $v = 0.26$, which are standard values for silicon. For analyzing the mechanical equilibrium deformation of a cantilever caused by a microdrop, we deposited a non-evaporating ionic liquid (1-butyl-3-methylimidazolium hexafluorophosphate from Merck KGaA, Darmstadt, Germany) onto a silicon cantilever. We imaged the face of the cantilever opposite to the one where the drop was sitting with a confocal profilometer (µSurf from NanoFocus AG, Oberhausen, Germany) and obtained a 3D image of the bent cantilever (Fig. 6). The upper half of the image represents the measurement, the lower half the simulation, the white circle the position of the drop. Due to the symmetry of the drop-cantilever system around the longitudinal axis, we simulate only half of the cantilever. The agreement between simulation and measurements is very satisfactory. In the lower left graph three transversal profiles of the cantilever, taken at the three places indicated by the arrows, are shown. Modeling this bending is beyond the capabilities of the analytic model. In the lower right graph the longitudinal profiles, experimental and simulated, are shown. The overall relative error is below 6% in this case, and below 10% on the average. This is an extremely good agreement, especially considering that all parameters of the simulation are fixed: they are determined from independent measurements (cantilever and drops dimensions) or are known from literature (silicon and ionic liquid properties).

From further parameter studies it resulted that the transversal bending and the clamping at the base cause negligible effects, so that the Poisson's ratio appears to be the dominating factor that causes the discrepancy between the analytic model and the experimental results [24].

Negative Inclination

Evaporation curves recorded with water drops deposited on clean hydrophilic cantilevers systematically differ from curves recorded with drops on clean hydrophobic cantilevers. Inclination and evaporation time have been nor-

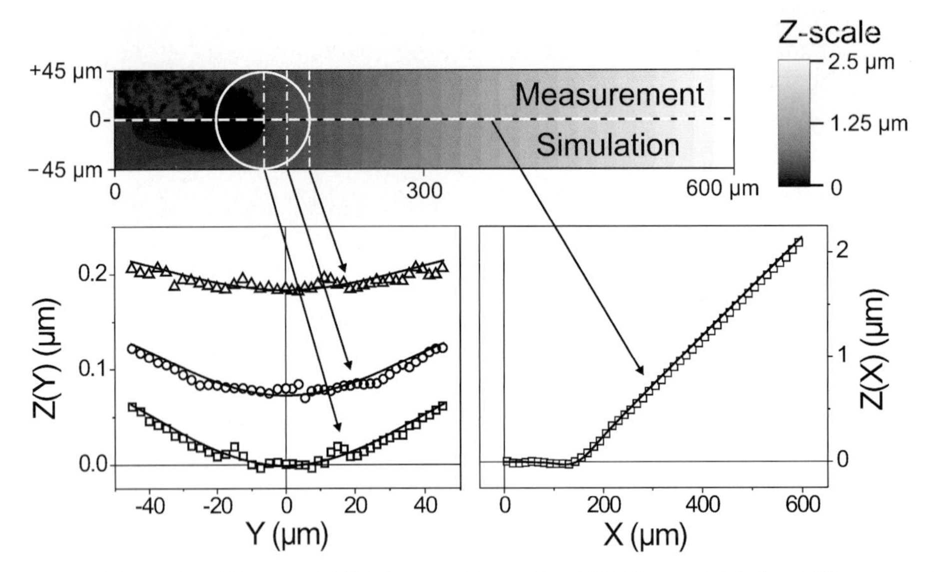

Fig. 6 Ionic liquid drop: $a = 45$ μm, $\Theta = 64°$, $\gamma_L = 0.045$ N/m; cantilever: $E = 180$ GPa, $v = 0.26$, $l_0 = 600$ μm, $w = 90$ μm, $d = 0.7$ μm. *Solid lines* represent simulations, *open symbols* experimental data points

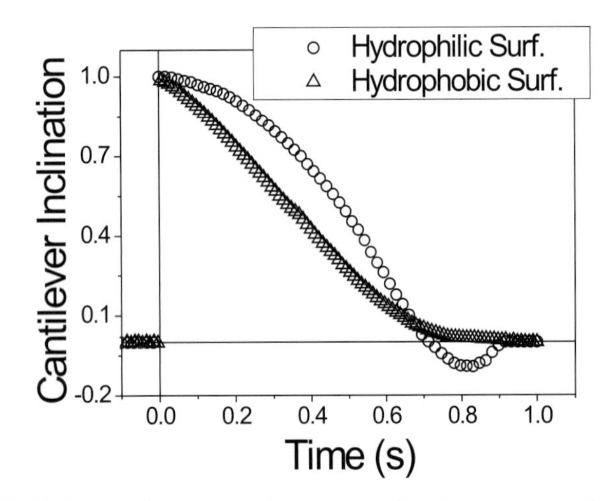

Fig. 7 Evaporation curves of two water microdrops on a cantilever with a hydrophilic (*circles*) and a hydrophobic (*triangles*) surface. $T \sim 25$ °C, RH $\sim 30\%$. Hydrophilic case: $a = 37$ μm, $\Theta = 70°$, $m_0 = 66$ ng. Hydrophobic case: $a = 34$ μm, $\Theta = 116°$, $m_0 = 187$ ng

malized in order to compare two measurements (Fig. 7). Two differences catch the eye:

(i) The shape of the two curves is different and the "hydrophilic curve" lies above the "hydrophobic curve" at the beginning of the evaporation.

(ii) The inclination becomes negative at the end of the "hydrophilic curve". It is not an artifact and is reproducible.

The first difference can be easily explained by the different initial drop volumina and by the different evaporation modi of the two drops. In fact, on the hydrophilic cantilever, the TPCL of the drop remained pinned for almost the entire evaporation, as far as we could conclude from video microscope images, while the contact angle of the drop gradually decreased. This corresponds to the CCR evaporation mode. Conversely, on the hydrophobic cantilever, both, contact radius and contact angle, changed during evaporation. The first part of both curves can be described by the model proposed above.

The negative inclination, on the other hand, can not be explained by the model, because Eq. 7 contains only positive terms:

(i) The first is the contribution from the Laplace pressure ΔP and from the vertical component of the surface tension $\gamma_L \sin\Theta$;

(ii) the second is basically Young's equation, which should be zero under the assumption that the drop is in thermodynamic quasi-equilibrium.

It is thus necessary to introduce an additional term, which must be negative and which describes the mechanical stress applied to the cantilever towards the end of evaporation of the drop. We call it $\Delta\sigma$, and it is not constant

$$\frac{dz}{dx} = \frac{3\pi a^3}{Ewd^3} \left[\gamma_L \sin\Theta + \frac{2d}{a}(\gamma_L \cos\Theta - \gamma_{SL} + \gamma_{SL}) + \Delta\sigma\right] . \tag{8}$$

This additional stress is for now just a free parameter, which helps to describe the experimental curves. We try to give a tentative explanation as to its physical origin.

(i) When the contact angle is "large enough", the drop pulls at the TPCL and the cantilever bends upwards (Fig. 8). The first term of the equation, $\gamma_L \sin\Theta$, dominates over the second term, $\gamma_L \cos\Theta - \gamma_{SL} + \gamma_{SL}$,

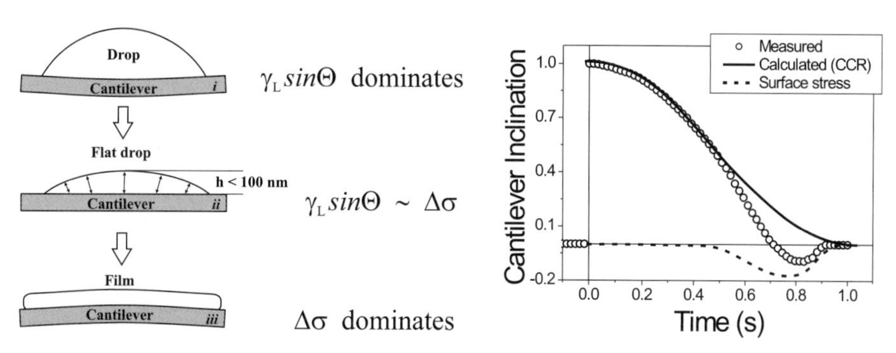

Fig. 8 *Left*: schematic of the model to explain the negative cantilever inclination, and range of action of the forces involved. *Right*: experimental evaporation curve (*circles*), calculated curve assuming CCR evaporation (*solid line*), and calculated surface stress (*dashed line*)

which is zero under the assumption of Young's equation being valid. The third term, $\Delta\sigma$, must also be zero.

(ii) When the drop becomes thinner than some 100 nm, the disjoining pressure inside the flat drop starts to play a role, favoring the formation of a thin film and acting to stabilize it. At this stage, ΔP, γ_L, and $\Delta\sigma$ have a similar magnitude and their effects on the bending of the cantilever cancel out each other. The cantilever crosses the "zero inclination" axis for the first time.

(iii) When the remainings of the drop wet the surface and form a thin film, which could span a larger area than the original drop contact area, $\Delta\sigma$ dominates over the other two terms, which become vanishingly small. The interfacial tension between film and cantilever is smaller than between air and cantilever, so that the cantilever bends away from the drop. The measured signal is negative.

It may also be possible to determine $\Delta\sigma$ experimentally from the inclination curve (Fig. 8, right). In fact, from video images we know that the drop evaporates in the CCR mode. We assume that the CCR mode holds until the end (last 100 ms). From the contact radius and contact angle data we calculate the inclination as if only ΔP and γ_L were acting on the cantilever (solid line). Subtracting these two curves yields the curve of $\Delta\sigma$ (dashed line). So, although we managed to measure it, we still have to find a quantitative model for $\Delta\sigma$.

In summary, using cantilevers as sensors we have discovered an effect arising with microscopic, pinned drops which to the best of our knowledge was never observed using other methods. Our tentative explanation is that a thin liquid film wets the surface, reduces the surface tension on the top side, and causes the cantilever to bend towards the bottom side.

Mass and Inclination

Surface forces exerted by the drop on the cantilever cause its bending. With the light lever technique we can measure the resulting inclination at the end of the cantilever. We call this signal "static", since it changes continuously, but slowly, during the evaporation of the drop which takes usually one second. Additionally to this use as surface stress sensors, cantilevers can also be employed as microbalances. Cantilevers are harmonic oscillators, whose resonance frequency depends, among other parameters, on their own mass and on their load. We can thus record the change of the resonance frequency caused by the mass change of the evaporating drop. We call this signal "dynamic" because a cantilever oscillation takes less than a millisecond.

Assuming that Young's equation is valid at all times, that the additional surface stress $\Delta\sigma$ is negligible, and that the evaporation takes place in the constant contact angle (CCA) mode (all three issues are practically achieved using a hydrophobized cantilever), the "static" equation simplifies to

$$\frac{dz}{dx} \approx \frac{3\pi a^3}{Ewd^3}\gamma \sin\Theta \,, \tag{9}$$

while the "dynamic" equation states that the mass added to the cantilever (load) is inversely proportional to the resonance frequency squared [25, 33].

$$m \propto \frac{1}{(2\pi f)^2} \,. \tag{10}$$

The two signals, inclination and resonance frequency, are acquired simultaneously, but are independent of each other. They yield the stress and the mass (Fig. 9A). We deposited a water drop on a silicon cantilever hydrophobized with a monolayer of hexamethyldisilazane (HMDS). The initial contact angle was $\sim 80°$. It decreased nearly linearly during evaporation, and was $\sim 70°$ at the end. The initial contact radius was ~ 33 μm, and decreased nearly linearly during evaporation. At the end it was below 10 μm. At present, we can record the inclination curve with a temporal resolution of ~ 0.1 ms between data points, and the frequency curve with ~ 5 ms. The mass calculated from the resonance frequency of the cantilever and from video microscope images is similar (Fig. 9B), although the time resolution (~ 5 ms) and the sensitivity (~ 50 pg) is much

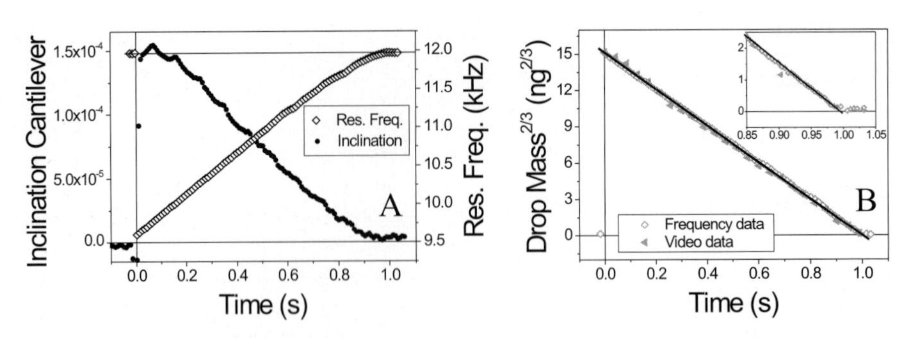

Fig. 9 **A** Simultaneously acquired inclination (*full circles*) and resonance frequency (*hollow diamonds*) of a cantilever versus time upon evaporation of a water microdrop on a hydrophobized silicon cantilever. $T \sim 25\,°\text{C}$, RH $\sim 30\%$. Drop data: $a = 33\,\mu\text{m}$, $\Theta = 80°$, $\gamma_L = 0.072\,\text{N/m}$, $m_0 = 60\,\text{ng}$. Cantilever data: $l_0 = 500\,\mu\text{m}$, $w = 90\,\mu\text{m}$, $d = 2.1\,\mu\text{m}$. **B** Drop mass$^{2/3}$ versus time from frequency (*hollow diamonds*) and video (*solid triangles*) data, and linear fit (*solid line*)

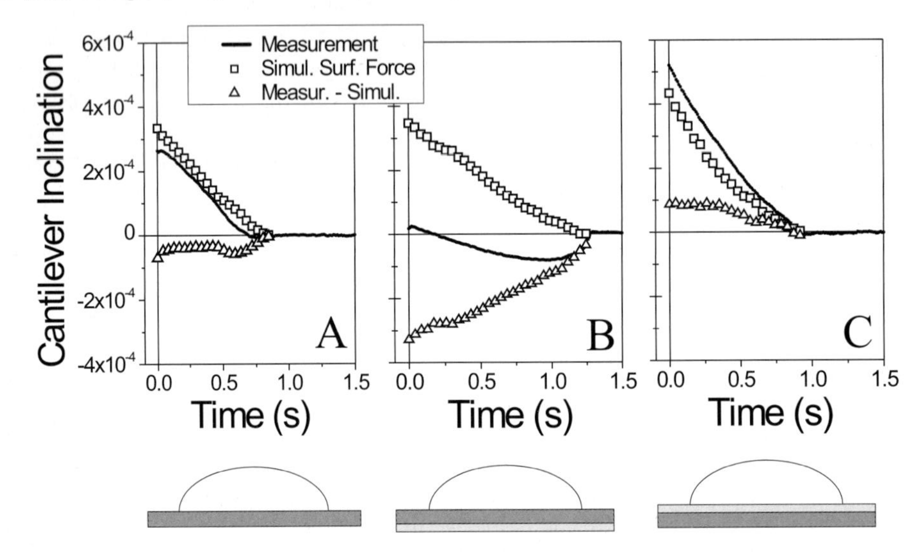

Fig. 10 Experimental (*solid line*), simulated (*hollow squares*), and difference (*hollow triangles*) inclination of silicon cantilevers versus time upon evaporation of water microdrops. $T \sim 25\,°\text{C}$, RH $\sim 30\%$. Drop data: various initial volumes, contact radii, and contact angles; $\gamma_L = 0.072\,\text{N/m}$. Cantilever data: $l_0 = 750\,\mu\text{m}$, $w = 90\,\mu\text{m}$, $d_A = 1.8\,\mu\text{m}$, $d_B = 1.5\,\mu\text{m}$, and $d_C = 1.7\,\mu\text{m}$; gold layer thickness $= 30\,\text{nm}$; Young's moduli: $E_{Si} = 180\,\text{GPa}$, $E_{Au} = 78\,\text{GPa}$; Poisson's ratios: $\upsilon_{Si} = 0.26$, $\upsilon_{Au} = 0.44$

higher for the frequency-derived mass. This offers the possibility of studying drop evaporation closer to its end (see inset). If we plot the mass as $m^{2/3}$ versus time we see that the evaporation law "$V^{2/3}$ linear versus time" valid for macroscopic drops can be extended to microscopic drops and is valid until the evaporation end.

In summary, using cantilevers as drop evaporation sensors we can simultaneously record the mass of the drop by tracking the resonance frequency of the cantilever, and the surface forces of the drop by tracking the inclination of the cantilever.

Vaporization Heat

Upon evaporation, a drop absorbs heat from its surroundings (air and cantilever). A drop of, e.g., water cools down by 1–2 °C [34], while benzene cools down by 15–20 °C [35]. The processes involved in heat dissipation in this case are mainly conduction and convection. If the cantilever is made of pure, crystalline silicon, the cooling does not affect its bending, as shown in the left graph in Fig. 10 [36]. The cantilever inclination measured upon deposition of a microdrop of water, which is the result of the combined action of surface forces and thermal effect, and the inclination simulated taking into account the surface forces only, from Eq. 7, are in good agreement. The difference between the two curves is smaller than 10% over all the evaporation. However, if the cantilever is, e.g., gold coated on one of its sides it behaves like a bimetal, since silicon and gold have different linear thermal expansion coefficients of respectively $\alpha_{Si} = 2.6 \times 10^{-6}\,\text{K}^{-1}$ and $\alpha_{Au} = 14.2 \times 10^{-6}\,\text{K}^{-1}$. The direction of the bending depends on which side the gold layer is with respect to the drop. The thermal effect alone (experimental inclination minus inclination simulated for surface forces) would cause the cantilever to bend downwards (negative inclination) if the gold layer is on the bottom (Fig. 10B), or upwards (positive inclination) if the gold layer is on the top (Fig. 10C). The three water drops used here had different initial volumina, and the three cantilevers had different thickness. The evaporation times and the inclinations are therefore not directly comparable. However, contact angle and radius of each drop were recorded during evaporation, and the cantilever properties are known. With these data we performed the simulations.

In summary, with the proper choice of the cantilever coating it is possible to sense the temperature of the evaporating drop, along with its mass and the effect of its surface tension.

Conclusions

I showed how the use of atomic force microscope (AFM) cantilevers as sensitive stress, mass, and temperature sensors can be employed to monitor the evaporation of microdrops of water. This technique has some advantages with respect to state-of-the-art techniques, like video microscopy, electronic or quartz crystal microbalances, since it allows measuring more drop parameters simultaneously for smaller drops sizes. This technique allowed detecting a difference between water microdrops evaporating from clean hydrophilic and hydrophobic surfaces. The difference is especially manifest close to the end of evaporation, and evidences arise for the formation of a thin water film on the hydrophilic surface. This is not the case for the hydrophobic surface. A tentative explanation is provided.

Acknowledgement Grateful acknowledgments are expressed to Dmytro Golovko, Ramon Pericet-Camara, Roberto Raiteri, Paolo Bonanno, Fabrizio Stefani, Karlheinz Graf, Thomas Haschke, Wolfgang Wiechert, and Hans-Jürgen Butt for fruitful discussions. I further express thanks to the board of the German Colloid Society for the assignment of the Richard-Zsigmondy Award.

References

1. Heilmann J, Lindqvist U (2000) J Imaging Sci Technol 44:491–494
2. Socol Y, Berenstein L, Melamed O, Zaban A, Nitzan B (2004) J Imaging Sci Technol 48:15–21
3. Kim EK, Ekerdt JG, Willson CG (2005) J Vac Sci Technol B 23:1515–1520
4. Tullo AH (2002) Chem Eng News 80:27–30
5. Chou FC, Gong SC, Chung CR, Wang MW, Chang CY (2004) J Appl Phys Part 1 Japan 43:5609–5613
6. Fabbri M, Jiang SJ, Dhir VK (2005) J Heat Mass Transf 127:38–48
7. Amon CH, Yao SC, Wu CF, Hsieh CC (2005) J Heat Transf Transact Asme 127:66–75
8. Shedd TA (2007) Heat Transf Engin 28:87–92
9. www.fogtec-international.com
10. Kawase T, Sirringhaus H, Friend RH, Shimoda T (2001) Adv Mater 13:1601–1605
11. Bonaccurso E, Butt HJ, Hankeln B, Niesenhaus B, Graf K (2005) Appl Phys Lett 86:124101
12. De Gans BJ, Hoeppener S, Schubert US (2006) Adv Mater 18:910–914
13. Karabasheva S, Baluschev S, Graf K (2006) Appl Phys Lett 89:031110
14. Ionescu RE, Marks RS, Gheber LA (2003) Nano Lett 3:1639–1642
15. Lefebvre AH (1989) Atomization and Sprays. Taylor & Francis, London
16. Lee ER (2003) Microdrop Generation. CRC Press, Taylor & Francis, Boca Raton
17. Edwards BF, Wilder JW, Scime EE (2001) Eur J Phys 22:113–118
18. Bourges-Monnier C, Shanahan MER (1995) Langmuir 11:2820–2829
19. Rowan SM, Newton MI, McHale G (1995) J Phys Chem 99:13268–13271
20. Birdi KS, Vu DT, Winter A (1989) J Phys Chem 93:3702–3703
21. Picknett RG, Bexon R (1977) J Colloid Interface Sci 61:336350
22. Pham NT, McHale G, Newton MI, Carroll BJ, Rowan SM (2004) Langmuir 20:841–847
23. Bonaccurso E, Butt HJ (2005) J Phys Chem B 109:253–263
24. Haschke T, Bonaccurso E, Butt H-J, Lautenschlager D, Schönfeld F, Wiechert W (2006) J Micromech Microeng 16:2273–2280
25. Golovko DS, Haschke T, Wiechert W, Bonaccurso E (2007) Rev Sci Instrum 78:043705
26. Obrien RN, Saville P (1987) Langmuir 3:41–45
27. Cordeiro RM, Pakula T (2005) J Phys Chem B 109:4152–4161
28. Soolaman DM, Yu HZ (2005) J Phys Chem B 109:17967–17973
29. Atkins P, De Paula J (2002) Physical Chemistry. Oxford University Press, Oxford
30. Butt HJ, Golovko DS, Bonaccurso E (2007) J Phys Chem B 111:5277–5283
31. Gasch R, Knothe K (1989) Strukturdynamik II. Kontinua und ihre Diskretisierung. Springer, Berlin
32. Landau LD, Lifshitz EM (2002) Theory of Elasticity. Butterworth-Heinemann, Oxford
33. Cleveland JP, Manne S, Bocek D, Hansma PK (1993) Rev Sci Instrum 64:403–405
34. David S, Sefiane K, Tadrist L (2007) Colloids Surf A 298:108–114
35. Blinov VI, Dobrynina VV (1971) J Eng Phys Thermophys 21:229–237
36. Golovko DS, Bonanno P, Lorenzoni S, Stefani F, Raiteri R, Bonaccurso E (2008) J Micromech Microeng, in press

Progr Colloid Polym Sci (2008) 134: 66–73
DOI 10.1007/2882_2008_076

Published online: 20 May 2008

G. G. Badolato
F. Aguilar
H. P. Schuchmann
T. Sobisch
D. Lerche

Evaluation of Long Term Stability of Model Emulsions by Multisample Analytical Centrifugation

G. G. Badolato (✉) · F. Aguilar ·
H. P. Schuchmann
Institute of Process Engineering in Life Sciences, Section of Food Process Engineering, University of Karlsruhe (T.H.), Kaiserstr. 12, 76131 Karlsruhe, Germany
e-mail:
gabriela.badolato@lvt.uni-karlsruhe.de

T. Sobisch · D. Lerche
L.U.M. GmbH, Rudower Chaussee 29 (OWZ), 12489 Berlin, Germany

Abstract Emulsion-based products are found within the chemical and agrochemical, cosmetic, pharmaceutical, and food industries. As emulsion structures are thermodynamically unstable, shelf-life stability is a main aspect in product and process development. The objective of this work was to evaluate multisample analytical centrifugation with STEP-technology as an accelerated test for predicting the long-term stability of emulsions. Therefore, model emulsions were designed that exhibited creaming, coalescence, flocculation, and Ostwald ripening as the dominating instability mechanism.

The results obtained using analytical centrifugation were compared with those obtained by conventional methods: visualization using graduated cylinders, viscosity, and droplet size measurements. Generally the results using conventional methods are in good accordance with those obtained by multisample analytical centrifugation. Multisample analytical centrifugation can thus be successfully applied to assess the stability of emulsion formulations.

Keywords Accelerated stability test · Droplet size distribution · Emulsion stability · Multisample analytical centrifugation · Shelf life · Viscosity

Introduction

Many products in the chemical and agrochemical, cosmetic, pharmaceutical, and food industries are emulsion-based. Their internal structure is composed of one or more fluids, with one being finely dispersed as droplets within the other one. The size distribution of the droplets mainly influences characteristic product properties as color, texture, flow- and spreadability, viscosity, mouthfeel, shelf-life stability, and release of active ingredients. It therefore has to be maintained for the life-time of a product. Due to the extremely high interfacial area in these systems, this microstructure is thermodynamically unstable. By applying emulsifiers and thickeners, emulsions are kinetically stabilized for a certain amount of time. However, shelf-life stability always is a big challenge in emulsion product and process development. An emulsion is physically stable if no change in its droplet size distribution occurs independently of time and place [1]. Mechanisms affecting emulsion stability are creaming, coalescence, flocculation, Ostwald ripening, and phase inversion (Fig. 1), which can occur separately or simultaneously.

Density differences between the dispersed and continuous phase result in *creaming* or *sedimentation* of the droplets. The droplet concentration within the product becomes inhomogeneous but the individual droplets preserve their size. In emulsions of low viscosity and smaller droplet size ($< 0.7\ \mu m$), separation under earth gravity will be counteracted by Brownian motion. Creaming and sedimentation processes in very dilute emulsions (volume fraction $\varphi < 0.01$) can be described using Stokes' law

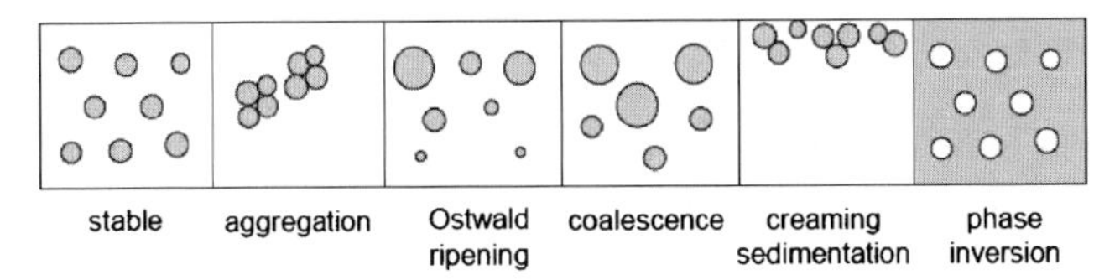

Fig. 1 Instability mechanisms that may occur in emulsions

(Eq. 1):

$$v_0 = \frac{(\rho_c - \rho_d)\, g x^2}{18 \eta_c} \qquad (1)$$

Where v_0 is the Stokes' velocity, ρ_c and ρ_d are the densities of the continuous and the dispersed phase, respectively, g is the gravitational force, x is the emulsion droplet size and η_c is the dynamic viscosity of the continuous phase. For higher dispersed phase concentrations one has to take into account hydrodynamic interactions between the droplets, which usually lead to a reduction in creaming velocity.

Aggregation, *flocculation*, and *coagulation* are terms used to describe the destabilization process when attractive forces interact between droplets only separated by a thin film of the continuous phase. For small interaction potentials, flocculation or aggregation will be reversed by rehomogenization similar to creaming and sedimentation.

During creaming/sedimentation and flocculation/aggregation/coagulation the original droplet size distribution of individual droplets is preserved. However, droplets also tend to grow, thus reducing the total interfacial energy stored within their interfaces. Thus the droplet size distribution will inevitably shift to larger sizes during longer storage (shelf-life). This may happen by *coalescence* when smaller droplets merge into larger ones resulting in a bigger mean droplet size or, in extreme cases, in total separation of the phases (break of emulsion structure) or by *Ostwald ripening* when larger droplets grow at the expense of the smaller ones due to diffusion of droplet phase molecules from smaller to bigger droplets, driven by their capillary pressure.

The kind and concentration of emulsifier molecules applied have a big influence on the coalescence process during the emulsification process itself and during prolonged shelf-life. In emulsification (droplet disruption processes), the droplets are disrupted into smaller ones. New interfacial area develops, being insufficiently covered by emulsifier molecules. Interfacial active molecules (emulsifiers) are transported by laminar and turbulent flow to the droplet subsurface, diffuse to the interface and adsorb and re-orientate at it. This stabilizes the new droplets formed. However, this process takes some time (milliseconds to minutes), depending on the emulsifier molecular structure. Droplets colliding with each other in the meantime will coalesce [2]. A detailed study about droplet coalescence within emulsification processes was performed by Danner (2001) [3].

Ostwald ripening is mediated by the diffusion of molecules of the dispersed phase through the continuous phase. Therefore, a determining aspect of Ostwald ripening is the solubility of the molecules of the dispersed phase in the continuous phase.

The rate of Ostwald ripening can be calculated using Eq. 2 [4]:

$$\frac{dx}{dt} = \frac{8 \nu_d^0 \gamma S^\infty D_d}{9RT} \,. \qquad (2)$$

Where ν_d^0 is the dispersed phase molar volume, γ is the interfacial tension, S^∞ is the dispersed phase solubility, D_d is the dispersed phase diffusion coefficient, t is the time, and T is the temperature.

In almost all cases, emulsions are not monodisperse and are often characterized by a wide droplet size distribution. Droplets of different sizes also have different capillary pressures. According to Eq. 3 the capillary pressure (p_{cap}) is increased with decreasing droplet size:

$$p_{cap} = \frac{4\gamma}{x} \,. \qquad (3)$$

On the other hand, the concentration of soluble components initially is the same in all droplets. Consequently there are no differences in the osmotic pressures. The osmotic pressure can be calculated using Eq. 4, where P_{osm} is the osmotic pressure, R is the universal gas constant and c is the concentration of dissolved components:

$$p_{osm} \approx RTc \,. \qquad (4)$$

Because of the higher capillary pressure of smaller droplets, they preferentially partition into the larger droplets. In the case of w/o emulsions, for example, if soluble substances are added to the aqueous dispersed phase (especially those of small molecular weight like salts or sugars) their concentration will increase with Ostwald ripening. Consequently the osmotic pressure in the smaller droplets will increase, counteracting the capillary pressure differences. Water diffusion from smaller to bigger droplets through the continuous phase will proceed until the system achieves a metastable equilibrium ($\Delta p_{cap} = \Delta p_{osm}$), when the emulsion will be stabilized against further Ostwald ripening.

Stability Tests Under Normal Storage Conditions

Under normal storage conditions (room temperature), the samples are stored on a shelf and are not submitted to any additional stress.

The easiest "real-time" stability test is visual observation of the product. Clear phase separation due to creaming or coalescence can thus be noted. Inhomogeneities of the emulsions can also be visualized by adding a dye into the dispersed phase [5]. Another possibility is the measurement of the emulsion characteristics, like droplet size distribution during storage time. Applying a photometer with high sensitivity, the measurement of emulsion turbidity or transmission in dependence of space and time [6–8] can also be used as stability test. Easy correlation is found, once it is proportional to disperse phase concentration and droplet size [9]. Creaming or sedimentation can be measured quantitatively using a graduated cylinder [10] or by measuring the emulsion electric conductivity at different positions of the emulsion container [11].

Tests under normal storage conditions can be efficient at predicting stability; however, they take too much time. This is especially the case when long term stability (over months to several years) is required. In order to solve this problem, accelerated stability tests are applied.

Accelerated Stability Tests

The simplest accelerated test consists of storing the emulsion at elevated temperature (at constant temperature; usually 45 or 50 °C, and/or using a temperature cycle). From the results, stability is usually predicted using empirical rules. For example, a stability of 3 months at 45 °C corresponds to at least 6 months stability at room temperature [11].

During a temperature cycle, emulsion stability can be evaluated using dynamic mechanical analysis (DMA) by means of viscosity measurements. Here, the sample is exposed to a small shear deformation with fixed amplitude and frequency and the complex dynamic shear modulus $G^* = G' + G''$ is determined. G' characterizes the elastic properties and G'' the viscous properties of the material. A change in the modulus from cycle to cycle indicates structural changes induced by the applied thermal stresses [12, 13].

Multisample analytical centrifugation is also used as an accelerated stability test. The principle of this method is similar to the photometrical measurement of turbidity, with the difference that the sample is centrifuged so that the measurements can be done in less time [14–18]. Creaming/sedimentation can be accelerated directly by using a centrifugal field instead of storing the samples. On the other hand, changes in the local distribution, structure, and droplet size distribution will change the creaming/sedimentation behavior. Therefore, analytical centrifugation should be able to trace destabilization independently of the underlying reason.

To evaluate the applicability of multisample analytical centrifugation as a tool to predict emulsion stability, model emulsions were designed with flocculation, coalescence, or Ostwald ripening as the dominating destabilization mechanism. The results of multisample analytical centrifugation were compared with those obtained during storage of the samples by measurement of viscosity and droplet size distribution. In addition, samples submitted to elevated storage temperatures and freeze–thaw cycling were analyzed by different methods.

Materials and Methods

For emulsion production, the components were first mixed using an agitator and then passed through a high pressure homogenizer (Microfluidizer, Microfluidcs Corp.; USA) at 500 bar of pressure. All the concentrations are given in % by mass.

Model Emulsion with Induced Flocculation

To study the effect of flocculation, an anionic emulsifier was used and a salt was added. An emulsion containing 40% silicon oil AK 100 (Wacker Chemie, Germany) as dispersed phase and 1% SDS as emulsifier was prepared. Deionized water or salt solution (NaCl) was added to this emulsion in order to obtain a final oil concentration of 30% and a salt concentration varying from 0 to 3000 mM.

Model Emulsions with Induced Coalescence

For inducing coalescence, o/w emulsions were prepared using the different emulsifiers Tween 20, Tween 80, and Tween 65. The emulsifiers were used at a concentration of 1% m/m (above the CMC). Vegetable oil (Floreal Haagen, Germany) was used as dispersed phase. The hydrophilic–lipophilic balance (HLB) values of each emulsifier, as well as the characteristics of the emulsions, are shown in Table 1.

Table 1 Emulsifiers used for stabilization of o/w emulsions against coalescence and emulsion characteristics

Emulsifier	HLB	Sauter diameter (μm)	Viscosity (mPa s^{-1})
Tween 20	16.7	0.66	3.7
Tween 65	10.5	1.35	3.1
Tween 80	15	0.62	3.1

Model Emulsions with Induced Ostwald Ripening

Ostwald ripening has a particular tendency to occur in w/o emulsions with an aqueous phase having a relatively high solubility in the continuous oil phase. Therefore, w/o emulsions with 10% disperse phase and a continuous phase containing clove and vegetable oil in different ratios were chosen. Clove oil has a distinctly higher solubility for water. As emulsifier, 10% of PGPR 90 (HLB around 1) was dissolved in the oil phase. Sucrose solution (1 M) was used as disperse phase in order to decrease the velocity of Ostwald ripening. The composition of the continuous phases and characteristics of the obtained emulsions are summarized in Table 2.

Table 2 Composition of the continuous phases of w/o emulsions to study Ostwald ripening and emulsion characteristics

Sample	Clove oil (%)	Vegetable oil (%)	Sauter diameter (μm)	Viscosity ($mPa\,s^{-1}$)
1	0	100	0.09	141.5
2	25	75	0.14	102
3	50	50	0.24	76
4	75	25	0.42	47

Visualization Tests, Droplet Size Analysis and Viscosity Measurements

Emulsions were stored at 4, 25, and 45 °C and also submitted to a number of freeze–thaw cycles (24 h freezing at −20 °C followed by 24 h thawing at room temperature). During the storage time, different methods were used to analyze emulsion stability. In the case of pronounced instability or destabilization during storage, the modification of the emulsion was also observed by the naked eye. For study of flocculation, emulsions were stored in graduated cylinders and the water separation was measured by visual observation [10].

Droplet size analysis was used for characterization of the emulsions just prepared and to detect alterations of emulsions during storage. The analysis was conducted using laser diffraction with polarization intensity differential scattering (PIDS) technology (Coulter LS 230, Beckman-Coulter, Krefeld, Germany).

Rheological measurements were realized in order to detect alterations on the emulsion structures during the storage time. They were carried out using a controlled stress rheometer (CSL-100 Rheometer, Carri-Med, Germany). The measurements were made at 20 °C. Shear stress varying from 0 to 2 $N\,m^{-2}$ was applied to each sample while the shear rate was measured.

Stability Tests Using Multisample Analytical Centrifugation

The multisample analytical centrifuge (LUMiFuge, LUM, Berlin Germany) used in this study employs the STEP technology, which allows measurement of the intensity of the transmitted light as a function of time and position over the entire sample length simultaneously. A schematic of the multisample analytical centrifuge is shown in Fig. 2.

Up to eight different samples can be analyzed at the same time at constant or variable centrifugal acceleration up to 2300 g. The separation behavior of the individual samples can be compared and analyzed in detail by tracing the variation in transmission at any part of the sample or by tracing the movement of any phase boundary. The variation of the transmission profiles contains the information on the nature, extent, and kinetics of the separation process(es) [17].

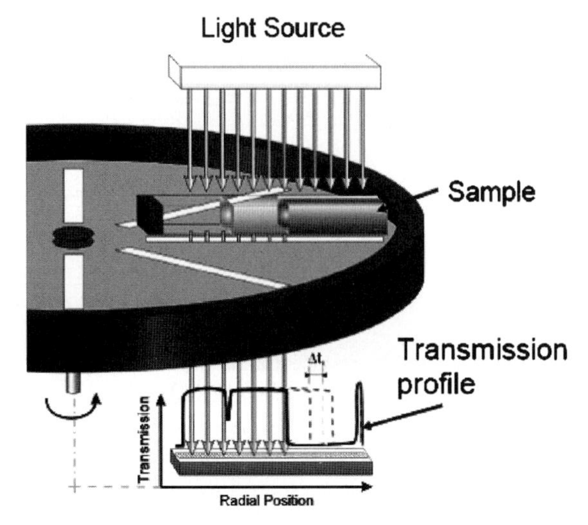

Fig. 2 Measurement scheme of the multisample analytical photocentrifuge. Parallel NIR-light is passed through the sample cells and the distribution of local transmission is recorded at preset time intervals over the entire sample length

All aging tests were conducted by storing the samples in the measuring cells and placing them directly into the centrifuge after various periods of storage or after a defined number of freeze–thaw cycles. Only the w/o emulsions were aged in separate containers, and samples were drawn after gentle rehomogenization because the emulsion with the highest clove oil content was difficult to handle due to phase separation.

Results and Discussion

Detection and Measurement of Flocculation Induced by Salt

No significant changes were detected in the droplet size distribution and viscosity of the emulsions with salt concentrations between 0 and 300 mM. An alteration of the droplet size distribution is not expected in weakly flocculated emulsions as the samples are agitated before the measurement, leading to redispersion of aggregates. All emulsions were very vulnerable to freeze–thaw cycling. Emulsions containing salt split after one cycle, whereas emulsions without salt split after two cycles.

Figure 3 shows the results of the observation (thickness of the separated layer of water during storage time) for the emulsions with high salt contents at normal storage conditions (under gravitational force 1 g) at 25 °C. As predictable by DLVO theory [19], the higher the salt content, the faster was the emulsion destabilization. The emulsion with 750 mM NaCl separated a water layer of about 20% within the first day of storage with only slight changes during the following days. A similar creaming profile was observed by Grotenhuis et al. 2003 [20] for emulsions containing casein micelles.

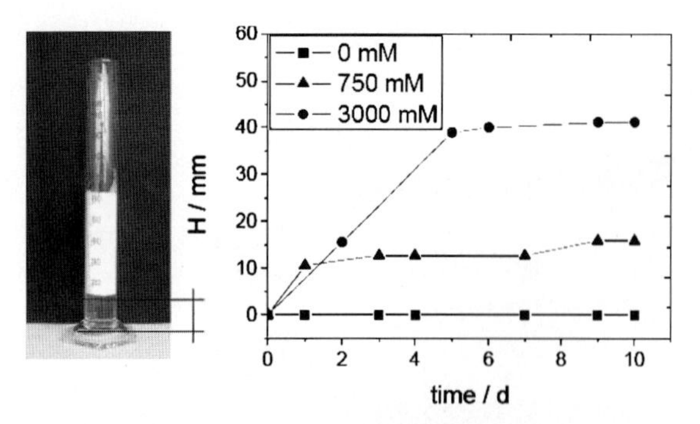

Fig. 3 Effect of electrolyte concentration on creaming velocity at 25 °C under normal gravitation conditions (1 g)

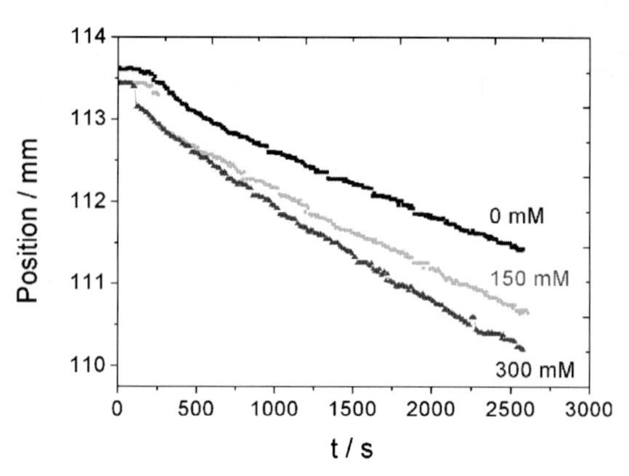

Fig. 5 Effect of electrolyte addition (flocculation with NaCl) on the stability of a silicon oil emulsion ($\varphi = 30\%$ m/m, emulsions, just prepared). Creaming kinetics at 1100 g displayed as movement of the boundary separating water and creaming emulsion

Multisample analytical centrifugation revealed a marked destabilization even for low electrolyte concentrations just after salt addition. Figure 4 shows the evolution of transmission profiles obtained during centrifugation of the o/w emulsion destabilized by addition of 300 mM NaCl. Creaming rates derived by software from the slope of the creaming curves (as shown in Fig. 5) were used to analyze and compare the stability of different emulsions during the storage time. After two weeks of storage visual separation had already occurred for 300 mM under gravity. Therefore, the sample was not analyzed further by centrifugation.

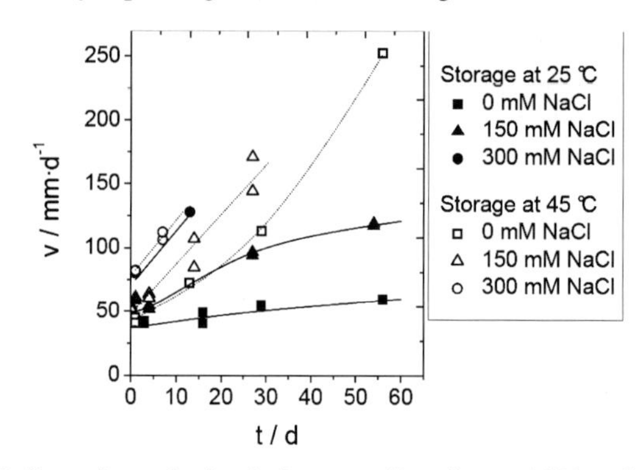

Fig. 6 Creaming velocity during centrifugation at 1100 g. Samples stored at 25 and 45 °C. The creaming velocity increases by increasing the amount of added salt, temperature, and duration of storage

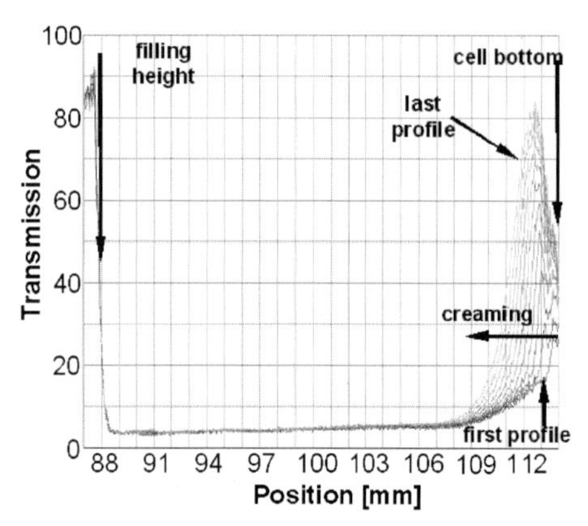

Fig. 4 Evolution of transmission profiles of an o/w emulsion ($\varphi = 30\%$ m/m) flocculated by addition of 300 mM NaCl during centrifugation at 1100 g for 43 min. First recorded profiles *bottom right*, last profiles *top right*

In Fig. 6 the creaming velocities at 1100 g are compared as a function of aging time during storage at 25 and 45 °C. As shown in Fig. 6 the creaming velocity for the emulsion without electrolyte increases only slightly during extended storage at 25 °C.

Stability and stability loss for the emulsion without electrolyte should be influenced by depletion flocculation by the anionic micelles [21]. As expected, creaming velocity increases by increasing the amount of added salt and increasing the storage temperature and duration. It should be mentioned that, due to the decreasing droplet concentration in the lower part of the samples during storage (lowering hindrance against creaming), creaming velocity is enhanced in addition to the effect of centrifugal acceleration, resulting in an increased sensitivity.

Detection and Measurement of Coalescence Influenced by Emulsifier Changes

Viscosity for emulsions prepared with Tween 65 and Tween 80 were identical. Viscosity for the emulsion with Tween 20 was slightly higher. Droplet size distribution was similar for emulsions with Tween 20 and Tween 80, and narrower and with smaller droplet Sauter diameter than for emulsions with Tween 65 (see Table 1). The bigger droplet size in the case of Tween 65 might be directly related to coalescence during emulsification. Due to its higher hydrophobicity and the high space requirement of the hydrophobic moiety, Tween 65 (polyoxyethylene-sorbitan tristearate) is probably less effective in stabilizing the droplets sterically after droplet break-up.

Under normal storage conditions, the emulsions with different emulsifiers showed no differences in emulsion stability over several weeks. No changes in droplet size and viscosity were observed during one month of storage independently of the storage temperature (25 and 45 °C). Emulsions with emulsifier Tween 65 exhibited a noticeable destabilization during freeze–thaw cycling. Freezing of the continuous water phase exerts pressure onto the droplets promoting coalescence [22]; however, no effect was seen in the case of emulsions containing Tween 20 and Tween 80.

However, multisample analytical centrifugation traced differences in the stability of all three emulsions just after preparation and destabilization during storage, even at 4 °C. The ranking of stability was proved by visual observation after 1 g-storage over several months. As expected from the differences in droplet size, emulsions stabilized with Tween 65 exhibited a considerably higher creaming rate during centrifugation at 1100 g compared to the other emulsions (Fig. 7). However, even at the beginning of storage, a distinctly higher creaming rate was also measured for the emulsion with Tween 80. Destabilization during storage increases with increasing temperature without changing the tendency for instability (Tween 65 > Tween 80 > Tween 20). As can be noted from Fig. 8, stability (creaming rate) changes only slightly

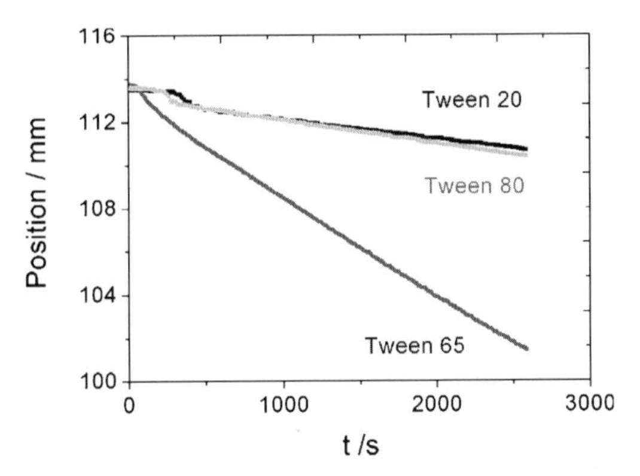

Fig. 7 Creaming kinetics of emulsions stabilized by different emulsifiers ($\varphi = 30\%$ m/m, emulsions, just prepared) in a centrifugal field at 1100 g displayed as movement of the boundary separating water and creaming emulsion

for emulsions stabilized with Tween 20 and 80, probably only by decreasing the droplet concentration in the lower part of the samples (hindrance against creaming). The pronounced destabilization with Tween 65 could be a combined effect of coalescence and decreasing droplet concentration due to creaming.

Detection and Measurement of Ostwald Ripening Influenced by Continuous Phase Composition

As seen in Table 2, by increasing the content of clove oil, the viscosity of the emulsion decreases and the emulsion droplet size increases. The same tendency for the relationship between continuous phase viscosity and resulting droplet size was observed by Tesch (2002) [23].

Differences in emulsion viscosities are principally due to the different viscosities of the emulsion continuous phase (oil phases). Although for some emulsions clear phase separation was visible to the naked eye, viscosity values revealed no significant changes during storage, even at higher temperature, due to the dominance of the continuous phase viscosity on the bulk viscosity.

Changes in droplet size distributions during storage are shown in Fig. 9 for sample 2. After 10 days of storage, the size distribution became bimodal. Smaller droplets than the original ones were observed. Furthermore, the fraction of big droplets increased. After 47 days, the fraction of small droplets apparently disappeared and the droplet size was shifted to higher values. One has to take into account that, during Ostwald ripening, the concentration of sugar in the smaller droplets increases, whereas it decreases in big droplets. This can cause an underestimation of the number of smaller droplets, since related changes in refractive index are neglected in the droplet size distribution calculation.

Increasing the clove oil content and storage temperature resulted in an increase of emulsion destabilization, as traced by droplet size and analytical centrifugation (Fig. 10).

Since Ostwald ripening is a diffusive process, it cannot be accelerated by centrifugation. Nevertheless, analytical centrifugation was able to trace emulsion destabilization

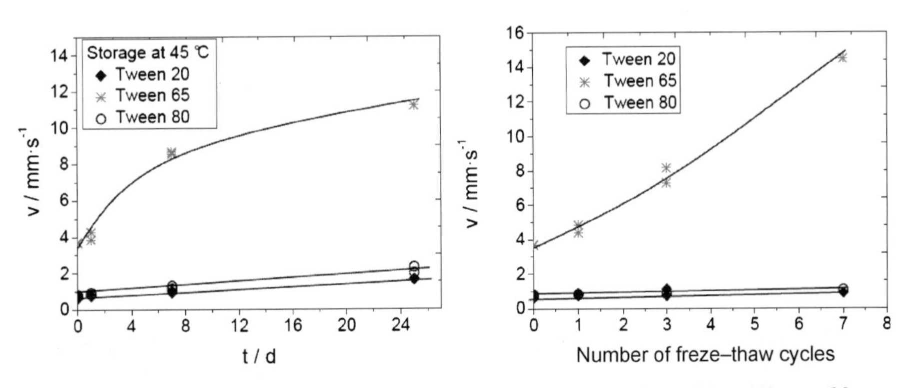

Fig. 8 Creaming during storage (*left*) and freeze–thaw (*right*). Emulsions stabilized by Tween 20 and Tween 80 are very stable, in contrast to emulsifier Tween 65

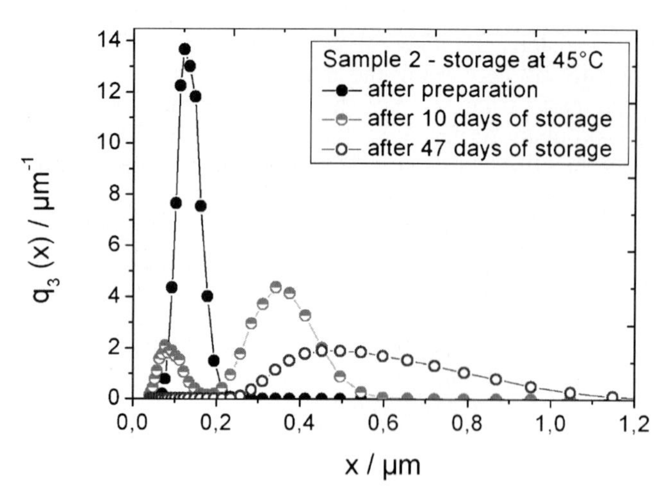

Fig. 9 Changes in droplet size distribution of sample 2 (25% m/m clove oil) during storage at 45 °C

by Ostwald ripening by taking measurements several times during storage, showing the same trend as the average droplet size. However, differences in stability due to sedimentation were also revealed because of the higher continuous phase viscosities of samples 1 and 2.

Conclusions

Fast prediction of emulsion stability is important for product development. As shown by this work, measurement of microstructure parameters such as droplet size distribution over storage time gives very precise information on starting and kinetics of instability and the mechanisms behind it. However, it takes time. Easy viscosity measurements (analysis of flow curves) were not sensitive enough to detect the first changes in the emulsion structure. Freeze–thaw cycles gave results that did not always correspond to shelf-life under "normal" conditions.

Multisample analytical centrifugation proved to be an efficient tool to predict emulsion instabilities due to flocculation, coalescence or Ostwald ripening. For emulsions tending to *flocculate*, results using 1 g-storage in graduated cylinders were also obtained by analytical centrifugation, which was faster and more sensitive. Flocculation could thus be detected even at very low salt concentrations. The influence of important parameters such as salt concentration and storage temperature was revealed by analytical centrifugation.

For emulsions tending to *coalesce* within the emulsification process (as induced by the use of different emulsifiers), multisample analytical centrifugation was able to trace differences in stability just after preparation and destabilization during storage even at 4 °C, even if no differences in the emulsion droplet size distribution and viscosity were observed during one month of storage independently of the storage temperature (25 and 45 °C). The ranking of stability was proved by visual observation after 1 g-storage over several months.

Ostwald ripening in emulsions was successfully followed, measuring droplet size distributions during storage. Being a diffusive process, Ostwald ripening can only be observed using multisample analytical centrifugation when the sample is measured several times during storage. The results obtained using analytical centrifugation were in accordance with those obtained by droplet size measurements. In addition, analytical centrifugation revealed differences in sedimentation stability due to differences in the continuous phase viscosity.

In summary, analytical centrifugation can be successfully used for screening purposes in emulsion product development. Different samples can be directly compared within a short time. To predict the shelf-life for emulsions aging not only by creaming or sedimentation, calibration with a known system has to be carried out.

Acknowledgement This study was supported by a grant from the Bundesministerium für Wirtschaft und Technologie (AIF PRO INNO II program, KA 0047301AWZ3).

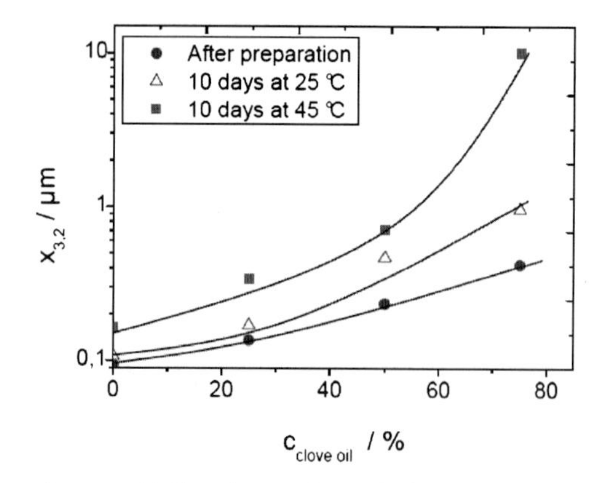

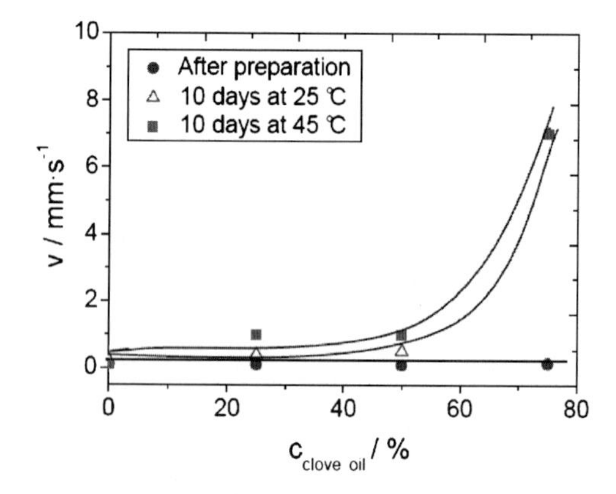

Fig. 10 Sauter diameter (*left*) and creaming velocity (*right*) of samples with different clove oil concentrations after preparation and during storage at 25 and 45 °C. Both diagrams reveal the same tendency for the different formulations concerning stability

References

1. Schubert H (ed) (2005) Emulgiertechnik, Grundlagen, Verfahren und Anwendung. Behr's, Hamburg (ISBN: 3-89947-086-9)
2. Kempa L, Schuchmann HP, Schubert H (2006) Chem Ing Tech 78/6:765
3. Danner T (2001) Tropfenkoaleszenz in Emulsionen. Dissertation, Institut für Lebensmittelverfahrenstechnik, Karlsruhe University. GCA, Waabs
4. Dickinson E (1992) An introduction to food colloids. Oxford, New York
5. Mühlebach S, Graf RB, Sommermeyer K (1987) Pharm Acta Helv 62:130
6. Oliczewski S, Daniels R (2004) Pulse laser – an innovative tool for the stability assessment of emulsions and creams. Proceedings international meeting on pharmaceutics, biopharmaceutics and pharmaceutical technology, Nuremberg, 15–18 March 2004. p 621
7. Formulaction (2003) Turbiscan. http://www.formulaction.com Last accessed: 23 April 2008
8. L.U.M. (2005) http://www.lum-gmbh.com Last accessed: 23 April 2008
9. Badolato GG, Freudig B, Idda P, Lambrich U, Schubert H, Schuchmann HP (2008) Membrane emulsification processes and characterisation methods. In: Güell C, Ferrando M, López F (eds): Monitoring and visualizing membrane-based processes. Wiley-VHC, Weinheim (in press) (ISBN: 3-527-32006-7)
10. Deutsches Institut für Normung (2002) ISO 6614: Bestimmung des Wasserabscheidevermögens von Mineralölen und synthetischen Flüssigkeiten. DIN, Berlin
11. Reng AK, Turowiski-Wanke A (2000) Stabilitätsprüfungen von Emulsionen und ihr Einsatz zur Entwicklung optimaler Emulgatorsysteme. Hochschulkurs Emulgiertechnik, Vortrag 20. GVT, Düsseldorf
12. André V, Willenbacher N, Fernandez P, Börger L, Debus H, Rieger J, Frechen T (2003) Cosmetics and toiletries manufacture worldwide. Aston Publishing Group, Florida, pp 102–109. Available at http://www.mvm.uni-karlsruhe.de/download/AME-No-Wi-reviewed-21.pdf Last accessed: 23 April 2008
13. Brummer R (2006) Rheology essentials of cosmetic and food emulsions. Springer, Berlin Heidelberg New York
14. Sobisch T, Lerche D (2000) Colloid Polym Sci 278:369
15. Sobisch T, Lerche D, Küchler S, Uhl A (2005) Rapid stability assessment for liquid and paste-like foods with multisample analytical centrifugation. In: Eklund T, Schwarz M, Steinhart H, Their HP, Winterhalter P (eds) Proceedings Euro Food Chem XIII: Macromolecules and their degradation products in food – physiological, analytical and technological aspects, Hamburg, 19–23 September 2005. Gesellschaft Deutscher Chemiker, Frankfurt, pp 672–675 (ISBN 3-936028-31-1)
16. Lerche D, Sobisch T, Detloff T (2006) Determination of stability, consolidation and particle size distribution of liquid or semi-liquid food products by multisample analytical centrifugation. In: Fischer P, Erni P, Windhab EJ (eds) ISFRS 2006, Proceedings 4th international symposium on food rheology and structure, Zürich, 19–23 Feb 2006. Kerschensteiner, Lapperdorf, pp 221–225 (ISBN 3-905609-25-8)
17. Sobisch T, Lerche D (2005) Rapid characterization of emulsions for emulsifier selection, quality control and evaluation of stability using multisample analytical centrifugation, SCI/RSC/SCS conference on cosmetics and colloids, London, 15 Feb 2005. http://www.soci.org/SCI/groups/col/2005/reports/pdf/gs3257_sob.pdf Last accessed: 23 April 2008
18. Küchler S, Schneider C, Lerche D, Sobisch T (2006) LabPlus Int 20/4:14
19. Schubert H (2005) Physikalisch-chemische Grundlagen der Stabilität von O/W- Emulsionen. In: Schubert H (ed) Emulgiertechnik, Grundlagen, Verfahren und Anwendung. Behr's, Hamburg, pp 207–231 (ISBN: 3-89947-086-9)
20. Grotenhuis E, Tuinier R, Kruif CG (2003) J Dairy Sci 86:764
21. Bibette J (1991) J Colloid Interf Sci 147:474
22. Sherman P (1968) General properties of emulsions and their constituents. In: Sherman P (ed) Emulsion science. Academic, London, p 131
23. Tesch S (2002) Charakterisieren mechanischer Emulgierverfahren: Herstellen und Stabilisieren von Tropfen als Teilschritte beim Formulieren von Emulsionen. Dissertation, Karlsruhe University. Shaker, Frankfurt (ISBN 3-8322-0194-4)

Progr Colloid Polym Sci (2008) 134: 74–79
DOI 10.1007/2882_2008_083

Published online: 16 July 2008

C. Oelschlaeger
N. Willenbacher
S. Neser

Multiple-Particle Tracking (MPT) Measurements of Heterogeneities in Acrylic Thickener Solutions

C. Oelschlaeger (✉) · N. Willenbacher
Institute of Mechanical Process Engineering and Mechanics, University Karlsruhe, Gotthard-Franz-Str. 3, 76128 Karlsruhe, Germany
e-mail: Claude.Oelschlaeger@mvm.uni-karlsruhe.de

S. Neser
Faculty of Mathematics and Science, University of Applied Sciences Darmstadt, Schöfferstrasse 3, 64295 Darmstadt, Germany

Abstract In this study the method of multiple-particle tracking (MPT) is used to quantify the degree of structural and mechanical microheterogeneity of two polymeric thickener solutions finally aiming at a better understanding of the contribution of microheterogeneities, which commonly occur in solutions of many synthetic as well as biopolymers, on bulk rheology. We have chosen the commercial polyacrylate ester Sterocoll FD and Sterocoll D (BASF Aktiengesellshaft) as model systems. For the Sterocoll FD solution the ensemble-averaged mean square displacement (MSD) is almost linear in time, as expected for such a weakly elastic fluid and relatively similar to that observed for a homogeneous aqueous glycerol solution, used as a reference system. However, the MSD distribution is broader than for the glycerol solution and their statistical analysis clearly reveals a heterogeneous structure on the μm length scale. For the Sterocoll D solution, the average MSD exhibits a subdiffusive behavior, typical for highly elastic solutions. Moreover, the displacements of microspheres at different locations within the solution display a wide range of amplitudes and time dependences. The MSD-distribution is very broad/bimodal and the statistical analysis indicates a degree of inhomogeneity slightly higher than for the Sterocoll FD solution.

Keywords Acrylic thickeners · Microheterogeneity · Microrheology · Multiple-particle tracking

Introduction

Synthetic acrylic polymers are frequently used as thickening agents in water-based coatings and adhesives or personal care products. Typically, these commercial alkali-swellable acrylates (as well as various other polymeric thickeners) form inhomogeneous partly aggregated or cross-linked solutions. Inter- and/or intramolecular aggregation is due to hydrophobic groups randomly distributed along the chains and can be varied through the solvent quality [1–3]. Crosslinking can be induced either thermally or by adding appropriate crosslinking agents during synthesis. Accordingly, such thickener solutions cover a wide range of rheological behavior, ranging from weakly elastic, almost Newtonian to highly elastic gel-like. Despite its high technical relevance, little is known so far about the contribution of the micro-scale inhomogeneities to the bulk viscoelastic properties [4]. Here we use the method of MPT, which was originally described by Apgar et al. [5] and Ma et al. [6], to quantify the degree of structural and mechanical microheterogeneity of such acrylic thickener solutions. Up to now MPT is frequently used for the study of microheterogeneity of actin filament network [7–10], living cells [11–15], proteins [16, 17], DNA solutions [18] or biological gels [19, 20]. The principle of MPT consists of monitoring the thermally driven motion of inert microspheres that are evenly distributed within the solutions and to statistically analyze the distribution of mean square displacements, from which information about the extent of heterogeneity can be extracted. In order to calibrate our MPT set-up we first investigated a mixture of glycerol/water solution, viscous liquid sample known as perfectly homogenous, even at length scales much smaller than the radius of the microspheres used for the MPT experiments. We also systematically compared the rheological properties derived from MPT measurements with those obtained from classical mechanical rheometry.

Materials and Methods

Samples

The acrylic thickeners were donated by the BASF-Aktiengesellschaft (Ludwigshafen, Germany). These acrylate esters are co-polymers with a typical composition of about 50% methacrylic or acrylic acid and 50% ethylacrylate and eventually a small amount of crosslinking agent [3]. They are made from emulsion polymerization and delivered as milky liquids with a solids content of 25–30% and pH $\gg 2.5$. Thickening properties in aqueous environment are recovered upon neutralization when the acid groups dissociate and the polymer chains get soluble. In our study the polymer concentration for both Sterocoll FD and Sterocoll D was 1 wt. %. Solutions were stirred at room temperature for 48 h and adjusted to pH $= 8$ slowly adding 1 N NaOH. Polymer concentration was determined thermo-gravimetrically after neutralization. Subsequently samples were equilibrated for at least 24 h prior to testing. No further change of pH and viscosity was observed within 3 months. A mixture of glycerol/water with 85% glycerol content and a zero shear viscosity of 0.1 Pa s at 20 °C was also investigated as a perfectly homogeneous reference system comparable in viscosity to our thickener solutions.

Mechanical Rheology Measurements

Steady and oscillatory shear measurements were performed at a temperature of 20 °C with a Rheo-Stress RS150 rheometer (Haake) using a Couette cell geometry for the glycerol and Sterocoll FD solutions and a cone-plate geometry (60 mm in diameter, 1° cone angle) for the Sterocoll D.

Multiple-Particle Tracking

In our study we have used green fluorescent polystyrene microspheres of 0.5 μm diameter (Bangs Laboratories: USA, lot Nr FC03F/7049) as tracer particles. The mixture (total volume: ∼ 20 μl) containing the sample solution including the tracers (volume fraction around 1%) was injected into a self build chamber, consisting of a coverslip and microscope glass slide. The sample thickeness was ∼ 150 μm and the microscope was focused roughly halfway into the sample to minimize wall effects. Images of the fluorescent beads were recorded onto a personal computer via a progressive scan camera (Allied Vision Technology: Pike F-100B, 2/3″ CCD, 1000 × 1000 square pixels (7.4 μm), up to 60 fps) mounted on an inverted fluorescence microscope (Axiovert 200, Zeiss), equipped with a C-Apochromate 40×, N.A. 1.2, water-immersion lens combined with a 2.5× optovar magnification changer. Movies of the fluctuating microspheres were analyzed by a custom MPT routine incorporated into the software Image Processing System (Visiometrics iPS). Statistical analysis of the trajectories was done using Enthought Microrheology Lab software (Austin, USA) based on the Crocker, Wirtz and Weitz calculations [21–23]. The displacements of the particle centers were simultaneously monitored in a 75 × 75 μm^2 field of view, for 100 or 200 s at a rate of 10 f/s. For each experiment a total of 100 to 150 particles was tracked.

Results and Discussion

Trajectories and MSDs

Typical trajectories of microspheres (0.5 μm diameter) dispersed in glycerol and in acrylic thickeners solutions are reported in Fig. 1. The extent of displacement is reduced by approximately a factor 2 for the Sterocoll FD solution (Fig. 1B) compared to the glycerol solution (Fig. 1A). This result corresponds to a higher viscosity of the Sterocoll FD solution. For the Sterocoll D sample, the magnitude of particle displacement is much lower (Fig. 1C) compared to both glycerol and Sterocoll FD solutions. This limited particle motion already indicates a higher viscosity but also suggests the presence of high elasticity in the system. From these trajectories the coordinates of the particles centroids were transformed into MSD traces [22]. For the glycerol and Sterocoll FD solutions the MSD traces adopt a power-law behavior as a function of time with a slope close to 1 $\left(\langle \Delta r^2(\tau) \rangle \propto \tau\right)$ throughout the probed time scales (Figs. 2A and B). We conclude that the motion of the beads in glycerol and Sterocoll FD is purely diffusive and that the microenvironment surrounding the particles responds like a viscous liquid. Interpreting the data leads to a measure of viscosity η using $\langle \Delta r^2(\tau) \rangle \propto 4D\tau$ in two dimension where the Stokes–Einstein equation gives $D = k_B T / 6\pi\eta a$ and a is the tracer particle radius. At long time scales (larger than 10 s), the profile of the MSD traces are affected by slight convection of the solutions, which is difficult to avoid for low viscosity fluids. In Sterocoll D the beads move significantly differently, as shown in Fig. 2C. Particle motion weakly depends on time for $\tau < 100$ s, indicating that particles are highly constrained by the surrounding solution. The motion is subdiffusive with a power-law having slopes well below one $\left(\langle \Delta r^2(\tau) \rangle \propto \tau^{\alpha}\right.$ with $\left.\alpha < 1\right)$

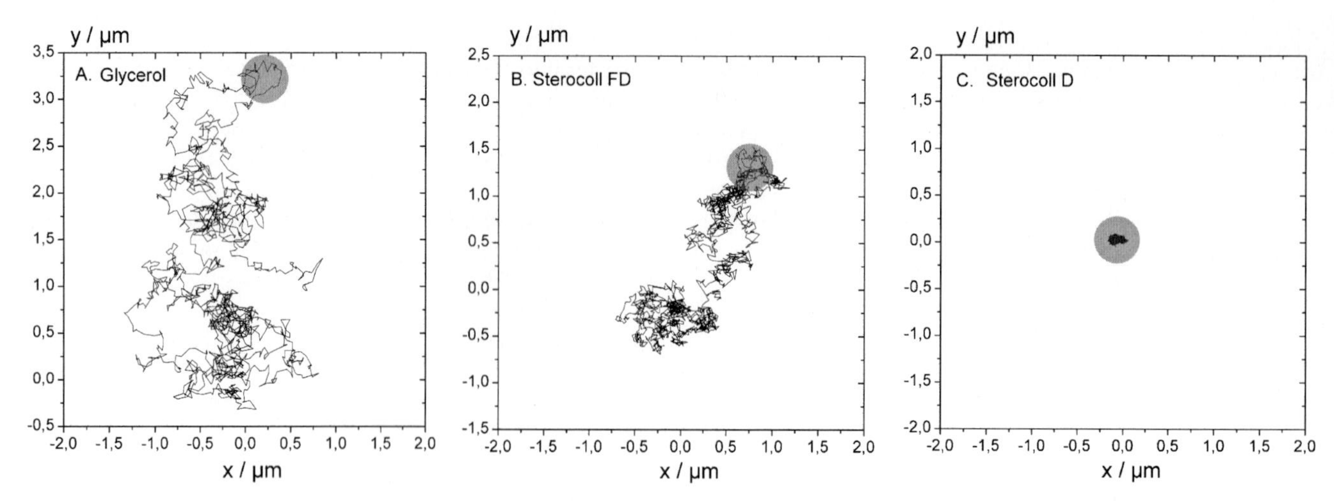

Fig. 1 Typical trajectories of a 0.5 µm diameter, fluorescent polystyrene microsphere (represented by a *sphere*) embedded in a solution of glycerol 85% (**A**), Sterocoll FD 1% (**B**) and Sterocoll D 1% (**C**). The total acquisition time was 100 s, with an interval of 100 ms between consecutive frames

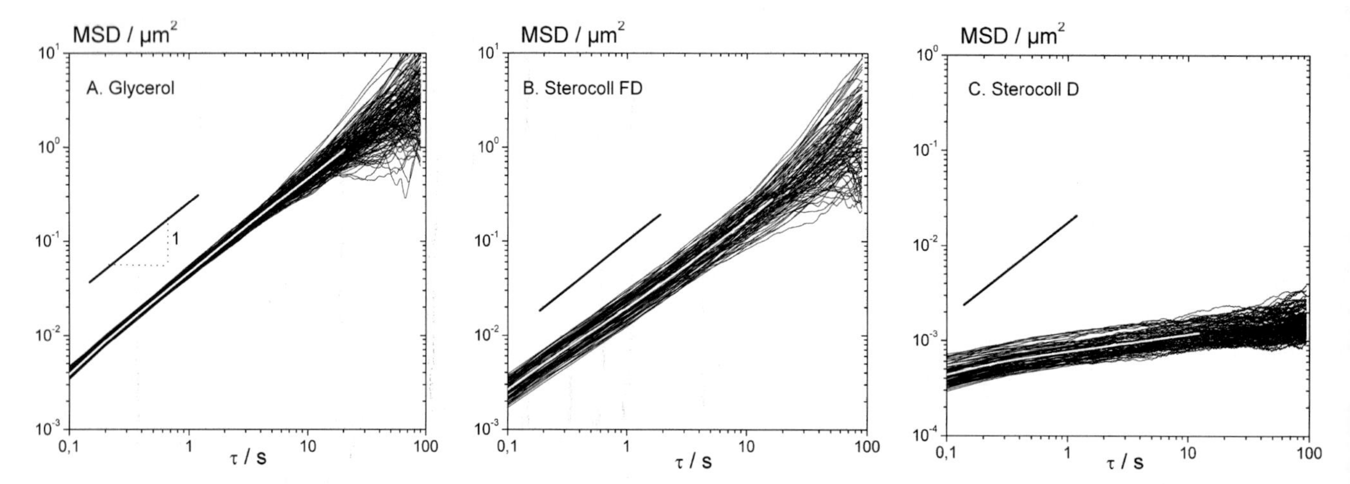

Fig. 2 Mean square displacement of individual microspheres dispersed in solution of glycerol 85% (**A**), Sterocoll FD 1% (**B**) and Sterocoll D 1% (**C**). The *white curve* is the ensemble-average MSD

that is the signature of elastic trapping of the beads by the mesh of the Sterocoll D solution. Another observation concerning the MSD traces is that the range of displacement at a given time scale is narrow for the glycerol/water mixture, but much wider for the two acrylic thickener solutions. To derive the macroscopic properties of each solutions, the ensemble-average MSD and ensemble-average diffusion coefficient were calculated from the individual MSDs. The ensemble-average diffusion coefficient decreases slightly from $1.17 \times 10^{-14}\,m^2/s$ (glycerol 85%) to $0.5 \times 10^{-14}\,m^2/s$ (Sterocoll FD). In comparison the same microsphere has a diffusion coefficient of $0.86 \times 10^{-12}\,m^2/s$ in water at 20 °C.

MSD Distributions and Statistical Analysis

To quantify the level of inhomogeneity in the different solutions we generated the MSD distributions from the MSD traces, normalized by the ensemble-average MSD. For the glycerol solution the MSD distribution is symmetric about the mean (Fig. 3A), as expected for a homogeneous liquid. For the Sterocoll FD solution, the distribution is also fairly symmetric (Fig. 3B), however it is significantly broader compared to the glycerol mixture. Finally, the MSD distribution further broadens and becomes even asymmetric (Fig. 3C) for the Sterocoll D solution.

By analyzing the contribution of the 10, 25, and 50% highest MSD values to the ensemble-average MSD at a time scale of 1 s (Fig. 4), we find that for the glycerol solution these parameters are close to those expected for a perfectly homogeneous liquid (10, 25, and 50%), for which all MSD values should theoretically be similar. For the Sterocoll FD, despite the fact that this solution behaves mostly like a liquid, it displays a much higher degree of heterogeneity than the glycerol solution (Fig. 4). The contributions of the 10, 25 and 50% highest MSD

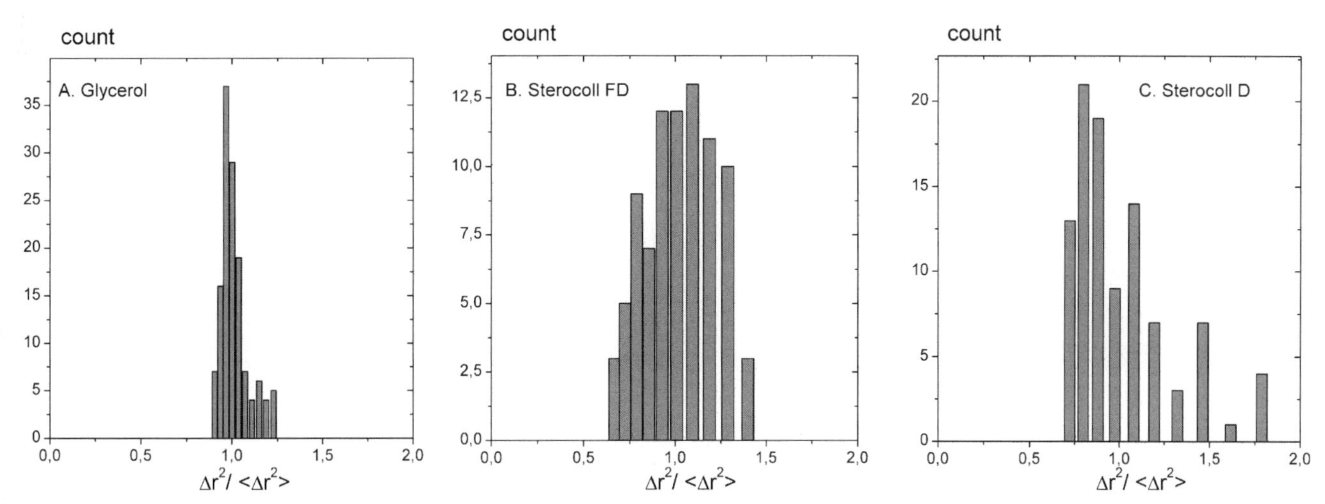

Fig. 3 MSD distributions measured at time lag of 0.1 s, normalized by the corresponding ensemble-average mean, for a solution of glycerol 85% (**A**), Sterocoll FD 1% (**B**) and Sterocoll D 1% (**C**)

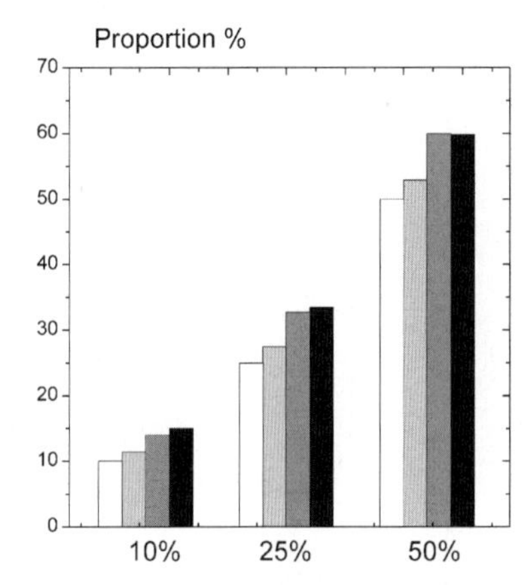

Fig. 4 Contributions (in %) of the 10, 25, and 50% highest MSD values to the ensemble-average MSD at a time scale of 1 s. Note that the contributions predicted for a perfectly homogeneous solution should be exactly 10, 25, and 50% respectively (*first column*). The *second column* represents the solution of glycerol 85%, the *third column* Sterocoll FD 1% and *fourth column* Sterocoll D 1%

values are (14, 32.7, 59.9%) respectively. For the Sterocoll D solution the contribution of the highest MSD values are slightly larger (15, 33.4, 59.8%) than those for Sterocoll FD (Fig. 4). This results indicate a slightly higher degree of heterogeneity for this system.

Van Hove Correlation Function

In order to perform the statistical analysis we examined the distribution of displacements at lag time τ, known as van Hove correlation function [25, 26]. The equation expression is given by:

$$G_s(r,\tau) = \frac{1}{N}\sum_{i=1}^{N}\langle\delta(r_i(\tau) - r_i(0) - r)\rangle = \frac{N(r,\tau)}{N},$$

where $r_i(\tau)$ is the distance traveled by a particle i in a time τ. $N(r,\tau)$ is the number of particles that move a distance between r and $(r + dr)$ in a time interval τ, and N is the number of particles. To a first approximation $G_s(r,\tau)$ has a Gaussian form but deviations from this form reflect the presence of heterogeneities. Such deviations can be characterized by the non-Gaussian parameter [27]

$$\alpha = \frac{\langle r(\tau)^4\rangle}{3\langle r^2(\tau)\rangle^2} - 1\,.$$

This quantity is zero for a Gaussian distribution, while broader distributions result in large values of α. The van Hove correlation functions for the glycerol, Sterocoll FD and Sterocoll D solutions are shown in Fig. 5A–C respectively. The solid line represents the Gaussian fit to the distribution. For the glycerol solution the ensemble-averaged data fits well to a Gaussian over several order of magnitude (Fig. 5A), as expected for a homogeneous, purely viscous fluid. The α value is 0.04, a strong indicator for the homogeneity of the solution. By contrast, in Sterocoll FD (Fig. 5B) and Sterocoll D (Fig. 5C) the α values are 0.225 and 0.283 respectively. This deviation of the ensemble-averaged distribution from Gaussian behavior indicates and confirms the heterogeneity of those systems.

Viscosity and Moduli

The variation of the viscoelastic moduli as a function of frequency has been determined as well as by MPT and standard oscillatory rheology. For MPT, the rheological properties were extracted from the MSD measurements following the method of Mason et al. [22]. After Laplace transformation of the MSD traces and the use of the generalized Stockes–Einstein relation (GSER), a complex modulus was calculated, from which frequency dependent elastic and viscous moduli, $G'(\omega)$ and $G''(\omega)$ respectively

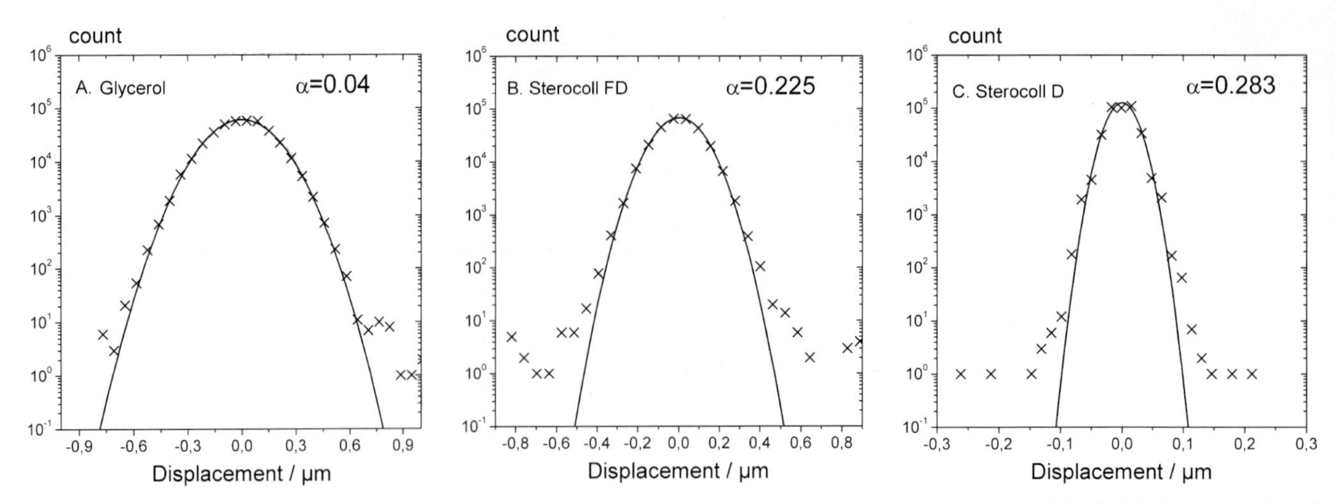

Fig. 5 Van Hove correlation functions for the ensemble-average particles, at time lag 1 s, moving in glycerol 85% (**A**), Sterocoll FD 1% (**B**), Sterocoll D 1% (**C**). The ensemble-average data is fit by a Gaussian distribution. α value is a measure of the fit of the curve to the data, with 0 being a perfect fit

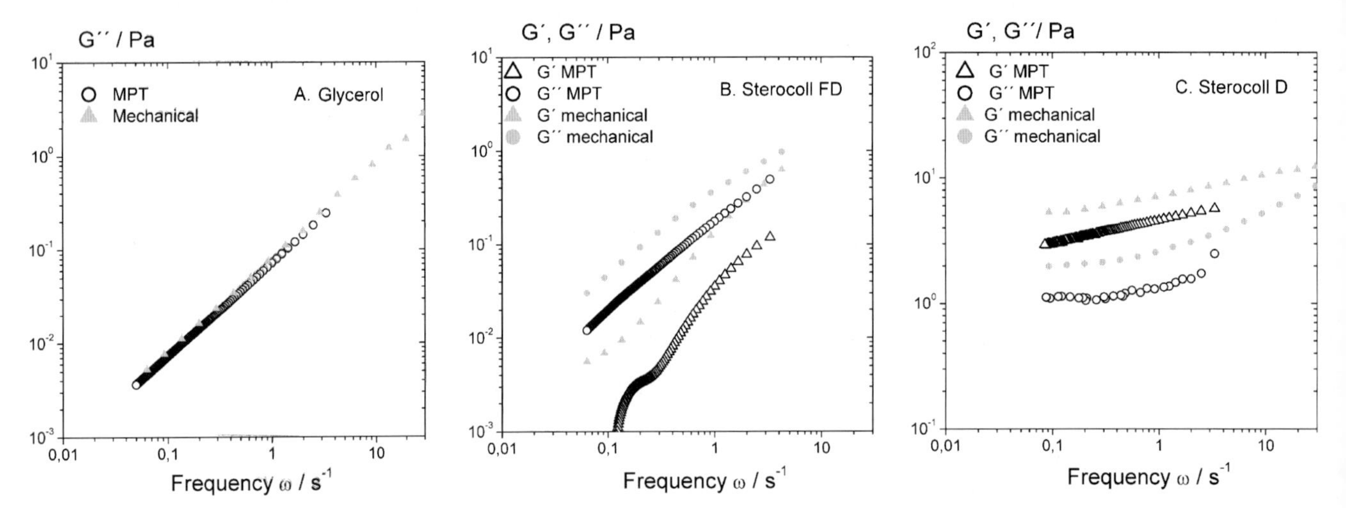

Fig. 6 Frequency-dependent elastic modulus G' and viscous modulus G'' of solution of glycerol 85% (**A**), Sterocoll FD 1% (**B**), Sterocoll D 1% (**C**). *Open symbols* were obtained with MPT, *solid symbols* with mechanical rheology

were extracted. For each sample the zero shear viscosity η_0 has been determined performing steady shear measurements. $\eta_0 = 0.1 \pm 0.01$ Pa s for glycerol/water mixture and 0.24 Pa s for the Sterocoll FD solution. The Sterocoll D solution is strongly shear thinning with a viscosity of 10 Pa s at a shear rate of 0.1 s^{-1}. The glycerol solution exhibits no measurable elasticity modulus G', but only a viscous modulus G'' that increases linearly with the frequency, as expected for this Newtonian liquid (Fig. 6A). The viscosity value determined from mechanical rheometry is $\eta = 0.084 \pm 0.005$ Pa s. From particle tracking measurement $\eta = 0.073 \pm 0.007$ Pa s using $\langle \Delta r^2(\tau) \rangle \propto 4D\tau$ with $D = k_B T / 6\pi\eta a$ and $D = 1.17 \times 10^{-14}$ m^2/s. The errors introduced here are the standard deviations of repeated measurements on the same sample and refer to the reproducibility of the measurements. The Sterocoll FD solution (Fig. 6B) shows both G' and G'' moduli with $G'(\omega) < G''(\omega)$ confirming a mostly viscous behavior. On the contrary, the Sterocoll D solution (Fig. 6C) shows a significant degree of elasticity with $G'(\omega) > G''(\omega)$ and constant modulus $G' = 10$ Pa in the whole frequency range investigated. The comparison between the rheological properties obtained from the two different techniques shows very good agreement for the glycerol solution, but in the case of the Sterocoll FD and Sterocoll D solutions the viscoelastic moduli are underestimated by a factor ~ 2. We believe this difference could be due to dynamical errors coming from the particle motion during the finite exposure time required for imaging. This error increases with the viscosity of the medium [24].

Conclusions

In this study, we have investigated the mechanical microenvironments and microheterogeneity of acrylic thickeners solutions using the MPT method and compared the results with those obtained for a glycerol/water mixture, characterized as a homogeneous fluid on the length scale of our tracer particles. For the Sterocoll FD solution, the MSD traces varies almost linearly with time, as expected for a predominantly viscous liquid. The MSD distribution exhibits a symmetric shape, however the histogram of MSD values as well as the van Hove analysis clearly reveal the heterogeneity of the system. For the Sterocoll D solution the MSD traces reach a quasi-plateau at time scales up to 100 s, consistent with an elastic trapping of tracer particles in a gel-like environment. For this system, the MSD distribution is highly asymmetric and the statistical analysis shows stronger heterogeneity than for the Sterocoll FD solution. Concerning the rheological quantities there is a fair agreement between MPT and mechanical measurements with in a factor of two, nevertheless improvement of our experimental set-up and calculation process is necessary for quantitative determination of bulk viscoelastic properties of solutions in the viscosity range > 0.1 Pa s.

Acknowledgement The authors acknowledge BASF-Aktiengesellschaft for providing the Sterocoll samples and Enthought company for free access to the analysis software. Many thanks to Jennifer Curtis and Heike Böhm from Heidelberg University (Germany) for interesting discussions and advice. We are indebted to Yaqi Xiong and Sachin Ghosh, who performed oscillatory shear and MPT measurements, respectively.

References

1. Wang C, Tam KC, Tan CB (2005) Langmuir 21:4191–4199
2. Ng WK, Tam KC, Jenkins RD (2001) Polymer 42:249–259
3. Kheirandish S, Guybaidullin I, Wohlleben W, Willenbacher N (2008) Rheologica Acta. Springer, Berlin. DOI: 10.1007/s00397-008-0292-1
4. English RJ, Raghavan SR, Jenkins RD, Khan SA (1999) J Rheol 43:1175–1194
5. Apgar J, Tseng Y, Fedorov E, Herwig MB, Almo SC, Wirtz D (2000) Biophys J 79:1095–1106
6. Ma L, Yamada S, Wirtz D, Coulombe PA (2001) Nat Cell Biol 3:503–506
7. Tseng Y, An KM, Wirtz D (2002) J Biol Chem 277:18143–18150
8. Tseng Y, Wirtz D (2001) Biophys J 81:1643–1656
9. Valentine MT, Kaplan PD, Thota D, Crocker JC, Gisler T, Prud'homme RK, Beck M, Weitz DA (2001) Phys Rev E 64:061506
10. Tseng Y, Kole PT, Lee JSH, Fedorov E, Almo SC, Schafer BW, Wirtz D (2005) Biochem Biophys Res Commun 334:183–192
11. Tseng Y, Kole PT, Wirtz D (2002) Biophys J 83(20):3162–3176
12. Tseng Y, Kole PT, Lee J, Wirtz D (2002) Curr Opin Colloid Interf Sci 7:210–217
13. Heidemann SR, Wirtz D (2004) Trends Cell Biol 14(4):160–166
14. Lee JSH, Panorchan P, Hale CM, Khatau SB, Kole TP, Tseng Y, Wirtz D (2006) J Cell Sci 119:1761–1768
15. Kole TP, Tseng Y, Jiang I, Katz JL, Wirtz D (2005) Mol Biol Cell 16:328–338
16. Dawson M, Wirtz D, Hanes J (2003) J Biol Chem 278(50):50393–50401
17. Xu J, Tseng Y, Carriere CJ, Wirtz D (2002) Biomacromolecules 3:92–99
18. Goodman A, Tseng Y, Wirtz D (2002) J Mol Biol 323:199–215
19. Panorchan P, Wirtz D, Tseng Y (2004) Phys Rev E 70:041906
20. Caggioni M, Spicer PT, Blair DL, Linberg SE, Weitz DA (2007) J Rheol 51(5):851–865
21. Crocker JC, Grier DG (1996) J Colloid Interf Sci 179:298–310
22. Mason TG, Ganesan K, Van Zanten JH, Wirtz D, Kuo SC (1997) Phys Rev Lett 79(17):3282–3285
23. Gardel ML, Valentine MT, Weitz DA (2005) Microrheology. In: Breuer K (ed) Microscale Diagnostic Techniques. Springer, Berlin
24. Savin T, Doyle PS (2005) Biophys J 88:623–638
25. Van Hove L (1954) Phys Rev 95:249–262
26. Hansen JP, McDonald IR (1986) Theory of Simple Liquids. Academic, London
27. Weeks ER, Crocker JC, Levitt AC, Schofield A, Weitz DA (2000) Science 287:627–631

Progr Colloid Polym Sci (2008) 134: 80–89
DOI 10.1007/2882_2008_090

Published online: 19 August 2008

Marcel Vrânceanu
Karin Winkler
Hermann Nirschl
Gero Leneweit

Influence of the monolayers composition on bilayer formation during oblique drop impact on liquids

Marcel Vrânceanu · Karin Winkler ·
Gero Leneweit (✉)
Carl Gustav Carus-Institute,
75223 Niefern-Öschelbronn, Germany
e-mail: gero.leneweit@carus-institut.de

Marcel Vrânceanu · Hermann Nirschl
Institute of Mechanical Process
Engineering and Mechanics, University of
Karlsruhe (TH), Germany

Abstract We study the dynamics of two phospholipid monolayers brought into contact by oblique drop impact on a liquid surface and bilayer/multilayer formation. Drop impact without monolayers shows that for low impact angles ($\alpha < 23°$) and low drop velocities the drop spreads as a thin sheet on the target liquid without immersion and mixing of the two liquids. When drop and target liquid surface are covered with monolayers, bilayer/multilayer formation is expected. The composition and mechanical properties of the monolayers can strongly influence the pattern of drop impact and bilayer/multilayer formation. Monolayers with either pure saturated or unsaturated phospholipids, and their mixtures with cholesterol were used. We show that under all conditions studied bilayer/multilayer synthesis takes place. Asymmetric bilayers can be produced by the coupling of drop and target monolayers. For some lipid mixtures the drop and target monolayer collapses during drop impact and symmetric bilayers/multilayers are formed.

Keywords Bilayer formation · Oblique drop impact · Phospholipid-monolayers

Introduction

Liquid interfaces with monolayers play an important role in many diverse industrial processes, creating colloidal dispersions such as emulsions, micelles or liposomes. When two monolayers come into contact, bilayers can be synthesized. Phospholipid bilayers form liposomes, which are used as vehicles for drug delivery, in cosmetics and in gene therapy [1].

In the present research, two monolayers are brought into contact by oblique drop impact on a liquid surface. The results presented extend a previous study on oblique drop impact without monolayers, focusing on the surface dynamics and mixing motions [2]. The latter shows that for low impact angles ($\alpha < 23°$) and low drop velocities (Weber number We < 140) the drop spreads as a thin sheet on the target liquid without immersion and mixing of the two liquids.

When the drop and target liquid are covered with monolayers, the behaviour of these monolayers apart from equilibrium and the role of their mechanical properties can be studied.

The surface rheological proprieties can be controlled by using monolayers of different lipid compositions. We used saturated and unsaturated lipids and their mixtures with different amounts of cholesterol. At the experimental temperature and film pressure the saturated lipids were in the liquid condensed or solid phase whereas the unsaturated lipids in the liquid expanded phase [3]. The mechanical properties of the monolayers can be tuned with addition of different amounts of cholesterol. The results are used for the proof of bilayer or multilayer synthesis and the conditions of their occurrence.

Materials and Methods

Materials

The target liquid consists of a glycerol/water mixture with 61% glycerol (w/w) and the density of $1158.6\,\text{kg/m}^3$. The drop bulk liquid also consists of a glycerol/water mixture, but with a higher glycerol content of 62.2% (w/w) and has a slightly higher density of $1161.2\,\text{kg/m}^3$. This density difference was chosen to make the drop bulk liquid to sink after the drop impact into the target liquid due to the density difference to visualize possible phospholipids in the contact zone of drop and target liquid. The surface tension of the glycerol/water mixture is 69 mN/m and the bulk viscosity 11.68 mPa s [4]. The glycerol/water mixture was chosen to simulate the impact of smaller water drops in the same Weber- and Reynolds-number range ($310 \leq \text{We} \leq 420$; $250 \leq \text{Re} \leq 490$) according to the laws of hydrodynamics similarity.

Monolayers with film pressures $\Pi = 30 \pm 1\,\text{mN/m}$ were formed with either pure 1,2-dioleoyl-sn-glycero-3-phosphatidylcholine (DOPC) or its mixture with 60 mol % cholesterol and with pure 1,2-dipalmitoyl-sn-glycero-3-phosphatidylcholine (DPPC) or its mixture with 33 and 60 mol % cholesterol. These concentrations were chosen to avoid phase separation of phospholipids and cholesterol, see [3]. *N*-(7-nitrobenz-2-oxa-1,3-diazol-4-yl)-1,2-dihexadecanoyl-sn-glycero-3-phosphoethanolamine, triethylammonium salt (NBD-PE) labelled on the head group, was used as fluorophore marker of the monolayers. At 3 mol % the brightest intensity of target liquid monolayer was found. To simulate the mechanical properties of NBD-PE, 3 mol % of 1,2-dipalmitoyl-sn-glycero-3-phosphoethanolamine (DPPE) was used for the monolayers without fluorophore. Calcein, a water-soluble fluorophore, was used for the visualization of the drop bulk liquid. Only one fluorophore was used at a time. The monolayers were deposited using a chloroform solution with a lipid concentration of 0.2 μmol/ml.

DOPC, DPPC and DPPE were obtained from Lipoid (Ludwigshafen, Germany), cholesterol and calcein from Sigma (Taufkirchen, Germany), chloroform > 99% and glycerol > 98% from Roth (Karlsruhe, Germany) and NBD-PE from Invitrogen (Karlsruhe, Germany). All substances were used without further purification. Bidistilled water with the quality for injectable drugs was used for the glycerol/water mixture.

Method

A schematic drawing of the experimental setup is presented in Fig. 1. The drop with a volume of 4 μl (corresponding to a drop diameter D of $\sim 2\,\text{mm}$) was produced at the end of a capillary with a pendent drop tensiometer PAT1 from Sinterface (Germany) [5]. To form the monolayer, approximately 0.2 μl of the chloroform phospholipid solution was injected at the drop surface with a μl-syringe as described in [6]. The tensiometer measures the drop volume and area, surface tension σ, film pressure Π, surface elasticity ε and surface dilational viscosity η of the drop monolayer. The target liquid was placed in a cuvette. Here, the monolayer was applied with a ml-syringe by releasing drops of the chloroform/phospholipid solution on different places of the liquid surface. The film pressure of this monolayer was measured with a Wilhelmy tensiometer.

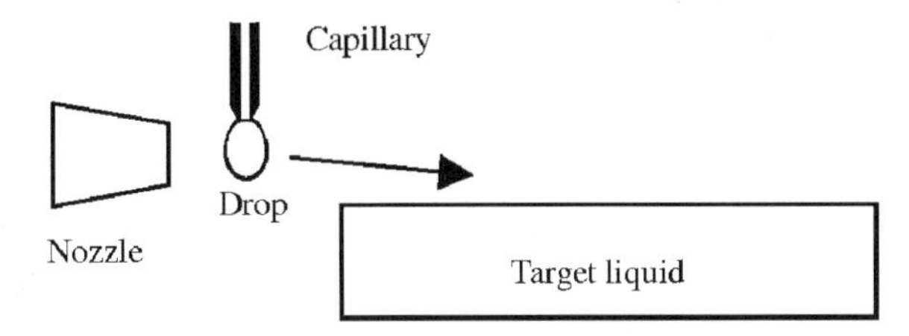

Fig. 1 Simplified sketch of the experimental setup

Once the properties of the drop and target liquid monolayers were determined, the drop was accelerated with a short pulse of compressed air from a nozzle.

The drop impact was recorded simultaneously in both vertical and lateral perspectives with two cameras: first, a high-speed rotating drum camera, which recorded the two perspectives of the drop impact on a 35 mm b/w film. This camera records the first 66 ms of the drop impact with a frame rate of 1666 Hz. The lateral perspective pictures give information about the impact angle $\alpha = 12 \pm 1°$ and the impact velocity $u = 2.3 \pm 0.1\,\text{m/s}$, from which the Weber number $\text{We} = \rho D u^2/\sigma = 380 \pm 20$ was deduced, where: ρ is the drop liquid density, D the drop diameter, u the drop impact velocity, and σ is the surface tension.

Second, a CCD camera connected to computer was used, which records 14 frames/s. This camera is equipped with a light filter and records only fluorophore emission, which is excited at 470 nm (maximum) by a xenon flashlight. The CCD camera detects either monolayer or drop bulk liquid distribution during and after the drop impact. Its lateral perspective shows possible submersions in the target bulk liquid.

All experiments were performed at $21.4 \pm 0.4\,°\text{C}$.

Results

Spreading of the Impacting Drop on the Target Liquid Surface

The spreading of the impacting drop is schematically presented in Fig. 2 as proved by Leneweit et al. [2], which studied the oblique drop impact without monolayers. For low impact angles ($\alpha < 23°$) and low drop velocities (Weber number $\text{We} < 140$) the drop spreads as a thin sheet on the target liquid without immersion and mixing of the two liquids [2].

Fig. 2 Schematic drawing in lateral perspective of drop spreading on the target liquid surface

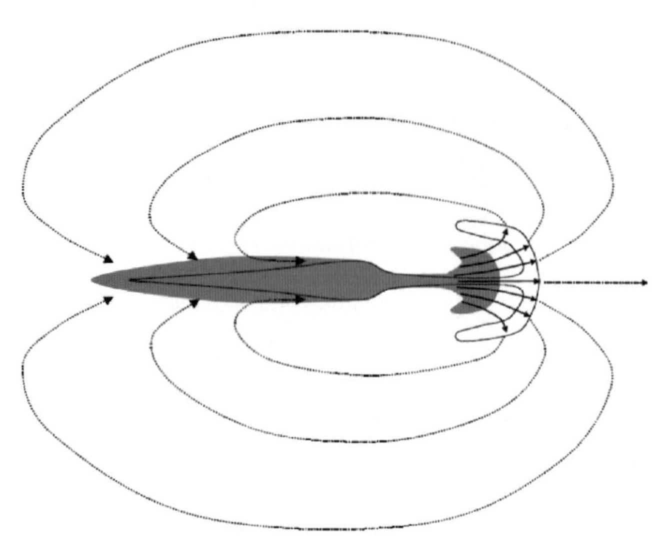

Fig. 3 Schematic drawing in vertical perspective of drop spreading on the target liquid surface [2]. The impacting drop coming from the left hand side, forms the *grey area* which moves further and forms the anchor pattern – the *area inside the solid black line*. The outlines of the drop patterns are extracted from experimental visualizations, the induced velocity field was drawn tentatively with arrows to give a qualitative impression

Figure 3 was taken from Leneweit et al. [2] and represents a sketch of the drop spreading pattern in vertical perspective. The impacting drop induces a velocity field in the target liquid, which expands the front part of the spread drop liquid to an anchor-like pattern and compress its rear part.

When lipid monolayers are deposited on the drop and on the target liquid the drop impact patterns depend on the monolayers rheological properties.

The Proof of Bilayer/Multilayer Formation Under the Drop Liquid

The main goal of this study is to prove the synthesis of bilayer structures with lipids from the two monolayers: drop and target monolayers. The place where the lipids from the drop monolayer could come in contact with the lipids from the target monolayer is under the thin liquid sheet of the drop impact pattern. If the insoluble lipids are captured under the drop bulk liquid they are in an aqueous medium and they form spontaneously bilayer and/or multilayer phases, like vesicles and liposomes, as shown in the literature [7–12]. Due to the fact that the new bilayer structures are formed with lipids from both drop and target monolayer, they are more or less asymmetric.

An experimental verification is necessary to prove whether lipids from the drop and/or target monolayer are under the drop liquid after impact. The impacting drop could push and displace the lipids from the two monolayers and the two bulk liquids could come in contact without separation lipid monolayers in between them. In this case there should be no lipids in the contact zone of the two aqueous media which means that no bilayer formation would take place.

To prove if there are lipids from the drop and from the target monolayer captured under the drop liquid after drop impact we made the drop bulk liquid to sink into the target liquid. This was done by using for the drop bulk liquid a water/glycerol mixture with 1.2% more glycerol than for the target bulk liquid, slightly increasing in this way the drop liquid density. After the impacting drop comes to rest in the anchor pattern, the drop liquid starts to sink into the target bulk liquid, due to the density difference, as shown schematically in Fig. 4.

To see the behaviour of the drop bulk liquid during and after impact, we labelled it fluorescently with calcein, as presented in Fig. 5.

The pictures in vertical perspective show the formation of the anchor form pattern, like described in Fig. 3 by Leneweit et al. [2]. After impact, the drop bulk liquid spreads as a thin liquid sheet on the target liquid surface in an anchor-form pattern and when the horizontal motion come to rest, it starts sinking into the target liquid due to the density difference, as seen in the lower line of Fig. 5 and presented schematically in Fig. 4. In the last three pictures from the vertical perspective the anchor form pattern cannot be clearly seen anymore because of the fluorescence of the sinking liquid.

Once we know the behaviour of the drop bulk liquid, which sink almost completely as shown in Fig. 5, we check if there are lipids captured under it. For this we labelled fluorescently the target or the drop monolayer. The sinking drop liquid submerges into the target liquid the fluorescently labelled lipids captured under it.

Figure 6A shows the lateral perspective pictures when the target liquid was fluorescently labelled. As can be observed there are lipids from the target monolayer withdrawn into the subphase. This means that the target monolayer was not displaced by the impacting drop, but was actually covered by the drop liquid. Due to the density difference the drop liquid is sinking into the target liquid withdrawing the target lipids captured under it. These lipids have to have the conformation of bilayer or multilayer structures as discussed above.

When the fluorophore substance is in the drop monolayer (Fig. 6B) it can be observed that the drop phospholipids are also submerged into the target bulk liquid once the drop liquid starts to sink. The quantity of the phospholipids from the drop, which are caught between the drop

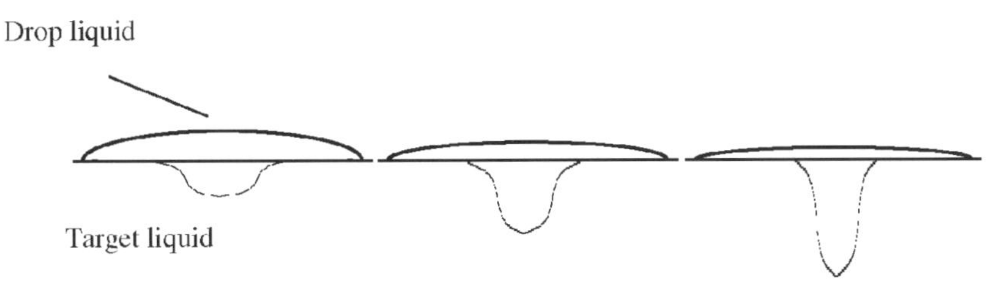

Fig. 4 The drop liquid sinks into the target liquid once the impacting drop comes to rest due to the density difference and interfacial instability

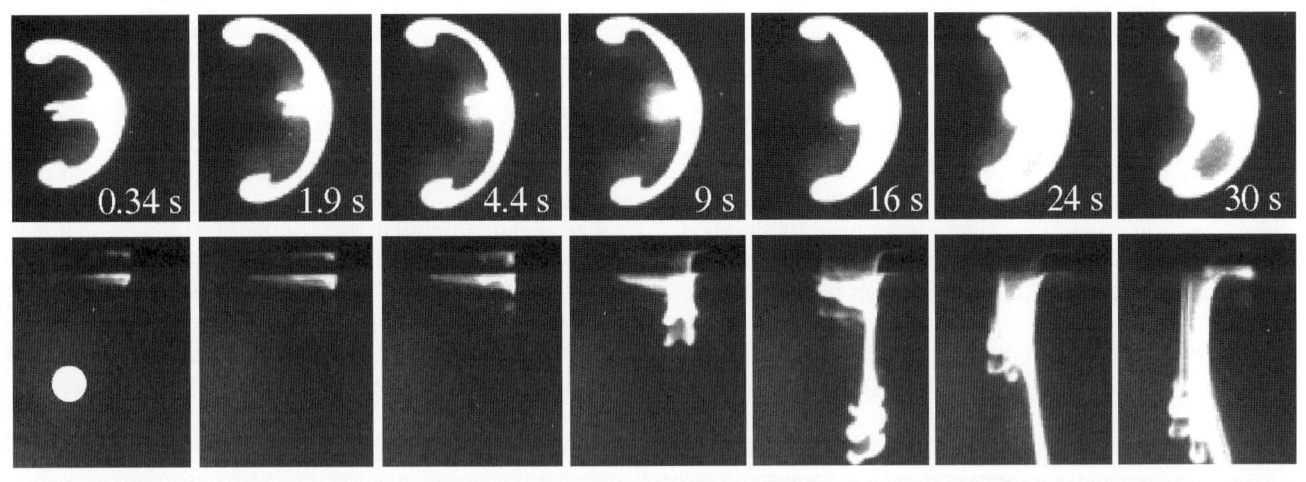

Fig. 5 Drop impact in vertical (*upper line*) and lateral (*lower line*) perspective with the drop bulk liquid fluorescently labelled with calcein. Drop and target monolayers: DOPC/cholesterol/DPPE = 64/33/3 (molar). The *white circle* from the first picture represent the drop before impact, $D = 1.8$ mm

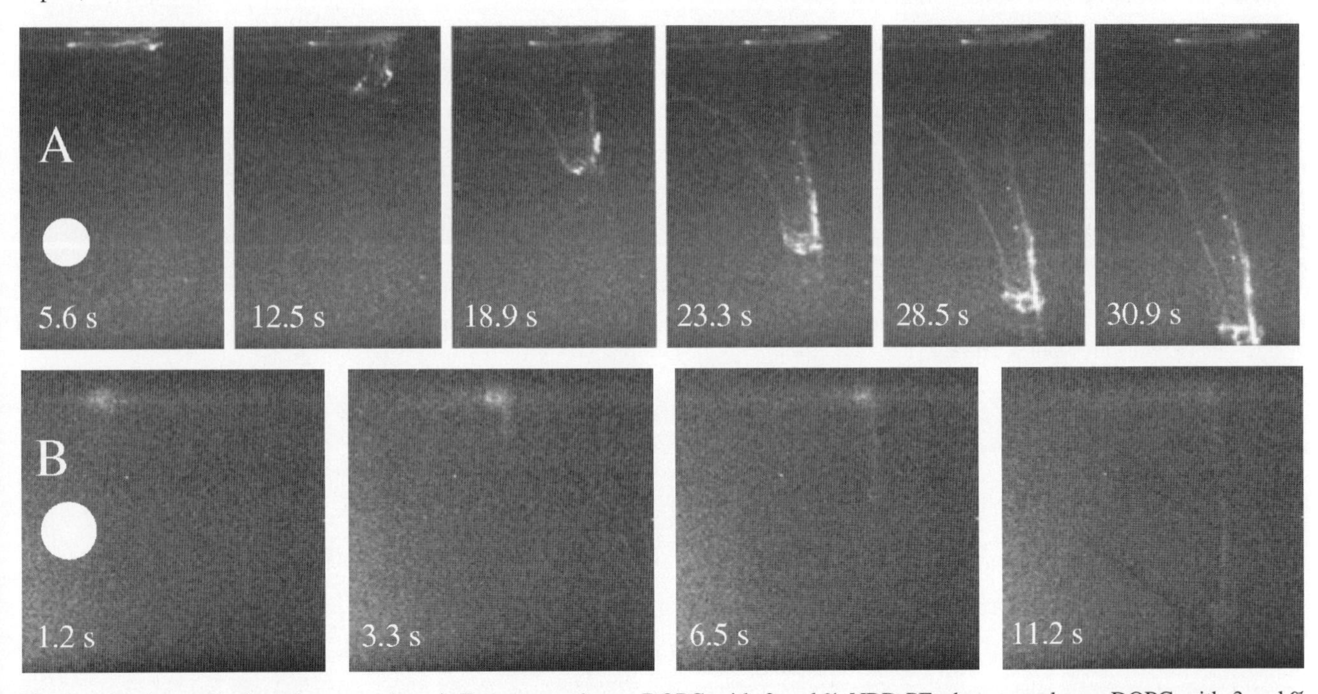

Fig. 6 Drop impact in lateral perspective: **A** Target monolayer: DOPC with 3 mol % NBD-PE; drop monolayer: DOPC with 3 mol % DPPE; **B** Target monolayer: DOPC/cholesterol/DPPE = 64/33/3 (molar); drop monolayer: DOPC/cholesterol/DPPE = 64/33/3 (molar). The sinking flow of the drop bulk liquid involves the phospholipids from the target and drop liquid monolayers, which were captured under the drop liquid

and the target liquid, seems to be much diminished compared to the phospholipids from the target liquid. This fact will be discussed in Sect. 4.

In this way we are able show that phospholipids from both drop and target monolayers are captured under the drop liquid in all experiments. As discussed above, lipids in a water medium means bilayer structures, like uni- or multi-lamellar liposomes.

To conclude we can say that bilayer structures can be formed by oblique drop impact on monolayers. The details of their generation will be shown in the following sections.

Drop Impact on Monolayers with Different Binary Mixtures of Either Unsaturated (DOPC) or Saturated (DPPC) Phospholipids with Cholesterol

The drop impact patterns, and the bilayer structures formed by drop impact, are highly influenced by the rheological properties of the drop and target monolayer. The rheological properties of monolayers can be modified by using different lipids, saturated or unsaturated, and their mixtures with different amounts of cholesterol [3]. The advantage of using cholesterol is that its presence stabilises the bilayer structures [13] and in high amounts inhibits the multilayer formation [14].

Drop Impact on Unsaturated Monolayers.

DOPC Monolayers Without Cholesterol. Figure 7 shows the vertical and lateral perspectives of a drop impact with pure DOPC monolayers. As seen as well in Fig. 6A, lipids from the target monolayer, captured under the drop liquid pattern, are submerged by the drop bulk liquid into the subphase.

To check if for DOPC monolayers, lipids from the drop monolayer are also captured under the drop liquid, we did experiments where the drop monolayer was fluorescently labelled, see Fig. 8. From these experiments we observe that fluorescently labelled lipids are also submerged into the subphase.

The anchor-formed pattern of the spreading drop liquid appears dark for the unlabelled drop monolayer in Fig. 7 and bright in Fig. 8 when the drop monolayer was fluorescently labelled. The anchor pattern is formed by the drop liquid, as shown in Figs. 3 and 5. The fact that the anchor form appears dark in Fig. 7 and bright in Fig. 8 means that

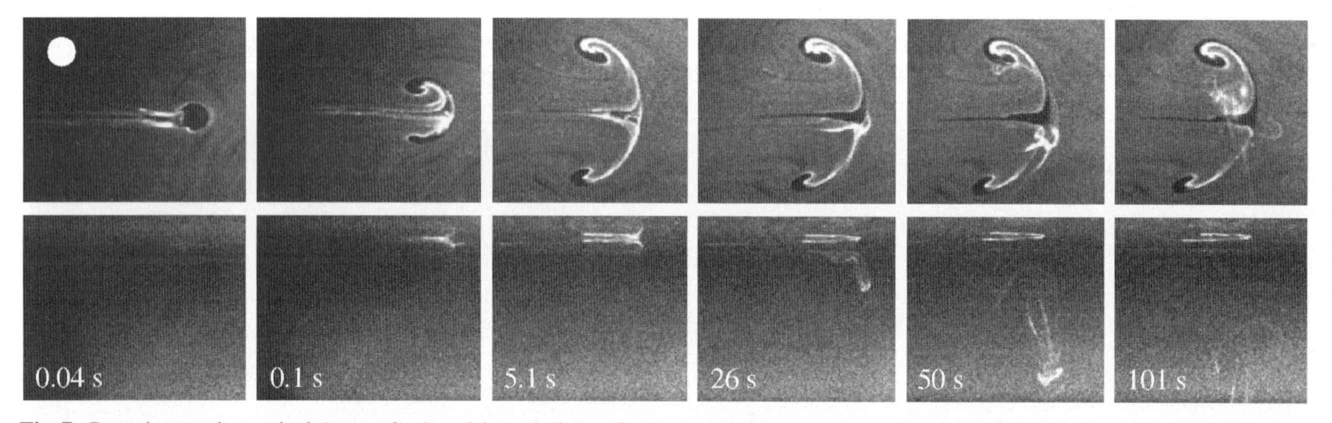

Fig. 7 Drop impact in vertical (*upper line*) and lateral (*lower line*) perspective, target monolayer: DOPC with 3 mol % NBD-PE; drop monolayer: DOPC with 3 mol % DPPE

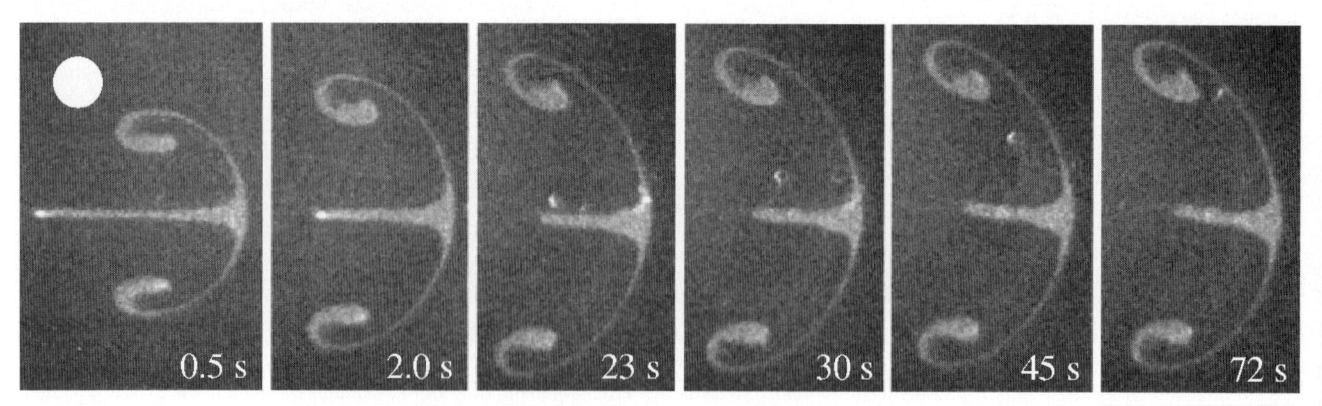

Fig. 8 Drop impact in the vertical perspective, target monolayer: DOPC with 3 mol % DPPE; drop monolayer: DOPC with 3 mol % NBD-PE

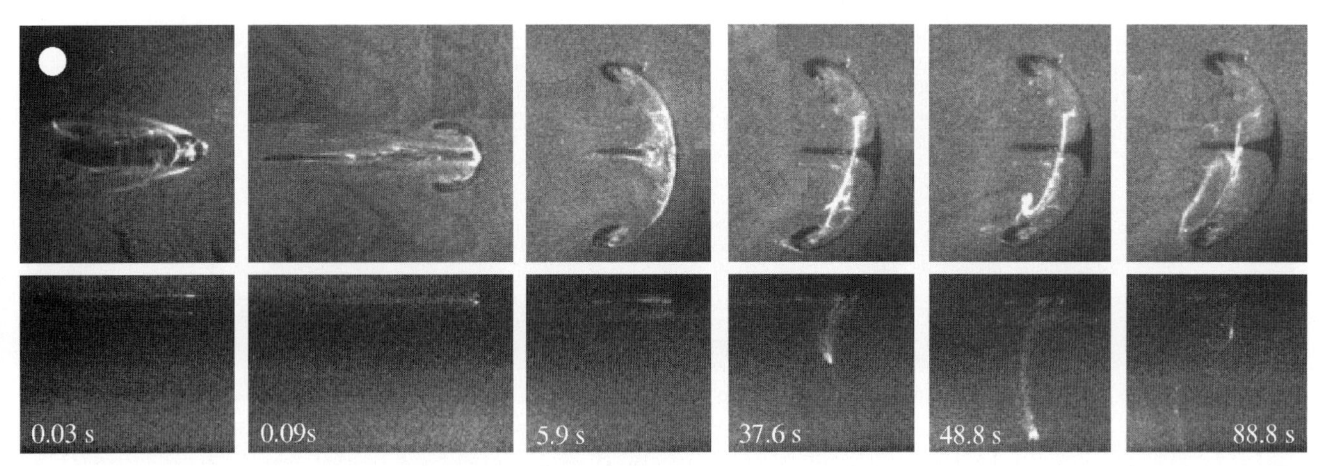

Fig. 9 Drop impact in vertical (*upper line*) and lateral (*lower line*) perspective, target monolayer: DOPC/cholesterol/NBD-PE = 37/60/3 (molar); drop monolayer: DOPC/cholesterol/DPPE = 37/60/3 (molar)

the drop liquid pattern is covered with lipids from the drop monolayer at its air-exposed surface. This monolayer from the top side of the anchor pattern is stable in time and the lipids forming it are not submerged into the subphase by the sinking of the drop bulk liquid.

For pure DOPC monolayers, if the target monolayer is fluorescently labelled (Fig. 7), extra-bright areas are formed in the first stage of drop impact, along the tail, and come to rest in the surrounding of the drop fluid at the backside of the anchor pattern. These areas are brighter than the fluorescently labelled target monolayer and remain on the target liquid surface. This is not the case when the drop monolayer is fluorescently labelled (Fig. 8), meaning that these areas contain lipids only from the target monolayer.

To conclude we can say that for drop impacts on DOPC monolayers, lipids from the drop and target monolayers are captured under the drop liquid pattern and extra-bright areas are formed at the back-side of the drop impact pattern with lipids from the target monolayer.

DOPC Monolayers with 60 mol % Cholesterol. Drop impact pictures on DOPC monolayers with 60 mol % cholesterol are shown in Fig. 9 where the target monolayer is fluorescently labelled.

During drop impact bright lipid structures are formed in the centre part of the anchor pattern and in the front part of the tail region. These bright structures are under the drop bulk liquid because once the drop bulk liquid sinks they are withdrawn into the subphase. No extra-bright areas are formed on the target monolayer like the case of pure DOPC (Fig. 7). The unlabelled drop monolayer remains on the target surface as a dark anchor pattern.

Drop Impact with Saturated Monolayers.

DPPC Monolayers Without Cholesterol. Drop impact pictures with pure DPPC monolayers are shown in Fig. 10, where the drop monolayer is fluorescently labelled.

We observe that the tail of the drop impact pattern is very long and does not form an anchor pattern. The first picture taken at 0.03 s after impact looks similar with the last one which was taken at more than 16 s after impact. From these pictures and from experiments where the target monolayer was fluorescently marked (results not shown), it can be seen that the target monolayer is stable and does not move at all. The impacting drop lands and stops quickly without inducing any movements on the target liquid.

From pictures taken in lateral perspective (results not shown), it can be observed that in the front part of the drop impact pattern, where the main mass of the drop liquid come to rest, the stable DPPC monolayer is broken and the drop liquid submerges into the target bulk liquid due to the density difference. This involves as well the drop and target monolayers lipids captured under the front part of the drop liquid.

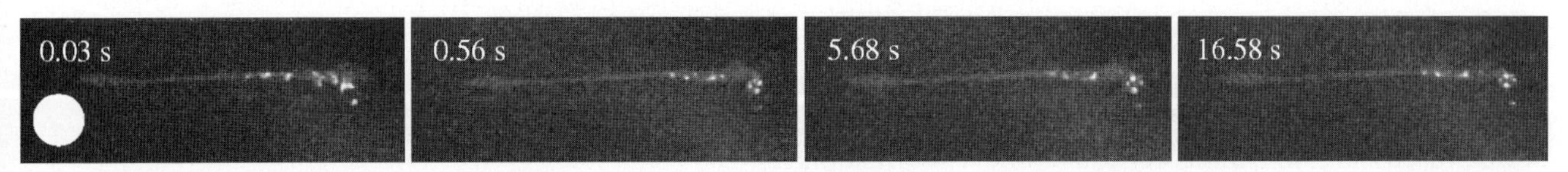

Fig. 10 Drop impact in vertical perspective, target monolayer: DPPC/DPPE = 97/3 (molar); drop monolayer: DPPC/NBD-PE = 97/3 (molar)

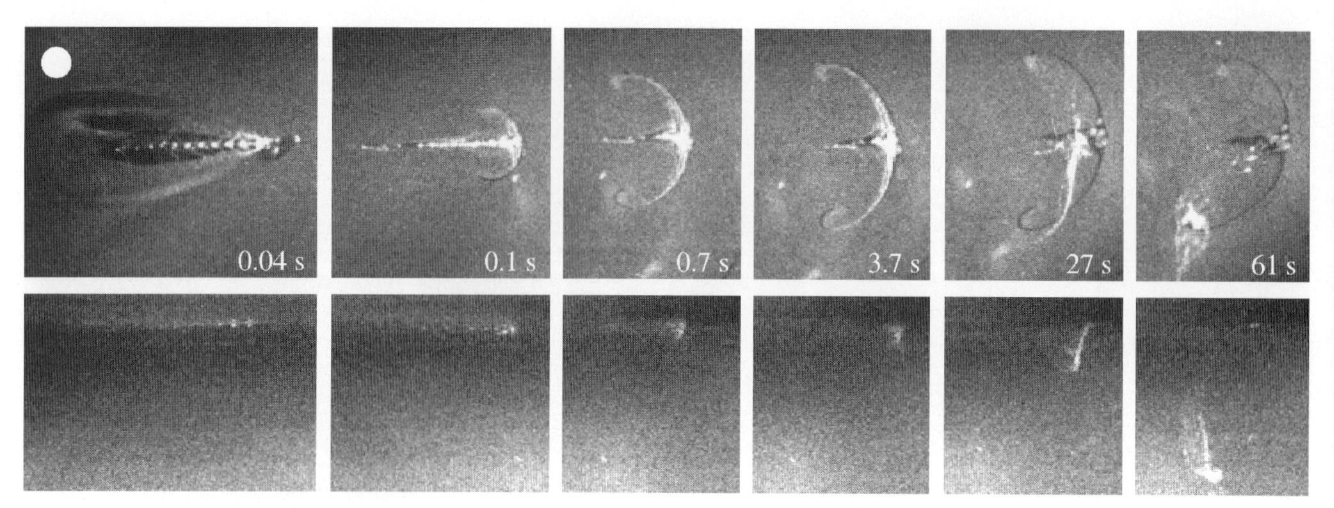

Fig. 11 Drop impact in vertical (*upper line*) and lateral (*lower line*) perspective, where the target liquid monolayer was formed by DPPC/cholesterol/NBD-PE (37/60/3 molar) and the drop monolayer by DPPC/cholesterol/DPPE (37/60/3 molar)

Extra-bright areas – appearing like spots, or fragments, are formed from both target (results not shown) and drop monolayers and remain at the target liquid surface.

DPPC Monolayers with 60 mol % Cholesterol. Drop impact pictures on DPPC monolayers with 60 mol % cholesterol are shown in Fig. 11 with the target monolayer being fluorescently labelled.

As can be seen there is a big difference between the drop impact on pure DPPC (Fig. 10) and on DPPC with 60 mol % cholesterol (Fig. 11). In Fig. 11 the drop impact pattern has an anchor pattern, like drop impacts on DOPC and DOPC-cholesterol monolayers.

From the vertical and lateral perspective, for the both cases: drop or target fluorescently labelled monolayers, it can be seen that the drop and target lipids are captured under the drop liquid and submerged into the subphase.

As can be observed in Fig. 11, during the drop impact some extra-bright areas, appearing like spots, are formed with lipids from the fluorescently labelled target monolayer. This happens as well when the drop monolayer is fluorescently labelled, results not shown. These extra-bright spots are not submerged by the drop bulk liquid into the subphase. They suddenly disappear, their lipids reintegrate back into the existing monolayer, and some of them remain at the liquid surface for more than 100 s. No extra-bright areas similar with the ones from the back-side of the anchor pattern of the pure DOPC monolayers are formed in this case.

Discussion

Monolayer Rheology

Before discussing our results regarding bilayer formation by oblique drop impact on different monolayers we have to take into account the differences between saturated and unsaturated monolayers and the effect of cholesterol on monolayer rheology.

DOPC monolayers, due to the unsaturation, i.e. kinks of the alkyl chains, are in the liquid expanded phase, which is a fluid phase at all film pressures Π [3, 13, 15]. At 21 °C and $\Pi > 25\,\mathrm{mN\,m^{-1}}$ DPPC monolayers are in the solid analogous phase [3, 13, 16], which is highly incompressible and condensed [13, 16]. Shah and Schulman [13] show that the effect of cholesterol on either saturated or unsaturated phospholipids is strikingly different. Cholesterol increases the surface elasticity, the dilational and the shear viscosity of unsaturated phospholipid monolayers [3, 13, 14, 17]. In saturated monolayers cholesterol disturbs the order between phospholipid molecules fluidifying the solid monolayer [13, 14, 18] and lowering its shear viscosity [18]. Pure cholesterol monolayers are liquid [13] and have very low surface shear viscosities which are hardly detectable [18].

In a previous study [3] we found that the surface elasticity and the surface dilational viscosity are higher for DPPC/cholesterol than for DOPC/cholesterol monolayers

Table 1 The surface elasticity and the surface dilational viscosity of DPPC-cholesterol and DOPC-cholesterol monolayers as determined in [3]

	Chol. (%)	ε (mN/m)	η (mN s/m)
DOPC-Chol.	0	120	40
	60	250	140
DPPC-Chol.	0	80	95
	33	180	200
	60	670	530

and both are increasing with the cholesterol content (see Table 1).

As a function of film pressure and cholesterol content, the DPPC-cholesterol monolayers are either in a solid, liquid/solid coexistence or in a liquid state, whereas DOPC-cholesterol monolayers are always in a liquid state [3, 15].

Bilayer Formation

As seen in the drop impact pictures (Figs. 6–11) lipids from the target and drop monolayers are captured under the drop bulk liquid and submerged into the target liquid. In Fig. 12 A we schematically present a possible mechanism of the drop and target monolayer dynamics, and bilayer formation under the drop bulk liquid. If the target monolayer is stiff enough, it is not displaced by the impacting drop, which rolls and spreads on it. In this way, asymmetric planar bilayers with lipids from the drop and target monolayers can be formed under the thin sheet of the drop liquid. Due to the sinking of the drop liquid the planar bilayer structures are bended and can form three dimensional bilayer structures as vesicles and liposomes. This is sustained by the observations of Ridsdale et al. [8], who showed that large bilayer structures, like folds, convert into more stable vesicular structures. For monolayers containing 60 mol % cholesterol it is expected that the vesicular structures are unilamellar because the high amount of cholesterol inhibits multilayer formation as was shown by Malcharek et al. [14].

We will discuss now the extra-bright areas formed during the drop impact and coming to rest at the back-side of the anchor pattern for pure DOPC monolayers. As presented in the results section, these extra-bright areas are formed only with lipids from the target monolayer and not from the drop monolayer. A possible mechanism of formation of these extra-bright lipid structures is presented in Fig. 12B and C. Pure DOPC monolayers are fluid and have low shear viscosity, low surface dilational viscosity and low surface dilational elasticity [3]. It is assumed that during the drop impact the target monolayer is compressed in the front part of the drop impact pattern. As the drop moves, this front compression

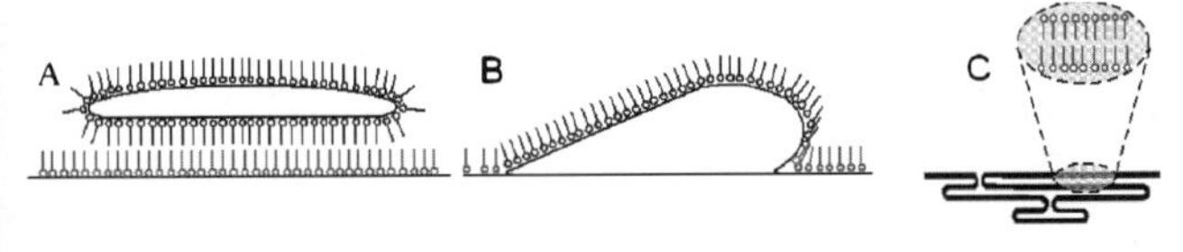

Fig. 12 Two mechanisms to describe monolayer dynamics during drop impact. **A** The target monolayer is stable and the drop liquid spreads on it. Asymmetric bilayers are formed under the drop liquid. **B** The target monolayer is displaced, compressed and expanded by the impacting drop. No bilayers are formed under the drop liquid. **C** As an enlargement of sketch (**B**) the compressed target monolayer is shown as it folds and forms symmetric bilayers or multilayers

forms the surrounding surface of the drop pattern. Compressed lipid monolayers are stable up to a maximum film pressure above which they collapse and form 3D bilayer structures [10–12]. The sketch C of Fig. 12 shows as an enlargement the collapsed target monolayer. Here the compressed target monolayer is shown as it folds and forms bilayers or multilayers. Gopal and Lee [10] show that monolayers in the liquid expanded phase, as is the case for DOPC monolayers, are not able to sustain large-scale folding, and collapse on a smaller length scale by forming vesicle-like structures. These structures are symmetric, being formed only with lipids from the target monolayer.

The fact that the extra-bright areas are 3D and not 2D structures anymore is sustained by the observation that there is no way to obtain such a extra-bright monolayer by increasing the monolayer film pressure (results not shown). A higher film pressures and higher fluorophore content does not give a brighter monolayer because of the quenching effect between the fluorophore molecules when they are restricted to form a monolayer.

As shown in Figs. 6–8 and discussed above, for drop impact on DOPC monolayers the first proposed mechanism (Fig. 12A) is as well correct. This means that for DOPC monolayers a combination of the two proposed mechanisms (Fig. 12A and B) seems to occur.

DOPC-cholesterol monolayers are liquid, but stiffer than pure DOPC monolayers, as discussed above, having higher dilational viscosity and dilational elasticity [3]. For drop impacts on DOPC monolayers containing 60 mol % cholesterol (Fig. 9), no extra-bright areas are formed on the target monolayer like in the case of pure DOPC monolayers (Fig. 7). In this case the impacting drop neither displaces, compresses or folds the target monolayer. The presence of cholesterol makes the target monolayer more stable than the pure DOPC monolayer.

We will discuss now the drop impact on saturated DPPC-cholesterol mixtures.

As is well known, at 21 °C and 30 mN/m film pressure a pure DPPC monolayer is solid and stiff, having high shear viscosity [18], but a relatively low surface dilational viscosity and elasticity [3]. The drop impact pattern on pure DPPC monolayers has a long tail and no anchor form (Fig. 10). The target monolayer is so stiff that the impacting drop cannot induce any surface movement in it as schematically drawn in Fig. 3.

During the drop impact some extra-bright small areas – appearing like discrete spots, or fragments, are formed from both target and drop monolayers and remain at the target liquid surface or sink into the subphase with the drop bulk liquid. The fact that these extra-bright spots are formed when the drop monolayer is fluorescently labelled, means that the drop monolayer fractures as well during the drop impact. This is not the case for pure DOPC, where only the target monolayer collapses and form continuous bright areas.

Using the imaging software ImageJ and its "Multi Cell Outliner" plugins we determined the areas of the drop impact patterns. These areas depend on the monolayers composition, being larger than the drop area before impact with approximately 140–160% for a pure DPPC monolayer and with approximately 220–250% for the others monolayers studied. There is a difference between the collapse of a liquid expanded monolayer like DOPC and that of a solid monolayer like DPPC. As discussed above, DOPC monolayers are not able to sustain large-scale folding, and collapse on a smaller length scale by forming vesicle-like structures [10], whereas the solid DPPC monolayers are apparently too brittle to bend, and collapse by fracture, as Lipp et al. [9] have shown. This means that in the case of DPPC the both drop and target monolayers fracture during drop impact and form asymmetric bilayer or multilayer structures which appear as extra-bright spots.

DPPC monolayers with 60 mol % cholesterol are in a fluid/fluid phase. Microscopically, they show domains of condensed complexes surrounded by a cholesterol rich phase [3, 19], but this µm-sized structures are at least two orders of magnitude smaller than the bright spots which we show in Fig. 11. As already shown in the literature [13, 18], cholesterol greatly reduces the shear viscosity of DPPC monolayers. The fluidifying effect of cholesterol in a saturated monolayer can be seen very well in our results by comparing the drop impact patterns of pure DPPC monolayer with the ones of DPPC with 60 mol % cholesterol. The last one present an anchor forming pattern similar to the drop impact pattern on the liquid DOPC and DOPC-cholesterol monolayers.

For monolayers of DPPC with 60 mol % cholesterol (Fig. 11) during the drop impact some big 3D structures, appearing like extra-bright spots, are formed with lipids from the drop and target monolayers. Earliest at 2.1 s after impact and at approximately one second after the drop liquid comes to rest in the anchor form, and after the drop bulk liquid starts to sink into the target liquid, the 3D structures start to disappear. We assume that their lipids reintegrate back in the existing drop or target monolayer because they transform into a larger bright round-area with approximately the same light intensity as the existing fluorescently labelled monolayer. Usually all extra-bright spots disappear in 12 to 33 s, whereas in some isolate cases the 3D structures are stable for more than 100 s.

We will discuss now the difference between the extra-bright 3D collapsed structures in the case of saturated DPPC monolayers with 60 mol % cholesterol and the ones of the pure DOPC monolayers. The former are formed from both drop and target monolayers and appears like extra-bright spots, unstable in time, the later ones are formed only from the target monolayer and appear like extra-bright large areas and are stable in time. We assume that this is due to the difference in structure of the two monolayers. The saturated monolayer with 60 mol % cholesterol contains two different phases: one of condensed complexes surrounded by a second phase of liquid cholesterol. Gopal et al. [10] and Leep et al. [9] show that the nucleation of the collapse takes place at the boundaries between the condensed domains and the fluid phase. This is in concordance with observations of Malcharek et al. [14] which found that the collapse of monolayers with high amount of cholesterol produces only isolated collapsed structures.

The unsaturated DOPC monolayer is in a liquid expanded homogeneous phase. Gopal and Lee [10] show that the liquid monolayers are not able to sustain large-scale folding and collapse on a smaller length scale by forming vesicle-like structures. These structures are stable in time and do not reintegrate into the existing monolayer like in the case of saturated DPPC with 60 mol % cholesterol.

Conclusions

Oblique drop impact was studied with phospholipid monolayers on both drop and target liquid surfaces. These experiments visualize the rheological properties of monolayers giving rise to complex pattern formation. During drop impact the monolayers composition and mechanical properties influence their dynamics and bilayer/multilayer formation. Asymmetric bilayer structures are formed by spontaneous aggregation of phospholipids captured under the thin sheet of the drop liquid. Symmetric bilayer structures are formed by collapse of the drop and target monolayers. Solid monolayers fracture during drop impact and bilayer/multilayer fragments are formed. For some monolayer compositions the 3D bilayer/multilayer structures reintegrate into the existing monolayer, as bilayer formation is reversible.

Acknowledgement Financial support by the DFG (LE 1119/3-1,2) is gratefully acknowledged.

References

1. Torchilin VP, Weissig V (2003) Liposomes – A Practical Approach. Oxford University Press, Oxford
2. Leneweit G, Koehler R, Roesner KG, Schäfer G (2005) Regimes of drop morphology in oblique impact on deep fluids. J Fluid Mech 543:303–331
3. Vrânceanu M, Winkler K, Nirschl H, Leneweit G (2008) Surface Rheology and Phase Transitions of Monolayers of Phospholipid/Cholesterol Mixtures. Biophys J 94:3924–3934
4. http://www.dow.com
5. Loglio G, Pandolfini P, Miller R, Makievski AV, Ravera R, Ferrari M, Liggieri L (2001) Drop and bubble shape analysis as a tool for dilational rheological studies of interfacial

layers. In: Möbius D, Miller R (eds) Novel Methods to Study Interfacial Layers. Studies in Interface Science. Elsevier, Amsterdam, pp 439–485

6. Vrânceanu M, Winkler K, Nirschl H, Leneweit G (2007) Surface rheology of monolayers of phospholipids and cholesterol measured with axisymmetric drop shape analysis. Colloids Surf A 311:140–153
7. Marrink SJ, Mark AE (2003) Molecular Dynamics Simulation of the Formation, Structure, and Dynamics of Small Phospholipid Vesicles. J Am Chem Soc 125:15233–15242
8. Ridsdale RA, Palaniyar N, Possmayer F, Harauz G (2001) Formation of folds and vesicles by dipalmitoylphosphatidylcholine monolayers spread in excess. J Membr Biol 180:21–32
9. Lipp MM, Lee KYC, Takamoto DY, Zasadzinski JA, Waring AJ (1998) Coexistence of Buckled and Flat Monolayers. Phys Rev Lett 81:1650–1653
10. Gopal A, Lee KYC (2001) Morphology and Collapse Transitions in Binary Phospholipid Monolayers. J Phys Chem B 105:10348–10354
11. Piknova B, Schram V, Hall SB (2002) Pulmonary surfactant: phase behavior and function. Curr Opin Struct Biol 12:487–494
12. Amrein M, V. Nahmen A, Sieber M (1997) A scanning force- and fluorescence light microscopy study of the structure and function of a model pulmonary surfactant. Eur Biophys J 26:349–357
13. Shah D, Schulman JH (1967) Influence of calcium, cholesterol, and unsaturation on lecithin monolayers. J Lipid Res 8:215–226
14. Malcharek S, Hinz A, Hilterhaus L, Galla H-J (2005) Multilayer Structures in Lipid Monolayer Films Containing Surfactant Protein C: Effects of Cholesterol and POPE. Biophys J 88:2638–2649
15. Nag K, Keough K (1993) Epifluorescence microscopic studies of monolayers containing mixtures of dioleoyl- and dipalmitoylphosphatidylcholines. Biophys J 65:1019–1026
16. Albrecht O, Gruler H, Sackmann E (1978) Polymorphism of Phospholipid Monolayers. J Phys (Paris) 39:301–313
17. Crane JM, Tamm LK (2004) Role of Cholesterol in the Formation and Nature of Lipid Rafts in Planar and Spherical Model Membranes. Biophys J 86:2965–2979
18. Evans RW, Williams MA, Tinoco J (1980) Surface viscosities of phospholipids alone and with cholesterol in monolayers at the air–water interface. Lipids 15:524–533
19. Worthman LAD, Nag K, Davis PJ, Keough KMW (1997) Cholesterol in condensed and fluid phosphatidylcholine monolayers studied by epifluorescence microscopy. Biophys J 72:2569–2580

Progr Colloid Polym Sci (2008) 134: 90–100
DOI 10.1007/2882_2008_074

Published online: 20 May 2008

E. Kettler
C. B. Müller
R. Klemp
M. Hloucha
T. Döring
W. von Rybinski
W. Richtering

Polymer-Stabilized Emulsions: Influence of Emulsion Components on Rheological Properties and Droplet Size

E. Kettler · C. B. Müller · R. Klemp · W. Richtering (✉)
Institute of Physical Chemistry, RWTH Aachen University, Landoltweg 2, 52056 Aachen, Germany
e-mail: richtering@rwth-aachen.de

M. Hloucha · T. Döring · W. von Rybinski
Research – Physical Chemistry, Henkel KGaA, Henkelstr. 67, 40191 Düsseldorf, Germany

Abstract We investigated o/w-emulsions containing polymeric thickener C_{10}–C_{30} acrylate (acrylate) and silicone-based emulsifier PEG-12 dimethicone (PEG-12) but without low molecular mass surfactants. Mechanical properties of emulsions were probed by oscillatory and continuous flow rheometry while droplet size was studied by flow particle image analysis (FPIA). By varying thickener and emulsifier content rheological properties and droplet size of emulsions changed significantly. Experimental results and a statistical analysis showed that the physical network, built up by acrylate in a concentration range from 0.1–1.0 wt. %, was the dominating factor for rheological properties and increased moduli and viscosity of emulsions. The development of droplet diameters revealed that a systematic control of droplet parameters was possible by increasing the PEG-12 concentration from 0.0–5.0 wt. %. In contrast, increasing acrylate concentration led to either large or small droplets. The influence of larger droplets in the emulsions was revealed when the arithmetic diameter and the Sauter diameter were compared and displayed huge differences. These differences resulted from a rather small amount of big droplets with diameters above 40 μm. An influence of oil droplets on emulsion elasticity was only observed for emulsions with low acrylate concentration (≤0.1 wt. %), because at higher concentrations the influence of oil droplets was superimposed by thickening properties of acrylate.

Keywords Droplet size · Emulsion · Polymer · Rheology

Introduction

Emulsions are mixtures of immiscible liquid phases in the presence of an emulsifier. In the case of classic oil in water (o/w-) emulsions the oil phase is dispersed into an aqueous phase by use of surfactants. Surfactants consist of a hydrophilic head group interacting with the polar phase and a hydrophobic tail which interacts with the non-polar phase. Thus, the usually used low molecular emulsifier adsorbs on the interface between the dispersed and continuous phase and reduces interfacial tension [1].

Microstructures of classic emulsions are influenced by various fabrication parameters such as energy input, mixer type (mechanically) [2–5], type of used oil, emulsifier (chemically) [6–10] and temperature protocol.

Temperature control is important in production of classic emulsions because non-ionic surfactants used as emulsifiers often form temperature dependent mesophases [11]. Variation in temperature control might lead to changes in morphology and therefore change viscoelastic properties, droplet size or appearance of emulsions.

In cosmetic applications optical properties of emulsions are rather relevant, they depend on droplet size, which is influenced by emulsifier.

To overcome temperature dependence of classic emulsions and to improve reproducible droplet sizes and skin compatibility, polymeric emulsifiers like dimethicones [poly(siloxanes)] [12], poly(acrylamidosulfonic) acids (AMP) [13] and hydrophobically modified poly(acrylates) with C_{12}-side chains [14] were used over the last years to substitute common surfactants. These polymer stabilized emulsions show good stability, high skin compatibility, even with sensitive skin types, high oil compatibility and a simplified production procedure [15].

Tamburic et al. [16] investigated polymeric thickener (acrylate) containing emulsions (0.2–0.4 wt. %) with high oil content (30 wt. %). The thickener is a poly(acrylic acid) with hydrophobic side chains (C_{10}–C_{30}) and cross linked by allylpentaerythritol. Continuous flow and oscillatory rheometry as well as texture analysis show that viscosity and elasticity of emulsions increase with higher acrylate concentrations, which is confirmed by our own measurements on emulsions with lower oil content (17 wt. %).

Usually pH has to be adjusted with a suitable base, since the acrylate is a polyacid, to get an optimal extended polymer conformation [16] which leads to thickening of emulsions. Final emulsion viscosity depends on acrylate concentration and is independent of temperature control during preparation in contrast to common emulsions. In emulsions acrylate adsorbs at the oil/water interface; at low concentrations of acrylate stabilization results from electro-steric effects and in the upper concentration range an associative thickening mechanism is postulated [17].

Goodrich [15] and Bremecker et al. [18] report medium droplet diameters of 20 to 25 μm, in the case of emulsions containing only acrylate (0.2–0.4 wt. %) and no emulsifier. Such large droplet sizes are often not acceptable in cosmetic applications.

To optimize droplet size and to enhance emulsifying properties in acrylate-stabilized emulsions use of non-ionic co-surfactants (sucrose esters) is proposed by Bobin et al. [19] who evaluated emulsions by means of rheology, microscopy and pH measurements.

Savic et al. [20] investigated emulsions stabilized by a combination of acrylate and conventional non-ionic surfactant keeping the oil concentration constant. Stability at different temperatures, droplet size and rheological properties of emulsions are analyzed. They find that use of non-ionic co-surfactants is necessary to decrease droplet size from 35 to 25 μm. Furthermore, non-ionic surfactants increase yield stress and viscosity of emulsions and improve aesthetic as well as applicative properties of emulsions.

The increase of moduli and viscosity depending on surfactant concentration is known from mixtures of hydrophobically modified poly(acrylates) and low molecular surfactants [21–25] in aqueous solution. Dimethicones, which have been in use for several years as co-surfactants in shampoo [26], lotions and sun-screen products [27], seem to be an appropriate substitute for low molecular co-surfactants. Suitthimeathegorn et al. [28] investigated the surface tension, controlled drug release and droplet size evaluation with time of anhydrous emulsions, which were stabilized by different dimethicones. A difference in the average particle sizes of 4 μm after one week storage compared to fresh emulsions was found.

The literature mentioned above on polymer-stabilized emulsions has, to our knowledge, dealt with emulsions containing only polymeric thickener or combinations of polymeric thickener and low molecular surfactants [21–29]. As a next step it seems rational to investigate systematically emulsions that combine the properties of polymeric thickener (acrylate) and polymeric emulsifier PEG-12. Knowledge on rheological properties of this type of emulsion might improve cosmetic applications, especially for sensitive skin types. Furthermore, it has to be clarified if there is a correlation between droplet size and elasticity in this type of emulsion.

Therefore, this contribution is focused on the influence of polymeric thickener (acrylate) and polymeric emulsifier PEG-12 on rheological properties (moduli, shear viscosity) and droplet size of emulsions. It will show how these properties can be systematically controlled depending on the combination of acrylate or PEG-12 concentration.

Investigations on moduli and shear viscosity are accomplished by oscillatory and continuous flow rheometry, while droplet size of emulsions is determined by flow particle image analysis (FPIA).

Various o/w-emulsions are examined containing 0.1–1.0 wt. % acrylate (constant PEG-12: 0.5 wt. %) and 0.0–5.0 wt. % PEG-12 (constant acrylate: 0.1 wt. %). Oil concentration is kept constant at an industrial commonly used value of 17 wt. % in emulsions where acrylate content is varied. In emulsions with different emulsifier content, water concentration was kept constant at 82.0 wt. %. Finally, a statistical analysis was applied which confirmed the possibility of systematic control of rheological parameters and droplet diameters by changing acrylate and PEG-12 concentration.

Experimental

Materials

All chemicals are used without further purification. Component names and suppliers are listed in Table 1.

Formulations of prepared emulsions are listed in Table 2. Sample names assemble as follows: SE47 – general name of investigated system, 1st number – acrylate content, 2nd number – emulsifier content, 3rd number – oil content (all in wt.%).

Table 1 Components used for emulsion preparation without further purification

Component	Function	INCI[a]	Supplier
Paraffin oil DAB 9	Oil, nonpolar	Paraffinum liquidum	Merck, Germany
PEG-12	Emulsifier	PEG-12 dimethicone	Dow Corning
Acrylate	Thickener, low modification	C_{10}–C_{30} acrylate	Noveon
Methylparabene	Preservative	Methylparabene, $C_8H_8O_3$	Merck, Germany
Propylparabene	Preservative	Propylparabene, $C_{10}H_{12}O_3$	Merck, Germany
Water	Solvent, polar	Water	Millipore

[a] INCI: **I**nt. **N**omenclature of **C**osmetic **I**ngredients

Table 2 Composition of investigated emulsions in wt.% [a]

Emulsion	Acrylate	PEG-12	Oil	Water
SE47-01-05-17	0.1	0.5	17.0	82.0
SE47-03-05-17	0.3	0.5	17.0	81.8
SE47-05-05-17	0.5	0.5	17.0	81.6
SE47-07-05-17	0.7	0.5	17.0	81.4
SE47-09-05-17	0.9	0.5	17.0	81.2
SE47-10-05-17	1.0	0.5	17.0	81.1
SE47-10-00-17	1.0	0.0	17.0	81.6
SE47-01-00-17	0.1	0.0	17.0	82.5
SE47-01-15-16	0.1	1.5	16.0	82.0
SE47-01-25-15	0.1	2.5	15.0	82.0
SE47-01-35-14	0.1	3.5	14.0	82.0
SE47-01-45-13	0.1	4.5	13.0	82.0
SE47-01-50-12	0.1	5.0	12.5	82.0

[a] Preservative concentration: 0.4 wt. % in each emulsion

Methods

Polymer-containing emulsions were prepared in a water bath by using an agitator type Eurostar digital (IKA Labortechnik, Germany) with a dispersing disk of 40 mm diameter. The oil phase consisting of paraffin oil, PEG-12 and propylparabene was preheated up to 50 °C under slow stirring (200 rpm). The aqueous phase containing water, acrylate and methylparabene was treated alike. Preheating was necessary to accelerate dissolving of acrylate in water. The aqueous phase was added to the oil phase under vigorous stirring (1700 rpm).

After 15 min emulsification the mixture was cooled down to RT by adding ice to the water bath. pH 6.5 was adjusted by using 0.1 M sodium hydroxide solution. Thickening of emulsions started when sodium hydroxide solution was added.

Rheological experiments were carried out on a stress-controlled rheometer type CVO-120 (Bohlin, Germany) with cone-plate geometry, cone-angle $\alpha = 1°$, cone radius $R = 20$ mm.

Investigations covered amplitude sweeps to specify the upper stress limit (σ_{max}) of the linear-viscoelastic (LVE) region. Stresses $\sigma < \sigma_{max}$ were used in frequency sweeps providing storage (G') and loss modulus (G''). The angular frequency (ω) was varied from 0.03 to 130 rad s^{-1}. Shear viscosity (η) of emulsions depending on shear stress ranging from 10 to 400 Pa was examined by measuring flow curves.

Droplet size and droplet size distribution were obtained with a Flow Particle Image Analyzer type FPIA-3000 (Sysmex/Malvern, UK) with a lens of 20-times magnification. The investigated size range covered 0.5 to 200 µm.

Results and Discussion

Viscoelastic Properties and Viscosity

Stress dependence of G' and G'' at 0.1 Hz is measured to determine the LVE-region and to investigate the influence of acrylate and PEG-12 concentration on yield stress of emulsions.

Increasing shear stress up to 20 Pa does not influence moduli of emulsions with acrylate concentrations higher than 0.1 wt. % (Fig. 1A). G'-values are constant and the LVE-region is not exceeded. The yield stress was always above the experimentally applied shear stress of 20 Pa. All emulsions exhibit a gel-like character ($G' > G''$). Increase of acrylate concentration leads to stepwise increase of G'-values (finally $G' \approx 600$ Pa at 1.0 wt. % acrylate). The fact that higher concentrations of hydrophobically modified thickeners in water like cellulose ethers [30–32] or poly(acrylic acids) [24, 26] lead to increased moduli and viscosities is generally explained with the agglomeration of hydrophobic side chains and a resulting network, which increases moduli and viscosity. Obviously this model is also valid in o/w-emulsions and the enhanced emulsion elasticity is caused by higher thickener content, whereas the influence of oil droplets on emulsion properties is subordinate.

An increasing PEG-12 concentration in emulsions with low acrylate content (0.1 wt. %) induces only small changes in G'-values (Fig. 1B). Shear stress dependence of storage modulus is evident as a clear decay of G' occurs above a certain σ_{max} in emulsions with increasing

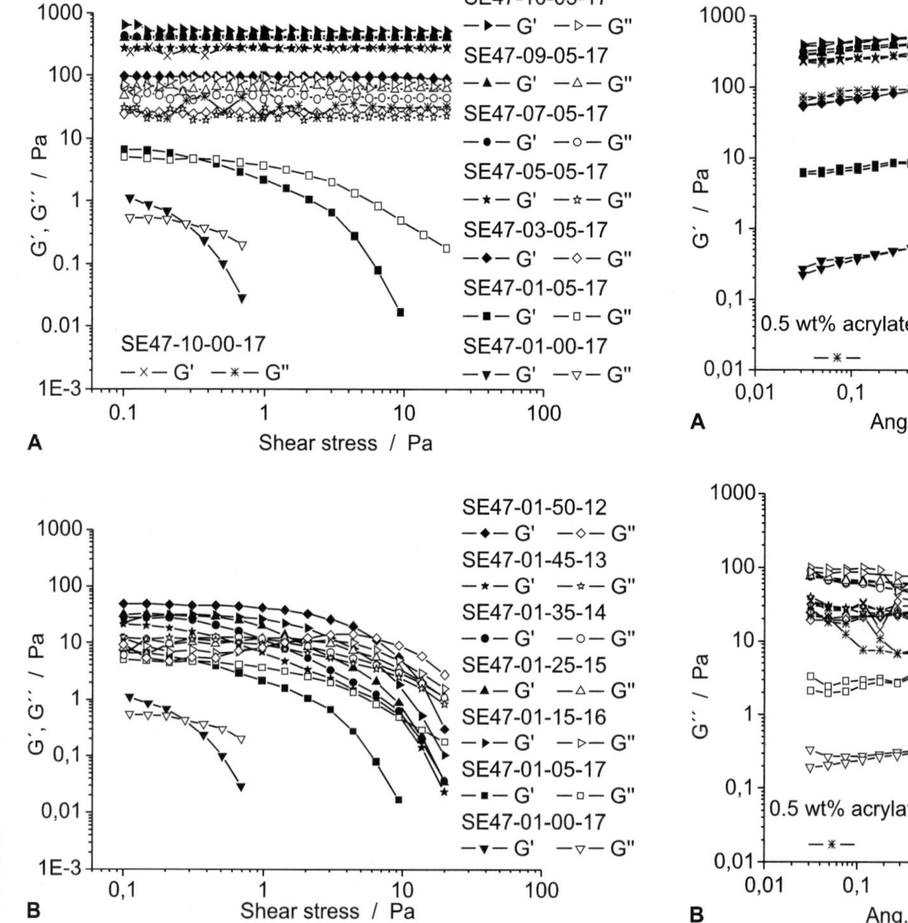

Fig. 1 **A** Stress dependence of G' and G'' at different acrylate concentrations (0.1–1.0 wt. %) and low PEG-12 concentration (0.5 wt. %). Applied frequency: $f = 0.1$ Hz. **B** Stress dependence of G' and G'' at different PEG-12 concentrations (0.0–5.0 wt. %) and low acrylate concentration (0.1 wt. %). Applied frequency: $f = 0.1$ Hz

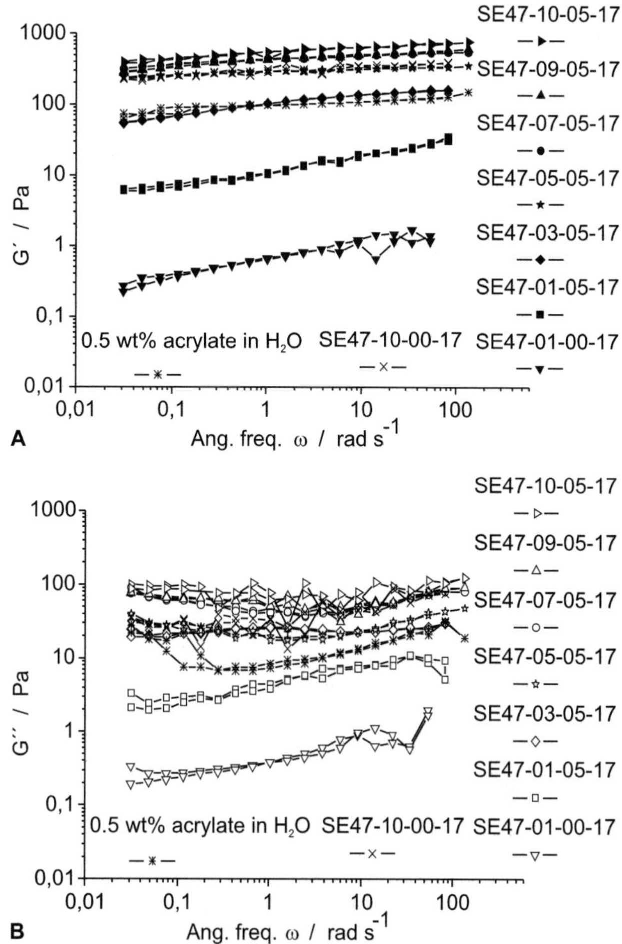

Fig. 2 **A** Frequency dependence of G' at different acrylate concentrations; applied stress SE47-01-00-17, SE47-01-05-17: $\sigma = 0.1$ Pa, other samples: $\sigma = 3.0$ Pa. **B** Frequency dependence of G'' at different acrylate concentrations; applied stress: SE47-01-00-17, SE47-01-05-17: $\sigma = 0.1$ Pa, other samples: $\sigma = 3.0$ Pa

PEG-12 concentration. This σ_{max} is shifted from < 0.1 Pa in SE47-01-00-17 (no PEG-12) to 1.0 Pa in SE47-01-50-12 (5.0 wt. % PEG-12), indicating improved elastic properties of emulsions. A similar behavior in yield stress values is reported by Savic et al. [20] for combinations of acrylate with classical low molecular surfactants.

Frequency dependence of storage (G') and loss modulus (G'') in emulsions with different acrylate and PEG-12 concentration is measured to evaluate influence of thickener and emulsifier concentration on the elasticity of emulsions.

In the whole range of measurement G'-values are higher than G''-values independent of acrylate or PEG-12 concentration (Figs. 2 and 3). The gel-like behavior shown here is also reported for multiple emulsion systems [33] and w/o-emulsions [34].

Lowest moduli are measured for emulsion SE47-01-00-17 (inverse triangles) with acrylate content of 0.1 wt. % and no emulsifier (Figs. 2 and 3). Frequency dependence of G' is weak but a steady increase of moduli towards high angular frequencies is observed. Addition of PEG-12 during emulsion preparation decreases frequency dependence of moduli. In SE47-01-50-12 (5.0 wt. % PEG-12, Fig. 3A: filled rhomb) frequency dependence is not as distinct as in emulsions with lower PEG-12 concentration and storage modulus is higher than in SE47-01-00-17.

The increase of acrylate concentration generally leads to higher and frequency independent moduli. Moreover, the presence of PEG-12 is unnecessary for frequency independence, if one compares G' of SE47-05-05-17 (filled stars) and SE47-10-00-17 (crosses) in Fig. 2A, which has the same value and progress. At an acrylate concentration of 0.7 wt. % (with 0.5 wt. % PEG-12 and 17 wt. % oil) G' becomes almost frequency independent around 500 Pa. This "plateau-value" is not raised significantly by further addition of acrylate, which indicates that an upper

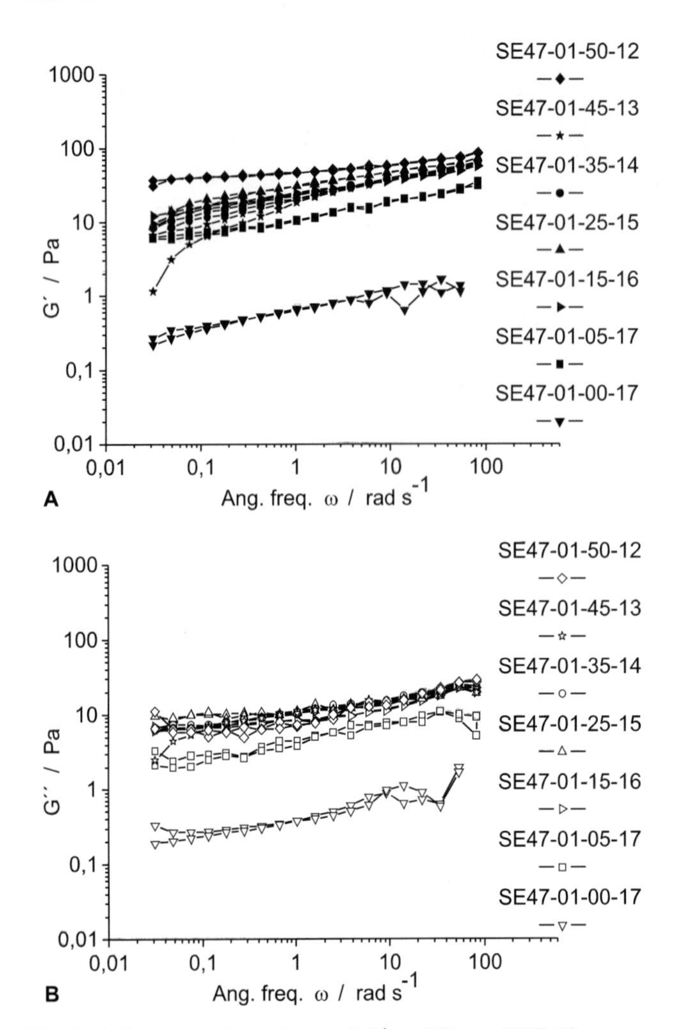

Fig. 3 **A** Frequency dependence of G' at different PEG-12 concentrations; applied stress SE47-01-00-17, SE47-01-05-17: $\sigma = 0.1$ Pa, other samples: $\sigma = 0.5$ Pa. **B** Frequency dependence of G'' at different PEG-12 concentrations; applied stress: SE47-01-00-17, SE47-01-05-17: $\sigma = 0.1$ Pa, other samples $\sigma = 0.5$ Pa

limit of emulsion elasticity is reached. A comparison of moduli from emulsions with and without emulsifier and a pure aqueous acrylate solution at an angular frequency of $\omega = 20\,\text{rad s}^{-1}$ will be described further below to answer the question whether or not the presence of oil droplets influences the elasticity of emulsions.

Flow curves demonstrate influence of changing PEG-12 concentration (0.0–5.0 wt. %) at a constant, low acrylate concentration (0.1 wt. %) on shear viscosity of emulsions.

All emulsions show reversible shear thinning behavior (Fig. 4), which indicates fully reversible structural changes within the scope of the experiment.

If PEG-12 concentration is increased from 0 to 1.5 wt. % the shear viscosity (η) at shear stresses of 10 Pa is increased about two orders of magnitude from 0.03 Pa s ($\eta_{(\text{SE47-01-00-17})}$) to 7 Pa s ($\eta_{(\text{SE47-01-15-16})}$). Further increase of PEG-12 concentration results in smaller increases of shear viscosity values. Similar results are reported by Sanchez et al. [35] for model emulsions containing a poly(ethylene glycol) derivative, sunflower oil and water.

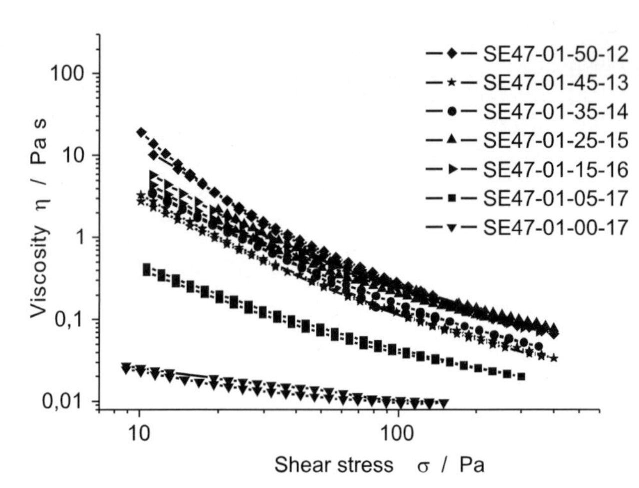

Fig. 4 *Flow curves*: shear stress dependence of shear viscosity in emulsions with low acrylate concentration (0.1 wt. %) and PEG-12 (concentration ranging from 0.0 to 5.0 wt. %). *Lines* are guidance for the eye

Droplet Size and Size Distribution

All emulsions were characterized with the FPIA-3000 in order to evaluate the influence of acrylate and PEG-12 concentration on droplet size and its distribution.

In the investigated concentration range of acrylate and PEG-12 droplet size decreases no matter which component concentration is increased. However, the influence of PEG-12 and acrylate on droplet size is significantly different.

The biggest average droplet diameter (d_{mean}), resulting from the emulsifying properties of acrylate, is 11 μm in emulsion SE47-01-00-17 (Table 3), which contains no emulsifier. Increasing acrylate concentration up to 1.0 wt. % leads to small droplets of $d_{\text{mean}} = 3.3$ μm, whereas addition of 0.5 wt. % PEG-12 in emulsion SE47-01-05-17 hardly decreases the droplet diameter at all.

Emulsifying properties of acrylate are very pronounced in this small concentration range and seem to outbalance the influence of PEG-12. By increasing acrylate concentration up to 1.0 wt. % (leaving PEG-12 constant at 0.5 wt. %) d_{mean} is reduced to 2.9 μm (Fig. 5, Table 3). A similar sharp diminishment of droplet diameter is not observed if PEG-12 concentration is increased, although the investigated concentration range is much broader than for acrylate (Table 2).

In emulsions with varied PEG-12 concentration a pronounced decrease of droplet size is observed when PEG-12 content exceeds 0.5 wt. % (Fig. 5). At the highest in-

Table 3 Influence of acrylate (0.1–1.0 wt. %) and PEG-12 concentration (0–5.0 wt. %) on droplet size. Droplet diameters d_S d_{90}, d_{mean}, d_{50} and d_{10} of emulsions in μm

Emulsion	Variation of acrylate concentration d_S	d_{90}	d_{mean}	d_{50}	d_{10}	Emulsion	Variation of PEG-12 concentration d_S	d_{90}	d_{mean}	d_{50}	d_{10}
SE47-01-05-17	36	31	10	4.3	1.7	SE47-01-00-17	63	30	11	3.8	1.5
SE47-03-05-17	40	7.1	3.6	2.3	1.4	SE47-01-05-17	36	31	10	4.3	1.7
SE47-05-05-17	20	6.2	3.4	2.4	1.4	SE47-01-15-16	56	11	5.5	2.9	1.5
SE47-07-05-17	16	5.2	3.1	2.1	1.4	SE47-01-25-15	36	11	5.4	3.1	1.6
SE47-09-05-17	13	5.4	3.0	2.2	1.4	SE47-01-35-14	15	7.9	3.9	2.6	1.5
SE47-10-05-17	11	4.9	2.9	2.2	1.4	SE47-01-45-13	22	4.6	2.7	2.0	1.4
SE47-10-00-17	12	5.8	3.3	2.1	1.4	SE47-01-50-12	44	5.8	3.5	2.8	1.6

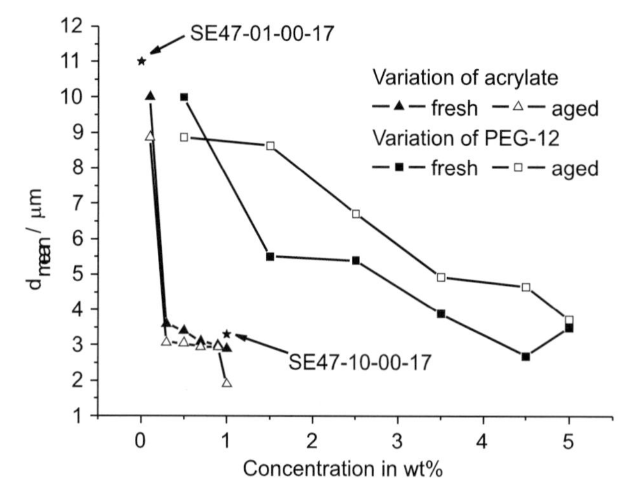

Fig. 5 Average droplet diameter (d_{mean}) of emulsions. *Triangles*: Variation of acrylate (0.1–1.0 wt. %), PEG-12 = 0.5 wt. %. *Squares*: Variation of PEG-12 (0.0–5.0 wt. %), acrylate = 0.1 wt. %. *Red stars*: SE47-01-00-17 (0.1 wt. % acrylate), SE47-10-00-17 (1.0 wt. % acrylate) no emulsifier, 17 wt. % oil. *Lines* are guidance for the eye

vestigated PEG-12 concentration of 5.0 wt. % (acrylate: 0.1 wt. %) the d_{mean} values become similar to that of high acrylate concentrations, but by increasing the PEG-12 concentration it is possible to decide between droplet sizes differing in diameter (approx. 1.2 μm), whereas by changing acrylate concentration the choice is between large and small droplets.

Emulsion SE47-01-00-17 comprises the lowest thickener and emulsifier content (Table 2). This emulsion has a high degree of polydispersity (Fig. 6A) in the range of 1 to 100 μm and shows a tendency towards a bimodal droplet size distribution. It is exemplary for emulsions with acrylate content of 0.1 wt. % and PEG-12 content up to 0.5 wt. %. Inhomogeneity indicates that the oil–water interface of oil droplets is less stabilized than in all other investigated emulsions, due to missing emulsifier or enough acrylate.

Emulsions with PEG-12 content of 1.5 to 2.5 wt. % (0.1 wt. % acrylate) still have a broad but monomodal droplet size distribution (Fig. 6C) and at PEG-12 concentrations ≥3.5 wt. % monomodal size distributions (Fig. 6F) with a lower degree of polydispersity appear. A similar behavior for model emulsions is reported in [35].

With low acrylate (0.1 wt. %) and PEG-12 concentrations (0.5 wt. %) a broad bimodal droplet size distribution is achieved (Fig. 7A). Further increase of acrylate content (0.7 wt. %) leads to a sharp decrease in width of droplet size distributions and the bimodal character vanishes. At concentrations ≥0.9 wt. % acrylate size distributions become even narrower. Emulsifying properties of acrylate, induced by hydrophobic side chains, are pronounced if acrylate concentration is high. However, it is only possible to choose between small and large droplets if only acrylate concentration is increased, because there is only one strong decrease of droplet size between 0.1 and 0.3 wt. %.

By increasing emulsifier content from 0.5 wt. % (in SE47-01-05-17) to 5.0 wt. % (in SE47-01-50-12) droplet diameters are reduced to $d_{90} = 5.8$ μm, $d_{50} = 2.8$ μm and $d_{10} = 1.6$ μm, respectively (Table 3). We can assume that more and smaller droplets are stabilized due to the higher emulsifier content.

The arithmetic droplet diameter (d_{mean}) is calculated by

$$d_{mean} = 2\left(\frac{\sum_i (ir_i)}{n}\right), \qquad (1)$$

with i: number of droplets of radius r_i and n: the total number of droplets. d_{90}, d_{50} and d_{10} are diameters which fractionate all measured droplets diameters in the way that 90, 50 and 10% of droplets are smaller than the values listed in Table 3. This intensifies the influence of smaller droplets on all of these diameters because the small droplets are numerous.

According to Sauter [36] the influence of larger droplets on the diameter can be stressed if an inhomogeneous emulsion is substituted by a model emulsion with droplets of uniform volume (Eq. 2) and surface (Eq. 3).

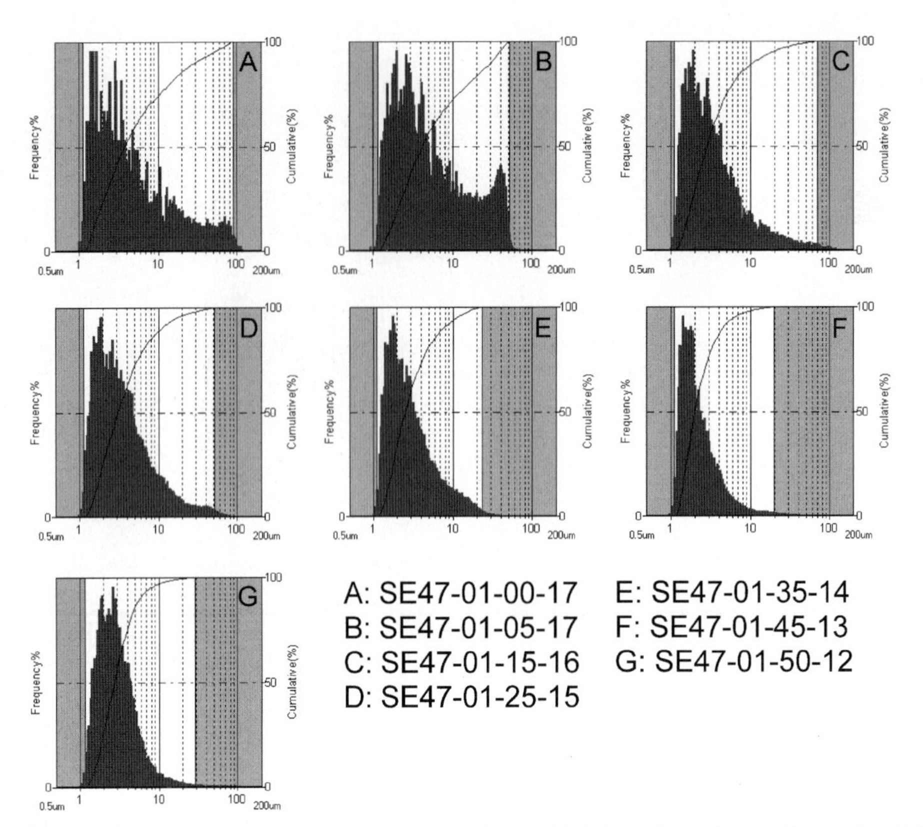

Fig. 6 Droplet size distributions in the range of 0.5 to 200 μm for emulsions with 0.1 wt. % acrylate and increasing PEG-12 concentration, measured with the FPIA. **A, B** typical for PEG-12 concentrations <0.5 wt. %. **C–E** typical for emulsions with PEG-12 content of 1.5 to 2.5 wt. %. **F, G** typical for emulsions with PEG-12 ≥3.5 wt. %

Furthermore, it is assumed that the model emulsion has the same total volume and droplet surface area as the original emulsion.

Droplet volume (V_{drop}) and droplet surface (S_{drop}) are calculated with

$$V_{\mathrm{drop}} = \frac{4}{3}\pi \sum_i \left(ir_i^3\right) \text{ and} \tag{2}$$

$$S_{\mathrm{drop}} = 4\pi \sum_i \left(ir_i^2\right) . \tag{3}$$

Afterwards the diameter d_{S} (Sauter diameter) is determined by

$$d_{\mathrm{S}} = 2\left(\frac{3V_{\mathrm{drop}}}{S_{\mathrm{drop}}}\right) = 2\left(\frac{\sum_i \left(ir_i^3\right)}{\sum_i \left(ir_i^2\right)}\right) . \tag{4}$$

This corresponds to the ratio of arithmetic mean volume and arithmetic mean droplet surface and comprises the droplet size distribution of the investigated emulsions. For a theoretical monomodal size distribution [36] the Sauter diameter is approx. two times higher than the arithmetic diameter, because the influence of larger but less numerous droplets is more pronounced.

The calculated d_{S} in our emulsions are four to ten times higher than d_{mean} values (Table 3), which resembles a broad monomodal or bimodal size distribution (Figs. 6 and 7). A small amount of droplets with diameters ≥40 μm or higher results in a sharp increase in d_{S} which explains the big differences between d_{S} and d_{mean} of emulsions SE47-01-00-17 to SE47-01-25-15. The polydispersity of emulsions SE47-01-35-14 to SE47-01-50-12 (Fig. 6F) is smaller and results in less pronounced differences between d_{mean} and d_{S}.

For maximum homogeneity a high emulsifier concentration is necessary for this kind of system. Nevertheless, formation of stable emulsions with quite small droplets can already be observed in emulsion with high acrylate content and no emulsifier (Table 3) and the observed size distributions are in the same order of magnitude as found by Jager-Lézer et al. [34].

Droplet diameters of emulsions stored for 12 months show smaller d_{mean} values than droplets of fresh emulsions in the case of acrylate variation (Fig. 5). For SE47-01-05-17 and SE47-10-05-17 the difference in d_{mean} is approx. 1 μm, whereas emulsions with concentrations between 0.3 and 0.9 wt. % acrylate have d_{mean} values which are 0.5 μm smaller compared to d_{mean} from fresh emulsions. Droplet size distributions of aged emulsions are less broad than

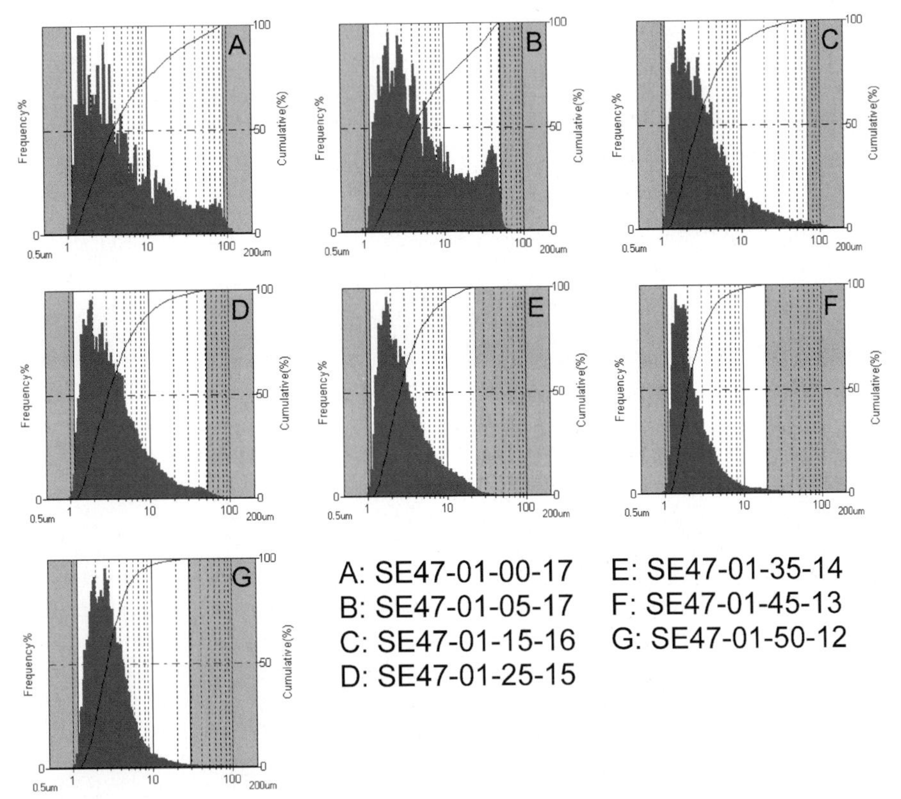

Fig. 7 Droplet size distributions in the range of 0.5 to 200 μm for emulsions with 0.5 wt. % PEG-12 and increasing acrylate concentration, measured with the FPIA. **A, B** typical for acrylate concentrations <0.3 wt. %. **C–E** typical for emulsions with acrylate content of 0.3–0.7 wt. %. **F, G** typical for emulsions with acrylate content ≥0.9 wt. %

in fresh emulsions and look like distributions shown in Fig. 6F. They show a smaller amount of big droplets which explains smaller d_{mean} values. The described emulsions with a combination of acrylate and PEG-12 are therefore quite stable.

For emulsions with low acrylate (0.1 wt. %) and PEG-12 concentrations up to 5.0 wt. % the d_{mean} values of aged emulsions are higher than in fresh emulsions. This is consistent with results from [18] and [28]. The droplet size distributions of the aged emulsions SE47-01-05-17 to SE47-01-50-12 reveal a polydispersity similar to that of fresh emulsions but the distributions are bimodal (compare Fig. 7A) with a clear increase of larger droplets with diameters of approx. 30–40 μm. This indicates droplet coalescence over the 12-month storage.

Comparison of Moduli and Droplet Size

To answer the question whether or not the presence of oil droplets influences the elasticity of emulsions the behavior of G' and d_{mean} with increasing acrylate or PEG-12 concentration have to be compared.

In the investigated concentration region it is possible to produce stable emulsions with droplets of $d_{mean} = 2.7\text{–}11\ \mu\text{m}$ (Fig. 5). The required amount of emulsifier is, however, a lot higher than the necessary amount of thickener. The comparison of moduli at $\omega = 20\ \text{rad s}^{-1}$ (Fig. 8) makes clear that changes in the magnitude of elasticity are easily achieved by varying acrylate concentration, whereas influence of PEG-12 concentration is quite low. PEG-12 seems to have the lower impact on rheological properties as well as on droplet diameters.

Emulsion SE47-01-00-17 contains 0.1 wt. % acrylate and no emulsifier. It has the lowest moduli and biggest droplet diameters. Addition of 0.5 wt. % PEG-12 increases the moduli by a factor of ten but droplet diameter is only reduced by 1 μm.

A solution of 0.5 wt. % acrylate in pure water, which matches approx. 0.4 wt. % acrylate in an emulsion, has almost the same moduli as emulsion SE47-03-05-17 (Figs. 2A and 8) with 0.3 wt. % acrylate. A stable emulsion without emulsifier but with 1.0 wt. % acrylate (SE47-10-00-17) has even higher G'-values and almost the same viscoelastic properties and droplet diameters as SE47-05-05-17 (0.5 wt. % acrylate). These examples show that acrylate, if it is dissolved in water, builds up a physical network, which dominates the rheological properties of emulsions.

At low acrylate concentrations (0.1 wt. %) an increase in PEG-12 content decreases the droplet size by a fac-

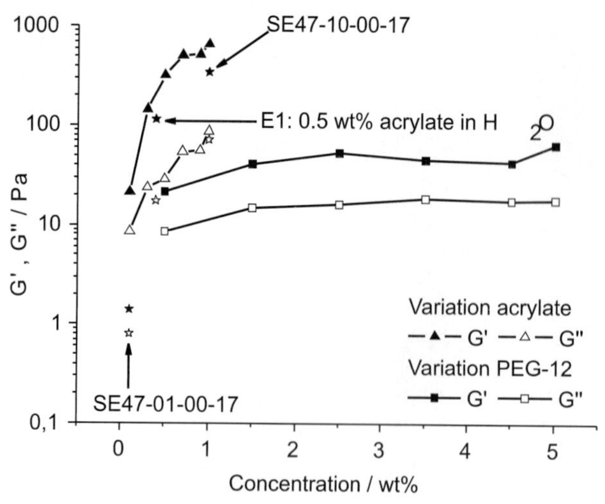

Fig. 8 G' (*closed symbols*) and G'' (*open symbols*) at $\omega = 20\,\mathrm{rad\,s^{-1}}$. *Triangles*: Influence of acrylate (0.1–1.0 wt. %), PEG-12 = 0.5 wt. % (x-axis: acrylate concentration). *Squares*: Influence of PEG-12 (0.0–5.0 wt. %), acrylate = 0.1 wt. % (x-axis: PEG-12 concentration). *Stars*: SE47-01-00-17 (0.1 wt. % acrylate), SE47-10-00-17 (1.0 wt. % acrylate) no emulsifier, 17 wt. % oil; E1: 0.5 wt. % acrylate in water—no oil, no emulsifier. *Lines* are guidance for the eye

tor of three. In contrast, the moduli increase from 20 to 65 Pa by a factor of three. The synergistic effect of acrylate and PEG-12 is even more pronounced in the change of viscosity (Fig. 4) by approx. two orders of magnitude. Formation of mixed micelles between thickening acrylate and polymeric PEG-12 is assumed. For other polymer/surfactant systems formation of mixed micelles is described in [21–25], explaining that aggregation of hydrophobic side chains of the polymer and surfactant is creating large intra- or intermolecular structures. The concentration of surfactant needed to build up this kind of structure is the critical aggregation concentration (cac), which is usually lower than the critical micelle concentration (cmc).

Correlation Parameters

To analyze the influence of emulsion components on rheological properties and droplet size a statistical analysis of form

$$f(A,B,C,D) = Ac_A + Bc_B + Cc_C + Dc_D + [AB]c_Ac_B + [AC]c_Ac_C + \ldots \quad (5)$$

A–D: correlation parameters (influence of single components),

$[AB]$, $[AC]$: cross correlation parameter of two components,

c_A, c_B etc.: component concentrations in wt.%

Table 4 Correlation parameters derived from Eq. 1

Correlation parameters	Component	G'	d_{mean}
A	Oil	−33	−1.9
B	acrylate	−13 175	+1388
C	PEG-12	+1125	+22
D	Aq. phase [c]	−1.7	0.0
AB	Oil/acrylate	+158	−14
AC	Oil/PEG-12	−12	−0.2
AD	Oil/aq. phase	+0.5	0.0
BC	Acrylate/DC193	−18	−12
BD	Acrylate/aq. phase	+137	−14
CD	PEG-12/aq. phase	−12	−0.3

[c] Aq. phase: water + preservative

is fitted to rheological parameters and droplet diameter (d_{mean}) of emulsions. The results are presented in Table 4.

The higher the absolute values of parameters G' and d_{mean} are the higher is the influence of a single component or combination of components on emulsion properties.

Acrylate influence on rheological properties of emulsions is approx. 10-times higher than the influence of PEG-12 (compare B and C for G' in Table 4). This statistical result corresponds to experimental data as increasing acrylate concentration has a stronger effect on moduli development than PEG-12 (Fig. 7).

The influence of acrylate on droplet size is even more pronounced and approx. 60-times higher than the influence of PEG-12 (compare B and C for d_{mean} in Table 4). This reflects the drastic d_{mean} decrease seen in Fig. 5 which is more pronounced than the decrease caused by PEG-12 content. These results show the strong impact of acrylate on rheological properties and droplet size in the low concentration regime (Figs. 5 and 7). However, differences in the influence of PEG-12 and acrylate on homogeneity of droplet sizes are not reproduced properly by the statistical analysis because d_{90}, d_{50} and d_{10}-values (Table 3) are neglected.

Conclusions

Emulsions combining properties of acrylate and PEG-12 are, due to good skin compatibility of the single components, especially useful in cosmetics for sensitive skin types. The experiments and the statistical analysis display that the physical network built up by acrylate in water is the dominating factor in the development of rheological properties in the investigated concentration range (Fig. 8). Regarding droplet diameters the only choice is between large ($d_{mean} = 10\,\mu\mathrm{m}$) or small ($d_{mean} = 3\,\mu\mathrm{m}$) droplets if acrylate concentration is higher than 0.1 wt. %. This is confirmed by droplet size distributions (Fig. 7) where a sharp decrease in the polydispersity is observed at acrylate concentrations above 0.1 wt. %.

In contrast, a systematic control of droplet diameter is possible if emulsifier concentration is increased. This is displayed by development of droplet diameters (Fig. 5) and the quite narrow size distributions for the bigger part of the droplets (Fig. 6). Nevertheless, existence and influence of large droplets in the emulsions is revealed when d_{mean} is compared to d_S. The big difference between these values (Table 3) is due to a small amount of large droplets seen in Fig. 6C.

A comparison of droplet diameters and moduli (Figs. 5, 8) shows that in the low concentration regime of acrylate an increase in PEG-12 concentration changes moduli and droplet diameters by a factor of three and viscosity by approx. two orders of magnitude. This behavior is due to formation of mixed micelles. If the acrylate concentration is higher than 0.1 wt. % these synergistic effects are superimposed by thickening properties of acrylate, which dominates the rheological behavior of the system.

This investigation shows that it is possible to produce emulsions with appealing optical properties (small droplet size) as well as needed rheological properties by combining a polymeric thickener with a polymeric emulsifier. The dominating role of acrylate on rheological properties is pronounced whereas an influence of oil droplets on emulsion elasticity is only seen for quite low concentrations of polymeric thickener.

Glossary

α	Cone-angle [°]
A–D	Correlation parameters (influence of single components)
$[AB]$, $[AC]$	Cross correlation parameter (two components)
c_A, c_B etc.	Component concentrations [wt.%]
d_{mean}	Arithmetic mean droplet diameter [μm]
d_s	Sauter diameter [μm]
d_{90}	90% of particles are smaller than d_{90}
d_{10}	10% of particles are smaller than d_{10}
G'	Storage modulus [Pa]
G''	Loss modulus [Pa]
f	Frequency [Hz]
LVE region	Linear viscoelastic region
M	Molar [mol/L]
min	Minute
R	Radius [mm]
r	Radius [cm]
rpm	Rotation per minute
RT	Room temperature
S_{drop}	Droplet surface [cm^2]
σ	Shear stress [Pa]
σ_{max}	Upper stress limit [Pa]
η	Shear viscosity [Pa s]
V_{drop}	Droplet volume [cm^3]
wt.%	Mass fraction
ω	Angular frequency [rad s^{-1}]

References

1. Aronson MP (1989) Langmuir 5:494–501
2. Schuchmann HP, Danner T (2004) Chem Ing Tech 4:364–375
3. Danner T (2001) Tropfenkoaleszenz in Emulsionen. GCA-Verlag, Karlsruhe
4. Rayner M, Trägardh G, Trägardh C (2005) Colloids Surf A 266:1–7
5. Vladisavljevic GT, Williams RA (2005) Adv Colloid Interf Sci 113:1–20
6. Bravin B, Peressini D, Sensidoni A (2004) Agric Food Chem 52:6448–6455
7. Wilde P, Mackie A, Husband F, Gunning P, Morris V (2004) Adv Colloid Interf Sci 108/109:63–71
8. Ma G, Gong F, Hu G, Hao D, Liu R, Wang R (2005) China Particuol 3:296–303
9. Cohen Stuart MA (2008) Colloid Polym Sci doi: 10.1007/s00396-008-1861-7
10. Balcan M, Anghel DF (2005) Colloid Polym Sci 283:982–986
11. Kavaliunas DR, Frank SG (1978) Colloid Interf Sci 66:586–588
12. Beck PH (1996) Patent No. WO 9614047, A1
13. Miller DJ, Morschhäuser R, Löffler M, Milbradt R, Stelter W (2004) Cosmet Toilet 119: 47–50
14. Perrin P, Monfreux N, Lafuma F (1999) Colloid Polym Sci 277:89–94
15. Goodrich Corp. (1994) Specialty Chemicals, Emulsification Properties. Bulletin 13
16. Tamburic S, Vuleta G, Simovic S, Milic J (1998) SÖFW-Journal 124:204–209
17. Lochhead RY, Rullson CJ (1994) Colloids Surf A 88:27–32
18. Bremecker KD, Koch B, Krause W, Neuenroth L (1992) Pharmazeut Ind 54:182–185
19. M-Bobin F, Michel V, Martini MC (1999) Colloids Surf A 152:53–58
20. Savic S, Milic J, Vuleta G, Primorac M (2002) STP Pharma Sci 12:321–327
21. Iliopoulos I, Wang TK, Audebert R (1991) Langmuir 7:617–619
22. Magny B, Iliopoulos I, Zana R, Audebert R (1994) Langmuir 10:3180–3187
23. Sarrazin-Cartalas A, Iliopoulos I, Audebert R, Olsson U (1994) Langmuir 10:1421–1426
24. Wang TK, Iliopoulos I, Audebert R (1988) Polymer Bull 20:577–582
25. Wang TK, Iliopoulos I, Audebert R (1991) Water Soluble Polymers ACS. Washington DC, pp 218–231
26. Kang KS, Park HH, Park DB, Cho YR (1999) Patent No. KR 182599, B1
27. Ogawa K, Ohashi K (2004) Patent No. JP 2002–259762, A2

28. Suitthimeathegorn O, Jaitely V, Florence AT (2005) Int Pharm J 298:367–371
29. Brugger B, Richtering W (2007) Adv Mater 19:2973–2978
30. Panmai S, Prud'homme RK, Peiffer DG (1999) Colloids Surf A 147:3–15
31. Piculell L, Egermayer M, Sjöström J (2003) Langmuir 7:617–619
32. Hoff E, Nyström B, Lindman B (2001) Langmuir 17:28–34
33. Michaut F, Perrin P, Hébraud P (2004) Langmuir 20:8576–8581
34. Jager-Lézer N, Tranchant J-F, Alard V, Vu C, Tchoreloff PC, Grossiord J-L (1998) Rheol Acta 37:129–138
35. Sanchez MC, Valencia C, Franco JM, Gallegos C (2001) J Colloid Interf Sci 241:226–232
36. Sauter J (1926) Forschungsarbeiten auf dem Gebiet des Ingenieurwesens, Heft 279, VDI-Verlag GmbH, Berlin

Progr Colloid Polym Sci (2008) 134: 101–110
DOI 10.1007/2882_2008_078

Published online: 13 June 2008

Omar A. El Seoud
Paulo A. R. Pires

FTIR and ^{1}H NMR Studies on the Structure of Water Solubilized by Reverse Aggregates of Dodecyltrimethylammonium Bromide; Didodecyldimethylammonium Bromide, and Their Mixtures in Organic Solvents

Abstract The structure of water solubilized by the reverse aggregates of dodecyltrimethylammonium bromide, $DoMe_3ABr$ in chloroform/*n*-heptane; didodecyldimethylammonium bromide, Do_2Me_2ABr, in *n*-heptane, and mixture of the two surfactants in the latter solvent has been probed by FTIR and ^{1}H NMR. The ν_{OD} band of solubilized HOD (4% D_2O in H_2O) has been recorded as a function of [water]/[surfactant] molar ratio, W/S. Curve fitting of this band showed the presence of a small peak at $(2375 \pm 12)\,cm^{-1}$ and a major one at $(2521 \pm 7)\,cm^{-1}$; the latter corresponds to $(92.5 \pm 1)\,\%$ of the total peak area. As a function of increasing W/S, ν_{OD} decreases, its full width at half-height increases and its area linearly increases over the W/S range investigated. Observed ^{1}H NMR chemical shift, δ_{obs}, of solubilized water, and the $C\underline{H}_2$–$N^+(CH_3)_3$, CH_2–$N^+(C\underline{H}_3)_3$ groups of $DoMe_3ABr$ change smoothly as a function of increasing W/S. Similar trends have been observed for Do_2Me_2ABr-solubilized water, and for water solubilized by a mixture of $DoMe_3ABr$ plus Do_2Me_2ABr. δ_{obs} for H_2O-D_2O mixtures solubilized by $DoMe_3ABr$ were measured as a function of the deuterium content of the aqueous nano-droplet. These data were employed to calculate the so called deuterium/protium "fractionation factor", φ_M, of the reverse aggregate-solubilized water. Plots of a function of δ_{obs} (for $\underline{H}$OD; $C\underline{H}_2$–$N^+(CH_3)_3$, and CH_2–$N^+(C\underline{H}_3)_3$) versus the atom fraction of deuterium in the aqueous nano-droplet were strictly linear, indicating that the value of φ_M is unity. The results of both techniques show that reverse aggregate-solubilized water does not seem to coexist in "layers" of different structures, as suggested, e.g., by the multi-state water solubilization model.

Keywords Cationic surfactants, reverse aggregate of · Interfacial water, structure of · Reverse micelles · Water-in-oil microemulsions

Omar A. El Seoud (✉) · Paulo A. R. Pires
Instituto de Química, Universidade de São Paulo, C.P. 26.077, 05599-970 São Paulo, Brazil
e-mail: elseoud@iq.usp.br

Introduction

Several surfactants aggregate in organic solvents of low polarity and relative permittivity, e.g., chloroform, and n-heptane (hereafter designated as heptane); traces of water are required for this aggregation. The molar ratio [water]/[surfactant], W/S, is usually used to designate the aggregates present in solution, reverse micelle, RM, or water-in-oil microemulsion, W/O μE. In RMs, the amount of solubilized water is ≤ the amount necessary to hydrate the surfactant head-group. Solubilization of water over and above this W/S threshold results in the formation of a W/O μE. Both aggregates have been used as "micro-reactors" for many classes of organic and inorganic reactions, including polymerization of water-soluble monomers and production of quasi mono-disperse inorganic particles [4, 12–15, 23, 24, 32, 35, 40, 42, 44, 79, 80].

The above-mentioned applications have provoked intense interest in studying the properties of reverse aggregate-solubilized water, e.g., its structure, microscopic polarity, and viscosity, as compared to those of bulk water and, for RMs of ionic surfactants, aqueous solutions of electrolytes. The majority of these studies have been on the anionic surfactant sodium bis(2-ethylhexyl)sulfosuccinate, AOT; less work has been carried out on reverse aggregates of cationic and nonionic surfactants. Figure 1 is a "snapshot" of a W/O μE formed by a mixture of single- and double-tail surfactants. It shows the types of water that have been claimed to exist within reverse aggregates; up to three "layers" [1, 2, 10, 19, 20, 34, 36, 37, 48, 50, 63, 66, 69, 88–91]. The first one, W_{bound}, is made of water molecules tightly bound to the surfactant head-group; the second, intermediate "layer", $W_{intermediate}$, refers to distorted H-bonded water species, e.g., cyclic dimers or higher aggregates with unfavorable H-bonds. The third, central "layer" contains bulk-like water, $W_{bulk\text{-}like}$; presence of the latter coincides with formation of the W/O μE. According to some authors, there is one more type of water, interfacial water, $W_{interfacial}$, which lies at the oil side of the interface. [37, 50]. This picture is based on the observation that several properties of RM-solubilized water show a pronounced change in wet RMs, followed by a smaller one in the W/O μE region. Note that the surfactant/water interface is not rigid, or static, but "undulates", a fraction of the counter-ions and surfactant monomers migrates from the interface into the water nano-droplet [24].

$DoMe_3ABr$ is soluble in chloroform/*n*-alkane mixtures, the resulting solution dissolves large quantities of water, up to W/S of ca. 24 for $CHCl_3$-heptane. On the

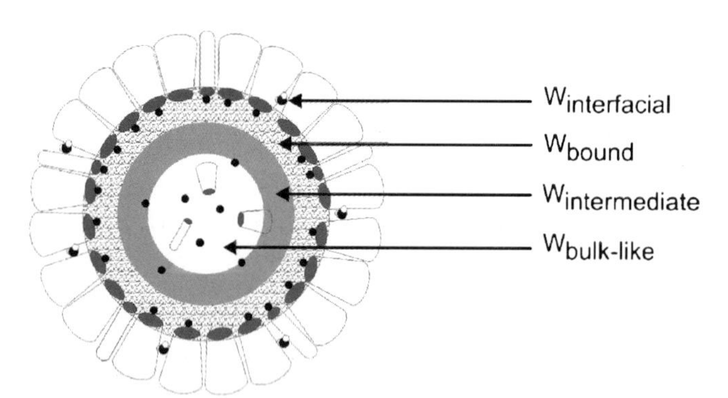

Fig. 1 Schematic representation of a W/O μE formed by a mixture of Do_2Me_2ABr and $DoMe_3ABr$ showing the types of water which have been claimed to be present in the system. The symbols refer to Do_2Me_2ABr ( ), $DoMe_3ABr$ ( ), the surfactant counter-ion (•), and water ( ). Species that migrated from the W/O interface into the aqueous nano-droplet are also shown, namely, a fraction of the surfactant counter-ion, and some surfactant molecules

other hand, Do_2Me_2ABr is soluble in heptane only in the presence of water. At 25 °C, the L2 μE phase is between W/S 18 to 26 [54]. Mixtures of $DoMe_3ABr$ and Do_2Me_2ABr in heptane dissolve much more water than either surfactant, e.g., up to W/S of 93 at 25 °C; [total surfactants] = 0.1 mol/L, $\chi_{DoMe_3ABr} = 0.28$, where χ refers to surfactant mole fraction, neglecting the solvent $\{\chi_{DoMe_3ABr} = [DoMe_3ABr]/([DoMe_3ABr] + [Do_2Me_2ABr])\}$ [11].

To our knowledge, there are no reports on the structure of water solubilized by dodecyltrimethylammonium halide RMs; by double-tailed cationic aggregates, or by mixed micelles of single- and double-tail cationic surfactants. Using FTIR and ^1H NMR, we have investigated the structure of water in the aqueous nano-droplet solubilized by reverse aggregates of $DoMe_3ABr$ in $CHCl_3$- or $CDCl_3$-heptane; aggregates of Do_2Me_2ABr in heptane, and mixed micelles of the two surfactants in the latter solvent. For simplicity, we use the terms W/S; "aqueous", and "water" to refer to the ratio of molar concentrations; the nano-droplet, and solubilized aqueous pseudo-phase, respectively. That is, these terms will be employed *irrespective* of the nature of the solubilizate, (pure) H_2O, or mixture of H_2O and D_2O, respectively. Note that W/S for the latter case is given by: W/S = $(n\ H_2O + n\ D_2O)/n$ (surfactant); *n* is the molarity of the species.

For the single-tail surfactant, curve fitting of the ν_{OD} band has resulted in a peak at $(2375 \pm 12)\,cm^{-1}$, and one at a higher frequency, $(2521 \pm 7)\,cm^{-1}$; the main peak corresponding to $(92.5 \pm 1)\,\%$ of the total peak area. As a function of increasing W/S, the area of the main peak increases linearly, without showing a break, expected due to the claimed formation of a second type of water. ^1H NMR has also been employed to study $DoMe_3ABr$-solubilized water. Values of the observed chemical shifts, δ_{obs}, of $\underline{H}OD$; $C\underline{H}_2\text{-}N^+(CH_3)_3$, and $CH_2\text{-}N^+(C\underline{H}_3)_3$ showed a smooth dependence on W/S. The same trend was observed for the system water-Do_2Me_2ABr/heptane, and for water solubilized by a mixture of $DoMe_3ABr$ plus Do_2Me_2ABr/heptane. Deuterium isotope effect on δ_{obs} of solubilized H_2O-D_2O mixtures, at constant values of W/S, has been employed to calculate the so-called deuterium/protium fractionation factor, φ_M, whose value bears on the structure of water within the aqueous nano-droplet [25–27, 59–61]. Results of both techniques indicate that Fig. 1 is an oversimplification, i.e., water solubilized by the above-mentioned cationic aggregates does not seem to exist in "layers" of different structures.

Experimental Section

Materials

The chemicals were purchased from Merck or SeAlopra Química (DF) and were purified as recommended elsewhere [3]. $CHCl_3$ and $CDCl_3$ were washed with water, dried, and kept in tightly closed brown bottles under nitrogen, in the dark, over activated type 4 Å molecular sieves. Before use, these solvents were tested for the presence of traces of phosgene by washing a sample with water, followed by the addition of acidified $AgNO_3$ solution to the aqueous phase. $DoMe_3ABr$ was recrystallized from acetone, and Do_2Me_2ABr was synthesized by the following scheme:

$$C_{12}H_{25}NH_2 + H_2C{=}O + HCO_2H \rightarrow C_{12}H_{25}N(CH_3)_2$$
$$C_{12}H_{25}N(CH_3)_2 + C_{12}H_{25}Br \rightarrow (C_{12}H_{25})N^+(CH_3)_2Br^- .$$

N,N-dimethyldodecylamine was prepared by the reductive methylation of 1-aminododecane [68], the product was purified by several fraction distillations, its purity was $> 99.5\%$, as determined by gas chromatography (Shimadzu 17A-2, equipped with a FID detector and Supelcowax 10 capillary column.). The reaction of 0.2 mole of the tertiary amine with 0.205 mole of purified 1-bromododecane in 50 mL acetonitrile was carried out in a Teflon-lined, stainless-steel reactor. The mixture was purged with N_2, and then kept for 8 h at 85 °C, under pressure (10 atmospheres). After cooling, the solvent was evaporated and the solid was recrystallized from acetone. Elemental analyses of both surfactants (as non-hygroscopic perchlorates) gave satisfactory results; Perkin-Elmer 2400 CHN apparatus; the Elemental Analyses Laboratory of this Institute.

Before use, each surfactant was dried at 50 °C, under reduced pressure, over P_4O_{10} until constant weight. Glass double-distilled H_2O was used throughout; D_2O was distilled before use. Its deuterium content was determined by ^{1}H NMR by using dioxane as an internal standard [72].

Methods

Surfactant stock solutions were prepared by weight. They contained the following solubilizates: (pure) H_2O, for recording the background of the FTIR experiment; 4% D_2O in H_2O, for measuring the effect of W/S on ν_{OD}; (pure) H_2O or (pure) D_2O, for the determination of φ_M. Effort has been made to ensure that these stock solutions contained the same amounts of surfactant and solvent, and equivalent amounts (within 0.1 wt %) of (pure) H_2O, (pure) D_2O, or a mixture of H_2O and D_2O, respectively. *This matching procedure has been followed in the preparation of each sample prepared therefrom.*

FTIR. A 0.2 mol/L solution of $DoMe_3ABr$ was prepared in chloroform/heptane (70 : 30 v/v). The following cells (Wilmad Glass, Buena) were used: CaF_2 (0.50 mm), ClearTran (0.20 and 0.10 mm) and NaCl (0.053 mm). The exact path length was calculated by the fringe method [18]. FTIR spectra were recorded with a Bruker Vector 22 FTIR spectrophotometer. The latter was fitted with a home-built, thermostated cell holder, whose temperature was controlled by circulating ethylene glycol. The temperature inside the IR cell was controlled at (25 ± 0.2) °C by using a digital thermometer (Novus SmartMeter, Porto Alegre). Transmission spectra were obtained by co-adding 32 spectra at 0.5 cm^{-1} resolution. The ν_{OD} spectral band is superimposed on a finite background. Provided that the reference and sample solutions were *carefully matched in composition*, this background can be approximated with the spectrum of $DoMe_3ABr$-solubilized H_2O, in the ν_{OD} spectral region [45,53,56,86]. Therefore the reference sample, *at each* W/S, was a surfactant solution containing (pure) H_2O. Deconvolution of the ν_{OD} band was carried out by commercial software (GRAMS/32 version 5.10, Galactic Industries, Salem).

^{1}H NMR. The "water" employed in the determination of the effects of W/S on δ_{obs} was a mixture of H_2O-D_2O, 25 : 75%, by volume. The following surfactant stock solutions were employed: $[DoMe_3ABr] = 0.2$ mol/L in $CDCl_3$-heptane; $[Do_2Me_2ABr] = 0.2$ mol/L in heptane; this solution contained solubilized water, W/S = 18.1; $[DoMe_3ABr] = 0.03$ mol/L plus $[Do_2Me_2ABr] = 0.17$ mol/L ($\chi_{DoMe_3ABr} = 0.15$) in heptane; this solution contained solubilized water, W/S = 30.6.

Values of φ_M were determined as follows: the desired atom fraction of deuterium in the aqueous nano-droplet was obtained by weighting, in 1 mL volumetric tubes, the appropriate volumes of *matched* surfactant stock solutions, one containing solubilized (pure) H_2O, the other solubilized (pure) D_2O. A Bruker DRX-500 NMR spectrometer was used. The spectra were recorded at 25.0 °C, at a digital resolution of 0.06 Hz/data point. The spectrometer probe temperature was periodically monitored, as recommended elsewhere [21]; values of δ_{obs} (after 10 min in the sample compartment for thermal equilibration, measured relative to internal TMS) were within the digital resolution limit.

Results

FTIR. Figure 2 shows typical ν_{OD} bands and corresponding band deconvolution. The results are for $DoMe_3ABr$ in $CHCl_3$-heptane (70 : 30, v/v); solubilized 4% D_2O in H_2O, at W/S = 4 and 16, respectively. The ν_{OD} peaks at other W/S (not shown) show a similar shape. Although curve fitting was carried out by considering contribution from Gaussian and Lorentzian components, our calculations showed that the bands are essentially Gaussian, in agreement with previous IR work on HOD in bulk aqueous phase [41,45,47,53,56,71,82–86], and in RMs and W/O μEs of anionic and cationic surfactants [58,59,61]. As shown, the agreement between the fitted and experimental curves is excellent.

The parameters that were calculated for the *main peak* of the ν_{OD} band are shown in Fig. 3. Part A shows

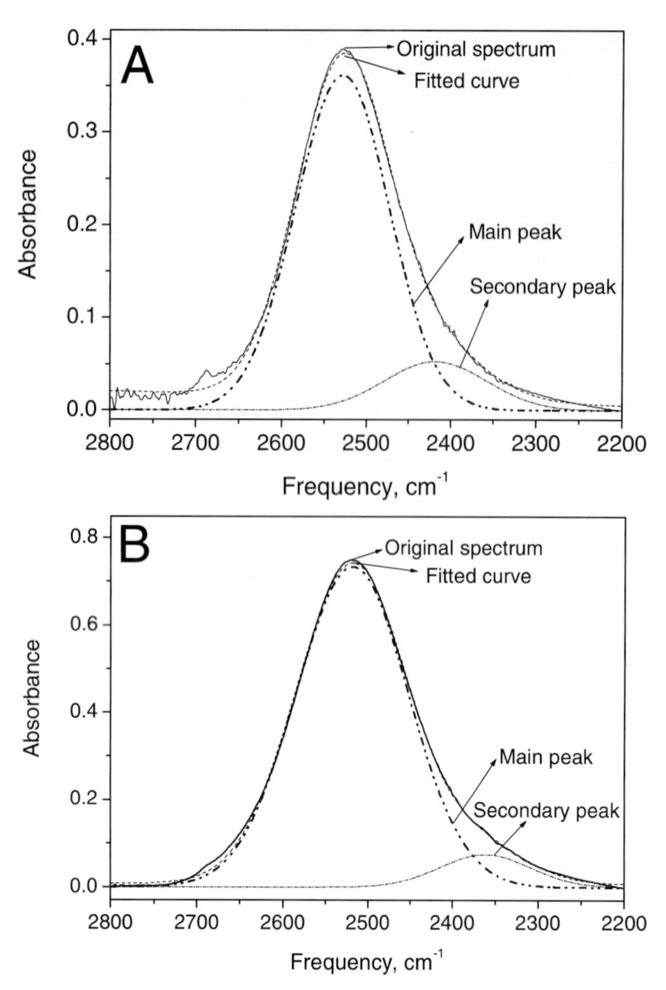

Fig. 2 Typical ν_{OD} bands and the corresponding band deconvolution. The results are for $DoMe_3ABr$, 0.2 mol/L in $CHCl_3$-heptane (70 : 30, v/v), at W/S = 4 (**A**) and 16 (**B**), at 25 °C. The aqueous nanodroplets are 4% D_2O in H_2O

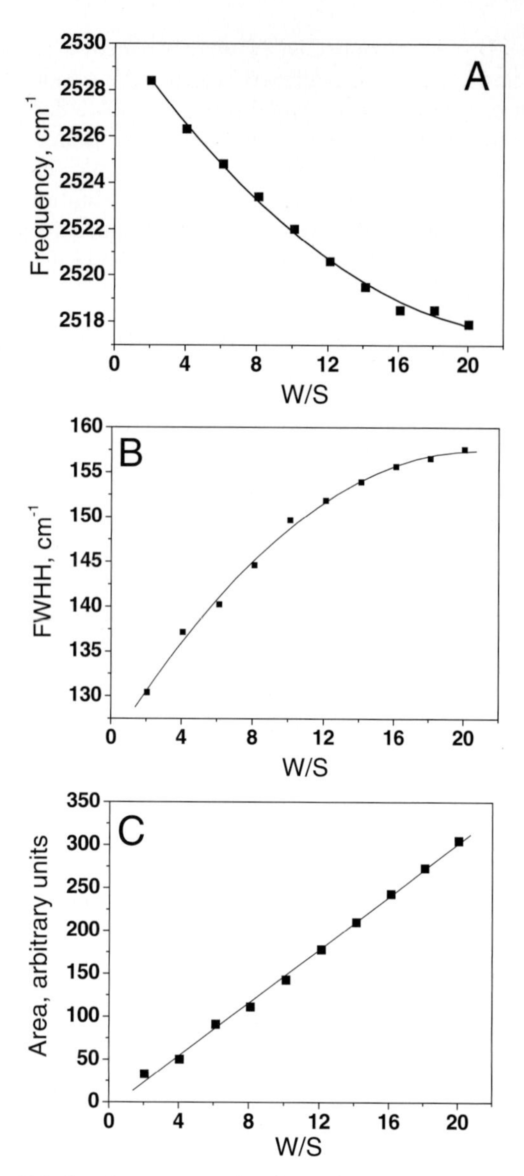

Fig. 3 Calculated parameters for the main peak of the ν_{OD} band. The results are for $DoMe_3ABr$, 0.2 mol/L in $CHCl_3$-heptane (70 : 30, v/v), 4% D_2O in H_2O; W/S = 2 to 20, 25 °C. The plots show the dependence on W/S of: The frequency (**A**); the full width at half height, FWHH (**B**); and the area (**C**)

a second-degree polynomial dependence of the peak frequency on W/S; part B shows a second-degree polynomial dependence of the peak full width at half height, FWHH, on W/S; part C depicts the linear correlation between peak area and W/S, correlation coefficient (r) and standard deviation (sd) are 0.9987 and 5.1516, respectively.

^1H NMR. Figures 4 to 6 show the dependence of δ_{obs}, for $\underline{H}OD$; $C\underline{H}_2$-$N^+(CH_3)_x$, and CH_2-$N^+(C\underline{H}_3)_x$ on W/S, where $x = 2$ or 3, for the single- and double-tail surfactant, respectively. The changes of δ_{obs} for $DoMe_3ABr$ are much larger because the starting W/S for Do_2Me_2ABr is large, > 18, i.e., well beyond the complete hydration of the head-ions [8, 52]. Consequently, more information about the structure of water can be gained by studying solubilization by $DoMe_3ABr$, where the effect of hydration of the head-ions on water structure is experimentally accessible. In Fig. 4, the δ_{obs} data for water are experimental and the solid curve was generated by the following exponential increase equation:

$$\delta_{obs} = A + Be^{\frac{(W/S)}{C}} . \quad (1)$$

Where A, B, and C are regression coefficients, whose values are: $A = 4.6436$, $B = -0.9320$, $C = -5.2869$, r^2 (regression coefficient for non-linear correlation) = 0.9978, for $DoMe_3ABr$, W/S = 3 to 24; $A = 4.7433$, $B = -0.3855$, $C = -10.4342$, $r^2 = 0.9999$, for Do_2Me_2ABr, W/S = 18 to 28. Likewise, an exponential decay equation

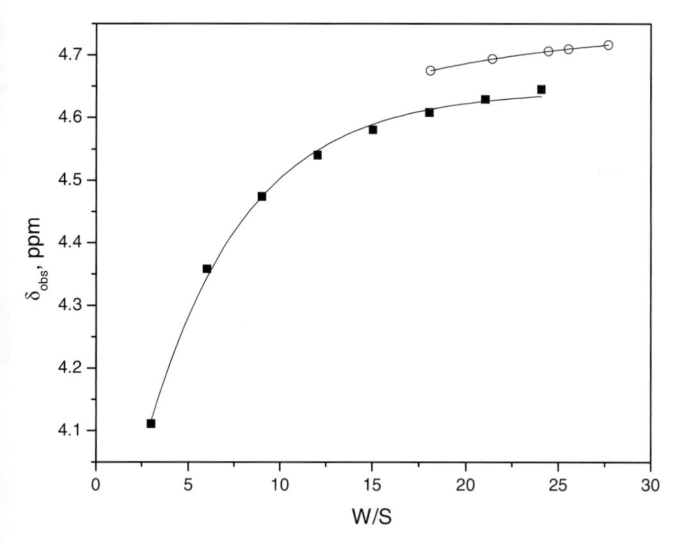

Fig. 4 Dependence of δ_{obs} of $\underline{H}OD$ on W/S for $DoMe_3ABr$ (■) and Do_2Me_2ABr (○). The *points* are experimental and the *solid curves* were generated by Eq. 1

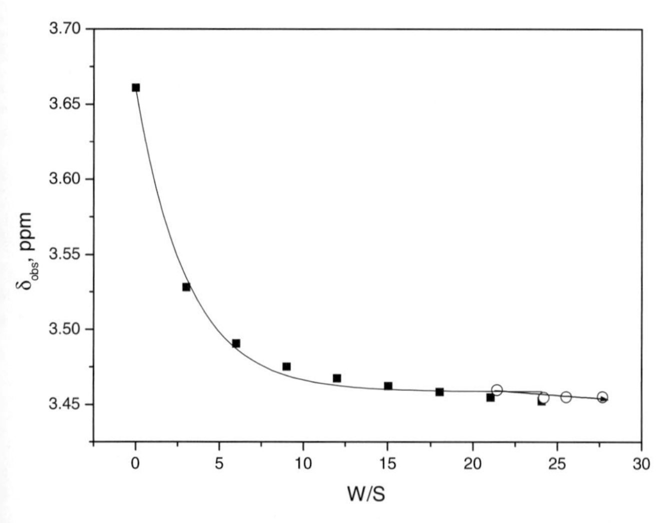

Fig. 5 Dependence of δ_{obs} of $C\underline{H}_2$-$N^+(CH_3)_x$ ($x = 3$ or 2) on W/S for $DoMe_3ABr$ (■) and Do_2Me_2ABr (○). The *points* are experimental and the *solid curves* were generated by Eq. 2

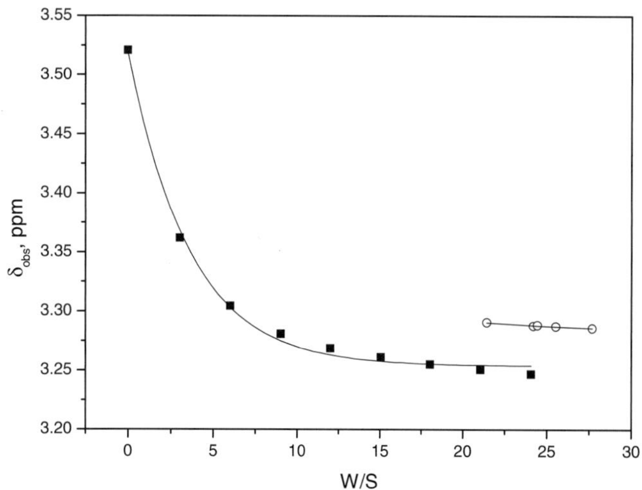

Fig. 6 Dependence of δ_{obs} of CH_2-$N^+(C\underline{H}_3)_x$ ($x = 3$ or 2) on W/S for $DoMe_3ABr$ (■) and Do_2Me_2ABr (○). The *points* are experimental and the *solid curves* were generated by Eq. 2

can be employed to describe the dependence of δ_{obs} on W/S for the surfactant head-group protons:

$$\delta_{obs} = A + Be^{-\frac{(W/S)}{C}}, \qquad (2)$$

The following values were calculated for the dependence of δ_{obs} of $C\underline{H}_2$-$N^+(CH_3)_x$ on W/S, where $x = 2$ or 3: $A = 3.4587$, $B = 0.2008$, $C = 3.0718$, $r^2 = 0.9952$, W/S $= 0$ to 24, $DoMe_3ABr$; $A = 3.4550$, $B = 1.2481$, $C = 0.2963$, $r^2 = 0.9939$, W/S $= 21.4$ to 27.7, Do_2Me_2ABr. The corresponding values for the dependence of δ_{obs} of CH_2-$N^+(C\underline{H}_3)_x$ on W/S are: $A = 3.2537$, $B = 0.2654$, $C = 3.5820$, $r^2 = 0.9970$, $DoMe_3ABr$; $A = 3.2801$, $B = 0.0782$, $C = 10.7106$, $r^2 = 0.9745$, Do_2Me_2ABr.

Discussion

FTIR. The observation times of IR and Raman spectroscopy are shorter (10^{-12}–10^{-14} s) than the time scales on which water molecules are expected to interchange with each other (between 10^{-7} and 10^{-11} s) [39, 73, 77, 78]. Therefore, both techniques are especially suited to detect different types of water at an interface. Thus, water molecules present in different environments should show up as separate bands in the IR or Raman spectrum, *provided that the difference in vibrational energies is suitably large*. Quantitative treatment of IR and Raman experimental data requires "a priori" hypothesis on the origin of the vibrational dynamics of the system under analysis. The model suggested should fit the data accurately (i.e., with the least possible error), and *agree with chemistry*. The latter proviso is central to curve fitting of spectroscopic data [49, 81, 87], and indeed to *any problem* whose solution relies on curve fitting [33]. Consequently, it is expected that examination of the same system by a certain spectroscopic technique should yield the same conclusions with respect to the number of water "layers" present within the reverse aggregate, and the relative concentration of each type. This, however, is not the case, as shown by the following IR and Raman data on RMs and W/O μEs of AOT and cetyltrimethylammonium bromide: (i) The number of water types present within the aggregate varies from one to three [1, 2, 10, 17, 19, 20, 34, 36, 37, 48, 50, 58, 59, 63, 64, 88, 89], (ii) More than one layer (or type) of water is claimed to be present at W/S ≤ 3 [19, 63, 65], a difficult-to-accept assumption because the sum of hydration numbers of the Na^+ and HSO_3^- ions (model for AOT head-ions) is > 5 [8, 52]; (iii) The reported dependence of W_{bound} on W/S is not uniform, being quadratic in one case [63], and complex (higher than fifth power dependence!) in other cases [20,37]; (iv) Curve fitting-based hydration numbers of AOT vary widely, 3.5 [2,19,63], 6.7 [34], and 12 [37]. In summary, examination of available literature shows that the question of presence of layers of water has not been settled.

We now address the results obtained. At the outset, the changes in frequency and FWHH observed in Fig. 3 are due to the hydration of the head-ions. There are two alternative interpretations of the results shown in Fig. 2: (i) The two peaks obtained by curve fitting correspond to different types of water in the nano-droplet, namely, W_{bound}, and $W_{bulk\text{-}like}$, respectively, (ii) There is one type of water present that gives rise to the observed main peak, the small peak need not be associated with HOD molecules present in a layer of different structure, as implied by the multi-state water solubilization model. If the former model is valid, one expects a correlation—*that agrees with chemistry*—between areas of the two peaks and W/S. The combined hydration numbers of Br^- and Me_4N^+ (model for the surfactant head-ions) is 5 ± 1 [29,51,52, 62]. Accordingly, the fraction of W_{bound} should level off at W/S of ca. 5, this is not observed in part (C) of Fig. 3. The fact that the ratio of the areas (main peak)/(total peak) is independent of W/S argues against any model based on discrete structures of the aggregate-solubilized water.

Therefore, the assumption made previously that the bands obtained by curve fitting of ν_{OH} (of solubilized *pure* H_2O) or ν_{OD} (of solubilized *pure* D_2O) are due to different types of water is an oversimplification; these bands may originate from coupled vibrations, and from a bending overtone often reported in the spectra of bulk H_2O or D_2O [16,17,22,55,64,70,74,75]. On the other hand, deconvolution of ν_{OH} or ν_{OD} vibrations of HOD is straightforward because both frequencies are essentially decoupled, provided that $[D_2O] \leq 10\%$, by volume [56,87]. This advantage has been recognized both for bulk aqueous phase [45,53,56,86,87], and for reverse aggregates [2,17, 58,59,61,65]. In curve fitting of the υ_{OD} peak, we have employed a small, second peak in order to get a better fit, Fig. 2. This use need not be associated with a second type of water within the aqueous nano-droplet, because the ν_{OD} peak is asymmetric, as shown elsewhere for bulk HOD [16,17,22,45,53,55,56,64,70,74,75,86]. That is, our IR data are best explained without resorting to the coexistence of structurally different water layers within the aqueous nano-droplet.

^{1}H NMR. We now examine the ^{1}H NMR data, especially Fig. 4, where δ_{obs} shows large dependence on W/S. The curves of water as well as those of the surfactant protons show no discontinuity at W/S of ca. 5, the threshold of formation of W/O μE. Such break is expected due to the formation of a new type of water, $W_{bulk\text{-}like}$. It is relevant that the dependence of δ_{obs} on W/S, for solubilized water, and for the surfactant head-groups of reverse aggregates of AOT and cetyltrimethylammonium bromide are also described by equations similar to Eqs. 1 and 2, *up to the phase separation limits* [25–27,61].

We have employed the dependence of δ_{obs} of solubilized H_2O-D_2O mixtures on their deuterium content to calculate the deuterium/protium fractionation factor, φ_M, for micelle solubilized water. The power of this technique is that calculation of φ_M does not require a *preconceived model* of structure of water in the nano-droplet. Calculation of φ_M from NMR data has been discussed elsewhere [25–27,38,72], so that only essential details will be covered. In case of a W/O μE, applying for simplicity the two-state water solubilization model and taking into account that water in the second layer is described as bulk-like water, φ_M is given by:

$$\varphi_M = (D/H)_{\text{bound water}} / (D/H)_{\text{bulk-like water}} \,. \tag{3}$$

The equation that is used to calculate φ_M is [25,38]:

$$(\delta_{HD} - \delta_H)/(\delta_D - \delta_H) = \{\varphi_M / [(1 - \chi_D) + \varphi_M \chi_D]\} \chi_D \,. \tag{4}$$

where δ_{HD}, δ_H, and δ_D refer to the observed chemical shift for aggregate-solubilized H_2O-D_2O, (pure) H_2O, and (pure) D_2O, respectively, and χ_D is the atom fraction of deuterium in the aqueous nano-droplet. A plot of the left-hand term of Eq. 4 versus χ_D should be linear for $\varphi_M = 1$, curve down for $\varphi_M < 1$, or curve up for $\varphi_M > 1$, this has been verified experimentally, outside the micellar domain [72].

Next we show how the value of φ_M is interpreted in terms of the structure of reverse aggregate-solubilized water, relative to bulk water. In a system where (^{2}H) and (^{1}H) equilibrate among a number of sites—bound and bulk-like water in the present case—deuterions accumulate, relative to protiums, at sites where they are most closely confined by potential barriers. Now, water in the first layer is expected to be more structured than that in the second layer. Therefore, (^{2}H) will tend to accumulate in bound water (i.e., that at the surfactant/water interface), relative to bulk-like water (in the center of the nano-droplet). Consequently, the fractionation factor for the aggregate-solubilized water is expected to be *greater than unity*, as can be deduced from Eq. 4.

Figures 7 and 8 show the result of application of Eq. 4 to the ^{1}H NMR data of $DoMe_3ABr$-water system, at W/S of 12 and 24, respectively. As discussed elsewhere for aqueous electrolyte solutions [38], and for AOT [26], Eq. 4 applies to *any* nuclei of both solvent and solute, respectively. The reason for the latter is that the solute (surfactant head-groups in the present case) is hydrated; this hydration is sensitive to the isotopic composition of the solvent, *provided that the structure of the solvent in the hydration shell is different from that in bulk solution.* Considering that the aqueous nano-droplet is akin to an electrolyte solution, we have applied Eq. 4 to solubilized water, and to

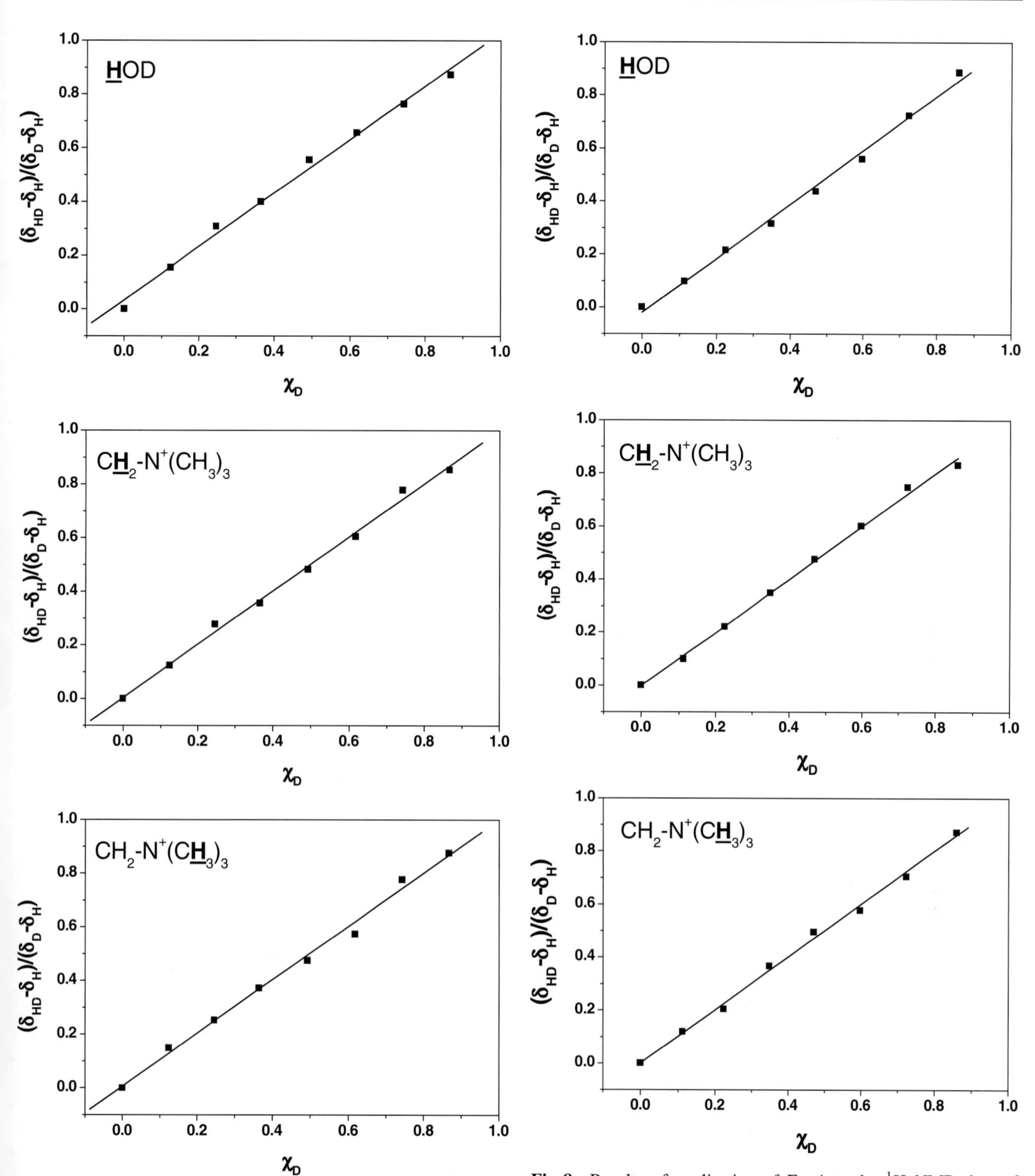

Fig. 7 Results of application of Eq. 4 to the ^{1}H NMR data of DoMe$_3$ABr-water system, at W/S = 12. The correlation coefficients (r) of the linear regressions are: 0.9972 (H̲OD), 0.9977 (CH̲$_2$–N$^+$(CH$_3$)$_3$) and 0.9968 (CH$_2$–N$^+$(CH̲$_3$)$_3$)

Fig. 8 Results of application of Eq. 4 to the ^{1}H NMR data of DoMe$_3$ABr-water system, at W/S = 24. Values of (r) of the linear regressions are: 0.9977 (H̲OD), 0.9987 (CH̲$_2$–N$^+$(CH$_3$)$_3$) and 0.9982 (CH$_2$–N$^+$(CH̲$_3$)$_3$)

the surfactant head-cation. As shown, *all plots are perfectly linear*, a result that does not agree with a multi-layer water solubilization model, where plots of $(\delta_{HD} - \delta_H)/(\delta_D - \delta_H)$ versus χ_D should have curved up, as discussed above.

Before addressing the previous result, the following points should be borne in mind: (i) As shown by Eq. 3, the fractionation factor is an equilibrium constant. That is, the relationship between the time-scales of NMR, and that of the diffusion of water molecules between bound- and bulk-like water, if present, *has no bearing* on the calculation of φ_M, (ii) Although fractionation factors like other secondary isotope effects are close to unity, they can be measured with high precision by NMR spectroscopy. For example, differences in φ as small as 1% have been reported [25, 38, 72]. Additionally, if we take W/S $= 5$ as the threshold of formation of a W/O microemulsion (vide supra), then the nano-droplets examined in the present work contained enough water to form a second, bulk-like water layer. Consequently, D/H fractionation between these layers should have occurred, *and experimentally observed*. Therefore, the unity fractionation factor obtained in the present work is neither due to a lack of sensitivity of the experimental technique nor to the presence of a negligible fraction of bulk-like water; (iii) We emphasize that the unity φ_M calculated for $DoMe_3ABr$-solubilized water does not mean that it is similar to bulk water (for which $\varphi = 1$, by definition). It means, however, that we did not detect the coexistence of two, or more water layers within the water pool that are structurally different (as implied by the two-state solubilization model) allowing D/H fractionation to occur, which should have resulted in $\varphi_M > 1$.

In summary, the results of both techniques indicate that treatment of experimental data in terms of the coexistence of structurally different water "layers" within the pool is probably an oversimplification. Water seems to be present as one pseudo-phase, whose properties change continuously as more water is solubilized. At high W/S these properties are akin, but not equal to those of water in electrolyte solutions. This conclusion agrees with IR and NMR studies of water within reverse aggregates of ionic and nonionic surfactants [17, 25–28, 58, 59, 64], fluorescence measurements in RMs [6, 7], NMR studies of concentrated salt solutions [5, 9], IR results of HOD in bulk aqueous phase [82–84], theoretical calculations on molecular dynamics of water [76], dielectric relaxation of water in hydrated phospholipid bilayers [30], and measurement of water chemical potential in the presence of phospholipid bilayer membranes [46].

The following factors contribute to an averaging of water structure over the *whole volume* of the aqueous nano-droplet: (i) Water within the pool is confined to such a small volume that it is highly likely that each molecule is simultaneously affected by several neighboring head-groups located at an undulating interface [43]. This may preclude the formation of water "layers" within the nano-droplet; (ii) Hydration of the species within the aqueous nano-droplet should be considered. These include a fraction of the surfactant counter-ion, and some surfactant monomers that migrate from the W/O interface [23]. Hydration of the latter species, and the possibility that the electrostatic fields (originated from opposite sides of the interface) do not decay to zero at the center aqueous nano-droplet [31, 67], contribute to an averaging out of water structure within the aqueous nano-droplet [57].

Conclusions

We used noninvasive techniques to investigate the state of solubilized water in RMs and W/O µEs of single- and double-tailed cationic surfactants. Although curve fitting of the ν_{OD} band required the use of two peaks, the relationship between individual peak areas and W/S does not support a multi-layer structural model. Calculated fractionation factor also points out in the same direction. Therefore the treatment of experimental data in terms of the coexistence of structurally different water layers within the aqueous nano-droplet is an oversimplification. The change in the slope of plots of certain physical properties as a function of increasing W/S, which is observed at the threshold of formation of the W/O µE may well reflect the expected decrease in water-surfactant interactions after completion of the hydration of the head-group. Factors responsible for averaging of water structure within the pool are discussed. The present NMR results are decisive because the information regarding the state of solubilized water (based on φ_M) *does not rely on a preconceived model of water structure.*

Acknowledgement We thank the FAPESP (State of São Paulo Research Foundation) for financial support. O. A. El Seoud thanks the CNPq (National Council for Scientific research) for Research Productivity fellowship. We thank Prof. Ali Bumajdad from Kuwait University for helpful discussions during the planning of this project, Cezar Guizzo and Gabriel O. El Seoud for the synthesis of Do_2Me_2ABr, and drawing Fig. 1, respectively.

References

1. Aliotta F, Migliardo P, Donato DI, Liveri VT, Bardez E, Larry B (1992) Prog Colloid Polym Sci 89:258
2. Amico P, D'Angelo M, Onori G, Santucci A (1995) Nuovo Cimento 17D:1053
3. Armarego WLF, Perrin DD (1998) Purification of Laboratory Chemicals. Butterworth-Heinemann, Oxford

4. Attwood D, Florence AT (1984) Surfactant Systems: Their Chemistry, Pharmacy, and Biology. Chapman and Hall, London
5. Baianu IC, Boden M, Lightowlers D, Mortimer M (1978) Chem Phys Lett 54:169
6. Belletête M, Droucher GJ (1989) J Colloid Interf Sci 134:289
7. Belletête M, Lachapelle M, Droucher G (1990) J Phys Chem 94:5337
8. Bockris JO, Reddy AKN (1973) Modern Electrochemistry. Plenum Press, New York
9. Boden N, Mortimer M (1978) J Chem Soc Faraday Trans 2 74:353
10. Boicelli CA, Giomini M, Giuliani AM (1984) Appl Spectrosc 38:537
11. Bumajdad A, Eastoe J, Griffiths P, Steytler DC, Heenan RK, Lu JR, Timmins P (1999) Langmuir 15:5271
12. Bumajdad A, Eastoe J, Zaki MI, Heenan RK, Pasupulety L (2007) J Colloid Interf Sci 312:68
13. Candau F, Leong YS, Fitch RM (1985) J Polym Sci 23:195
14. Candau F, Leong YS, Poyet G, Candau SJ (1984) J Colloid Interf Sci 101:167
15. Candau F, Zekhini Z, Durand JP (1986) J Colloid Interf Sci 114:398
16. Carey DM, Korenowski GM (1998) J Chem Phys 108:2669
17. Christopher DJ, Yarwood J, Belton PS, Hills B (1992) J Colloid Interf Sci 152:465
18. Compton SV, Compton DAC, Coleman PB (eds) (1993) Practical Sampling Techniques in Infrared Analysis. CRC Press, Boca Raton, p 217
19. D'Angelo M, Onori G, Santucci A (1994) J Phys Chem 98:3189–3193
20. D'Aprano A, Lizzio A, Liveri VT, Aliotta F, Vasi C, Migliardo P (1988) J Phys Chem 92:4436
21. Derome A (1987) Modern NMR Techniques for Chemistry Research. Pergamon Press, Oxford
22. Efimov YY, Naberukhin YI (1978) Mol Phys 36:973
23. Eicke H-F, Kvita P (1984) In: Luisi LP, Straub BE (eds) Reverse Micelles: Biological and Technological Relevance of Amphiphilic Structures in Apolar Media. Plenum Press, New York, p 21
24. El Seoud OA, Hinze W (eds) (1994) Organized Assemblies in Chemical Analysis. JAI Press, Greenwich, p 1
25. El Seoud OA (1997) J Mol Liq 72:85
26. El Seoud OA, El Seoud M, Mickiewicz JA (1994) J Colloid Interf Sci 163:87
27. El Seoud OA, Novaki LP (1998) Prog Colloid Polym Sci 109:42
28. El Seoud OA, Okano LT, Novaki LP, Barlow GK (1996) Ber Bunsenges Phys Chem 100:1147
29. Enderby JE, Neilson GW (1979) Water—A Comprehensive Treatise. Plenum Press, New York
30. Enders H, Nimtz G (1984) Ber Bunsenges Phys Chem 88:512
31. Feitosa E, Agostinho Neto A, Chaimovich H (1993) Langmuir 9:702–707
32. Fendler JH (1982) Membrane Mimetic Chemistry. Wiley, New York
33. Gandour RD, Coyne M, Stella VJ, Schowen RL (1980) J Org Chem 45:1733
34. Giammona G, Goffredi F, Liveri VT, Vassallo G (1992) J Colloid Interf Sci 152:465
35. Goba RD, Kon-no K, Kandori K, Kitahara A (1983) J Colloid Interf Sci 93:293
36. Goto A, Yoshioka H, Kishimoto H, Fujita T (1992) Langmuir 8:441
37. Jain TK, Varshney M, Maitra A (1989) J Phys Chem 93:7409
38. Jarret RM, Saunders M (1985) J Am Chem Soc 107:2648
39. Jeffrey GA (1997) An Introduction to Hydrogen Bonding. Oxford University Press, New York
40. Kon-no K, Kitatahara A, El Seoud OA, Schick M (eds) (1987) Nonionic Surfactants: Physical Chemistry. Marcel Dekker, New York, p 185
41. Kristiansson O, Eriksson A, Lindberg J (1984) Acta Chem Scand A38:609
42. Langevin D (1984) In: Luisi LP, Straub BE (eds) Reverse Micelles: Biological and Technological Relevance of Amphiphilic Structures in Apolar Media. Plenum Press, New York, p 287
43. Leonidis EB, Hatton TA (1989) Langmuir 5:741
44. Lianos P, Thomas JK (1986) J Colloid Interf Sci 117:505
45. Lindgren J, Hermansson K, Wójcik MJ (1993) J Phys Chem 97:5254
46. Lis LJ, McAlister M, Fuller N, Rand RP, Parsegian VA (1982) Biophys J 37:657
47. Lucas M, De Trobriand A, Ceccaldi M (1975) J Phys Chem 79:913
48. MacDonald H, Bedwell B, Gulari E (1986) Langmuir 2:704
49. Maddams WF (1980) Appl Spectrosc 34:2451
50. Maitra A, Jain TK, Shervani Z (1990) Colloid Surf 47:255
51. Marcus Y (1994) Biophys Chem 51:111
52. Marcus Y (1997) Ion Properties. Marcel Dekker, New York
53. Mikenda W (1986) Monatsh Chem 117:977
54. Monduzzi M, Caboi F, Larché F, Olsson U (1997) Langmuir 13:2184–2190
55. Monosmith WB, Walrafen GE (1984) J Chem Phys 81:669
56. Mundy WC, Gutierrez L, Spedding FH (1973) J Chem Phys 59:2173
57. Nakayama H, Yamanobe M, Baba K (1991) Bull Chem Soc Japan 64:3023
58. Novaki LP, El Seoud OA (1997) Ber Bunsenges Phys Chem 101:1928
59. Novaki LP, El Seoud OA (1998) J Colloid Interf Sci 202:391
60. Novaki LP, El Seoud OA, Lopes JCD (1997) Ber Bunsenges Phys Chem 101:1928
61. Novaki LP, Pires PAR, El Seoud OA (2000) Colloid Polym Sci 278:143
62. Ohtaki H, Radani T (1993) Chem Rev 93:1157
63. Onori G, Santucci A (1993) J Phys Chem 97:5430
64. Pacynko WF, Yarwood J, Tiddy GJT (1987) Liq Cryst 2:201
65. Piletic IR, Moilanen DE, Spry DB, Levinger NE, Fayer MD (2006) J Phys Chem A 110:4985
66. Profio PD, Germani R, Onori G, Santucci A, Savelli G, Bunton CA (1998) Langmuir 14:768
67. Rabie HR, Vera JH (1997) J Phys Chem B 101:10295
68. Reck RA, Horwood HJ, Ralston AM (1947) J Org Chem 12:517
69. Rosenfeld DE, Schmuttenmaer CA (2006) J Phys Chem B 110:14304
70. Scherer JR (1978) In: Clark RJH, Hester RE (eds) Advances in Infrared and Raman Spectroscopy, Vol 5. Wiley, New York, p 149

71. Schiffer J, Hornig DF (1968) J Chem Phys 49:4150
72. Schowen KB, Gandour RD, Schowen RL (eds) (1978) Transition States for Biochemical Processes. Plenum Press, New York, p 225
73. Senior WA, Verrall RE (1969) J Phys Chem 73:4242
74. Sokolowska A (1991) J Raman Spectrosc 22:31
75. Sokolowska A, Kecki Z (1986) J Raman Spectrosc 17:29
76. Teleman O, Joensson B, Engstroem S (1987) Mol Phys 60:193
77. Tiddy GJT (1979) Nucl Magn Reson 8:174
78. Tiddy GJT (1980) Phys Rep 57:1
79. Uskokovic V, Drofenik M (2006) J Magn Magn Mater 303:214–220
80. Uskokovic V, Drofenik M (2007) Adv Colloid Interf Sci 133:23
81. Vandeginste BGM, De Galan L (1975) Anal Chem 47:2124
82. Waldron RD (1957) J Chem Phys 26:809
83. Wall TT, Hornig DF (1965) J Chem Phys 43:2079
84. Wall TT, Hornig DF (1967) J Chem Phys 47:784
85. Walrafen GE (1968) J Chem Phys 48:244
86. Wiafe-Akenten J, Bansil R (1983) J Chem Phys 78:7132
87. Wills HA, Van der Maas JH, Miller RGJ (1987) Laboratory Methods in Vibrational Spectroscopy. Wiley, New York
88. Yoshioka H, Kazama S (1983) J Colloid Interf Sci 95:240
89. Yoshioka H (1983) J Colloid Interf Sci 95:81
90. Zhou G-W, Li G-Z, Chen W-J (2002) Langmuir 18:4566
91. Zhou N, Li Q, Wu J, Chen J, Weng S, Xu G (2001) Langmuir 17:4505

Progr Colloid Polym Sci (2008) 134: 111–119
DOI 10.1007/2882_2008_080

Published online: 22 May 2008

Aixin Song
K. Reizlein
H. Hoffmann

Swelling of Aqueous L_α-Phases by Matching the Refractive Index of the Bilayers with that of the Mixed Solvent

Aixin Song
Key Lab for Colloid and Interface,
Chemistry of Education Ministry,
Shan Da Nan Lu, 250100 Jinan, China

K. Reizlein
Bayer CropScience AG, D-FT, Bldg. 6820,
Alfred-Nobel-Str. 50, 40789 Monheim,
Germany

H. Hoffmann (✉)
University of Bayreuth, BZKG,
Gottlieb-Keim-Str. 60, 95448 Bayreuth,
Germany
e-mail: heinz.hoffmann@uni-bayreuth.de

Abstract It is shown that L_α-phases in two-phase L_1/L_α-regions from surfactants or phospholipids swell to single phase regions when part of the water in the system is replaced by hydrophilic co-solvents, like glycerol or dimethylsulfoxide (DMSO), or by the addition of sugar. The behavior is demonstrated on aqueous phases from non-ionic surfactants and co-surfactants and on lecithin in water. Single L_α-phases are reached in the mixed solvents when the co-solvents content reaches between 50 and 60% and the sugar content about 40%. These L_α-phases are completely transparent, birefringent, and they have a rather low viscosity with a finite structural relaxation time. The loss modulus is always larger than the storage modulus in the whole frequency region between 0.001 and 20 Hz.

Turbid, but homogeneously looking phases are obtained and they separate slowly with time when macroscopically separated two-phase samples are mixed again by shaking. The turbidity in the homogeneously looking phases decreases with increasing content of the co-surfactant or sugar. These metastable samples become transparent and stable when the refractive index of the bilayers in the L_α-phase becomes the same as the refractive index of the mixed solvent. The experimental results demonstrate that the matching of the refractive index of the bilayers and the solvent is relevant both for the turbidity and for the swelling of the bilayers in the L_α-phase. The interlamellar spacing between the bilayers is determined by a balance of repulsive forces given by undulations and attractive forces determined by van der Waals forces. The swelling of the L_α-phase is caused by a decrease of the Hamaker constant that controls the attractive forces, and is given by the difference of the refractive indices of the mixed solvent and the bilayers in the L_α-phase. Thus, it is concluded that the change of turbidity and the swelling of the L_α-phase have the same origin, namely the matching of the refractive index of the bilayer system with that of the mixed solvent.

Keywords Swelling of L_α-phases · Refractive index · Hamaker constant

Introduction

Many aqueous surfactant or phospholipid solutions contain L_α-phases either as a single phase or in equilibrium with other phases [1]. Typical examples are dispersions of phospholipids or non-ionic surfactants in combination with various co-surfactants [2, 3]. Freshly prepared samples in a multi-phase state are usually turbid phases which, with time, separate into two macroscopically separated phases, a single L_α-phase and a L_1-phase. Depending on

Fig. 1 Phase behavior of samples with 5% LA070 and different concentration of octanol at room temperature. The concentration of octanol was 0, 50, 80, 100, 150, 200, 250, 300, and 350 mM, respectively. Note that the samples with 50 and 80 mM octanol are two-phase samples while the samples with 100, 150 and 200 mM octanol are one-phase samples

the density difference between the L_α- and the L_1-phase, the L_α-phase can be on top or on the bottom of the sample. Under the polarization microscope the L_α-phase is a birefringent phase and can easily be recognized by its birefringence pattern, the so-called oily streaks or, if it contains multilamellar vesicles, by the spherolites [4, 5]. The bilayers in the L_α-phases are separated from each other by a well-defined interlamellar distance that is controlled by attractive and repulsive forces between the bilayers. In equilibrium, these forces balance each other. The attractive forces are usually caused by van der Waals forces and the repulsive forces from undulations and electrostatic forces coming from ionic charges on the bilayers [6–8]. In many situations the attractive forces are much larger than the repulsive forces and the bilayers collapse on to each other until there is only a thin solvation layer left between the head-groups of the bilayers [9]. This is f.i. the situation with aqueous dispersions of phospholipids for which interlamellar distance between the bilayers from the phospholipids does not swell in water. The other extreme situation occurs in iridescent phases in which the bilayers can be swollen up to several 1000 Å [10]. Such large interlamellar distances are usually caused by electrostatic repulsions between the bilayers. In this paper, it will be shown how the L_α-phases in two-phase L_1/L_α-regions can be made to swell until they become a one-phase region without changing the composition of the bilayers. The swelling of the L_α-phase is accomplished only by changing the composition of the aqueous solvent. We first observed such situations when we established phase diagrams of ternary and quaternary systems and then realized the general validity of the made observations, and finally found the explanation for the unusual behavior. Similar results with the L_α-phases of the blockcopolymer $(EO)_{15}$-$(PDMS)_{15}$-$(EO)_{15}$ have already been published [11].

Results

The Phase Diagram of the Non-Ionic Surfactant Genapol 070 (LA070) and Octanol

In Fig. 1, samples of phases are shown with 5% of a commercial non-ionic surfactant LA070 with increasing concentration of octanol. The samples were photographed with and without polarizers. With increasing co-surfactant concentration the phase sequence that is typical for nonionic surfactants and co-surfactants is observed: first, a single transparent L_1-phase, then a two-phase L_1/L_α-situation with the L_α-phase on top of the sample, then a single L_α-phase, and finally, at high co-surfactant concentration, a two-phase situation, probably from a L_α/L_2-system. The two-phase regions exist between a co-surfactant concentration of 30 mM and a co-surfactant concentration of 90 mM octanol. In this region, the volume fraction of the L_α-phase increases from zero to one. The sequence of the phases with increasing co-surfactant concentration can simply be explained by the concept of the packing parameter that was introduced by Ninham and Israelachvili [12]. The head-group of octanol is very small and the mean size of the polar head-groups decreases with increasing octanol concentration.

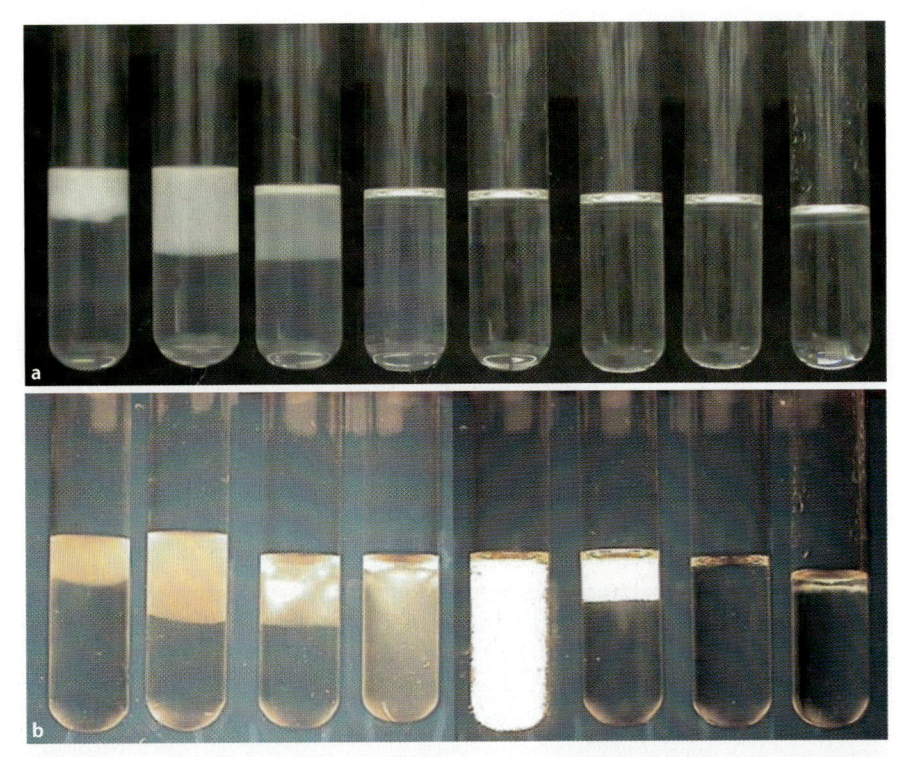

Fig. 2 5% LA070/50 mM octanol system (two-phase) in glycerol/H_2O mixed solvents at room temperature when the concentration of glycerol is 0, 20, 45, 50, 60, 70, 80, and 100%, respectively. Note that the first three samples are in a two-phase situation while the next two samples are in a one-phase situation

The Swelling of the L_α-Phase in the Two-Phase L_1/L_α-System by Glycerol

In Fig. 2, the two-phase systems with 50 mM octanol are shown when increasing amounts of water are replaced by glycerol. The result is that the volume of the L_α-phase becomes larger, and finally, with 50% glycerol the L_α-phase, fills out the whole volume of the sample. The single L_α-phase exists between a glycerol content of 50% and 60%. Even with 70% glycerol, a L_α-phase still exists, but now in a two-phase region. The two-phase region for this condition could be a result of attraction between the bilayers. The turbidity in samples within the single L_α-phase region decreases with increasing glycerol content in spite of the fact that the samples contain the same amount of surfactant and co-surfactant and have the same morphology. The difference of the turbidity comes from the differences of the refractive index of the bilayers and the refractive index of the solvent. The swelling of the L_α-phase indicates that with increasing glycerol the interlamellar distance between the bilayers increases until the whole sample is filled up with the L_α-phase. The complete phase separation in the two-phase region takes many hours after mixing. When the samples are mixed freshly they look like homogenous isotropic samples with a high turbidity. Such samples are shown in Fig. 3. The turbidity of these samples is decreasing with increasing glycerol content. The turbidity of the samples is thus correlated with the swelling of the L_α-phase in the two-phase L_1/L_α-region.

The freshly mixed samples are all optically isotropic and show no birefringence, even for the situation where the samples in Fig. 2 show a strong birefringence. This is somewhat surprising, but it can easily be explained. It is known that L_α-phases under shear are transformed into small unilamellar vesicles (SUV) and multilamellar vesicles (MLV). The higher the shear rate, the smaller the number of shells in the MLV's [13]. Therefore, it is likely that the L_α-phase and the vesicles in the L_α-phases can be completely transformed into SUV's by shaking of the samples. Solutions with small vesicles do not show birefringence when the solutions are at rest. It is only under flow when the vesicles are deformed that they show birefringence. The turbidity of the homogenously looking phases comes from the vesicles. It is explained later that the decrease of the turbidity comes from the matching of the refractive index of the bilayers with the refractive index of the solvent. This matching obviously must also lead to the swelling of the L_α-phase. In order to demonstrate the general validity of the given results, we shall now show additional results with the same surfactant/co-surfactant systems in which the matching of the refractive index is done with other co-solvents.

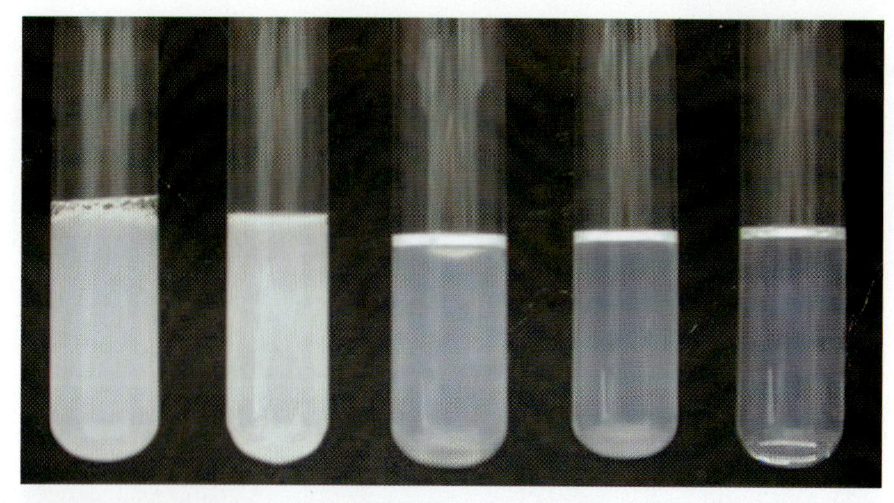

Fig. 3 The freshly mixed samples of 5% LA070/50 mM octanol system with increasing glycerol concentration of 0, 20, 45, 50, and 60%, respectively

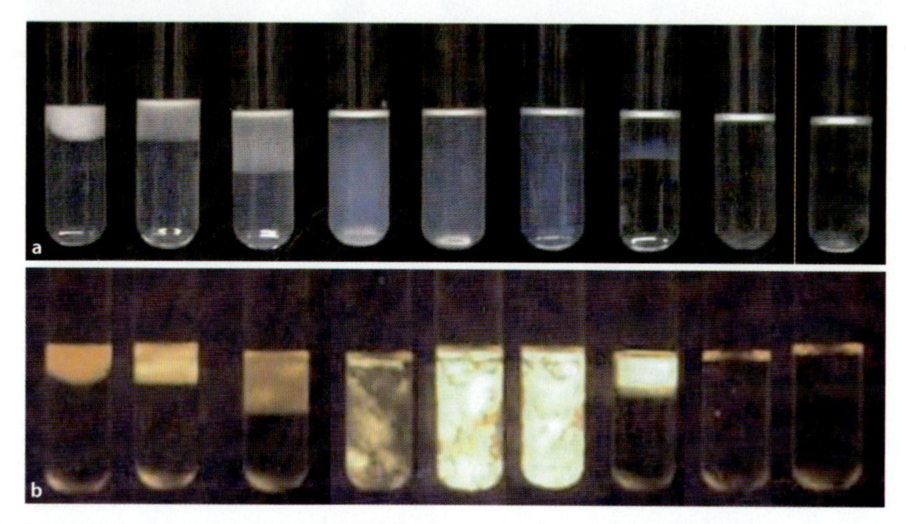

Fig. 4 5% LA070/50 mM octanol system (two-phase) in DMSO/H_2O mixed solvents at room temperature and the concentration of DMSO 0, 20, 25, 30, 40, 45, 50, 60, and 100%, respectively. Note that the first three samples are in a two-phase state while the next three samples are in a one-phase state

The Swelling of the L_α-Phase in the Two-Phase L_1/L_α-System by Dimethylsulfoxide

The samples in Fig. 4 show phases of 5% LA070 with 50 mM octanol in DMSO/water mixtures. Up to a concentration of 25% DMSO, the samples are in a two-phase L_1/ L_α-state, and from 30% to 45% DMSO, the samples are in a single L_α-phase. With 50% DMSO, the samples are in a two-phase situation. With even higher DMSO concentration, all samples are in a low viscous isotropic state. It is likely that at such high DMSO concentration DMSO has increased the critical micelle concentration (CMC) of the surfactant and co-surfactant concentration so much that micellar aggregates are no longer stable and the phases are molecular solutions. It is obvious that the concentration of 30% DMSO can swell the phases even more than a 30% glycerol concentration. It is conceivable that DMSO also affects the system by adsorbing on the bilayers and making the bilayers more flexible. As for samples with glycerol, the turbidity in the L_α-phases decreases with increasing content of DMSO. In this state, the samples look very similar as the L_α-phases with glycerol.

The Swelling of the L_α-Phase in the Two-Phase L_1/L_α-System by the Addition of Sugar

Normal household sugar was used as additive in the samples in Fig. 5. In the shown samples increasing amounts of sugar were added to a 5 ml sample of 5% LA 070 and 50 mM octanol. It is clearly seen that in spite of the increasing volume of the sample, and hence a dilution of

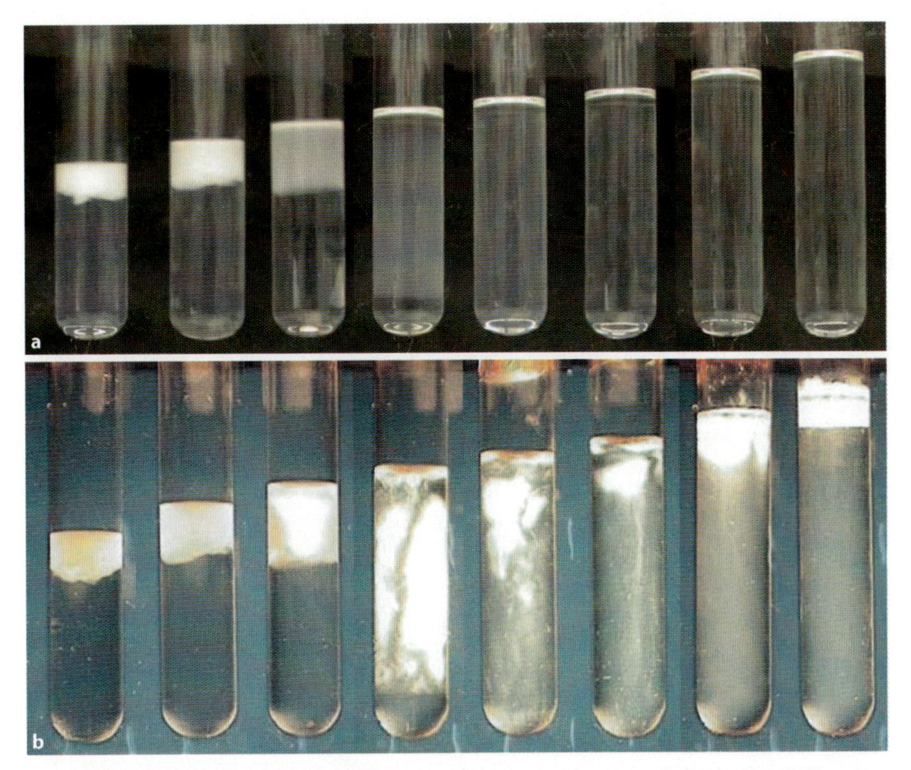

Fig. 5 5% LA070/50 mM $C_8H_{18}O/H_2O$ samples with different amounts of sugar at room temperature. The total amounts of the original samples are all 5 g like the first one. The amounts of sugar are 0, 1, 2, 3, 3.5, 5, 6, and 7 g, respectively. Note that the first four samples are in a two-phase state while the next four samples are in a one-phase state

the surfactant, the volume fraction of the L_α-phase grows, and, with 3.5 g of sugar, reaches a single phase. Sugar is sometimes used in surfactant phase studies to increase the viscosity in the aqueous solvent [14]. It is usually assumed that it is a very hydrophilic additive and it does not change the CMC of the used surfactant. In single phase regions, sugar usually does not change the morphology of the phases. However, as the results show here, sugar changes the swelling behavior of the L_α-phase in two-phase L_1/L_α-systems. In the three investigated systems with glycerol, DMSO, and sugar, a L_1-phase was in equilibrium with the L_α-phase and the L_1-phase had a different composition than the L_α-phase. Therefore, it could be argued that the added co-solvent changed the composition of the two phases, and the swelling was a result of the composition changes. It is assumed, however, that such changes, if they occur, are not relevant for the large swelling. In order to prove this point a binary system was studied in which the composition of the bilayer cannot change when a co-solvent is added.

The Swelling of the Lecithin Phase by the Addition of Glycerol

The results for lecithin dispersions are shown in Fig. 6 when water was replaced by glycerol. Even for this system, a swelling of the L_α-phase is observed, and for concentration of 90% to 100% glycerol, a single L_α-phase is observed with as little as 5% of lecithin.

The Rheological Properties of the L_α-Phases

All the L_α-phases that have been produced by swelling with a co-solvent are viscous shear thinning solutions. They behave similarly as shear thinning L_1-phases with a structural relaxation time [15]. A rheogram of such a phase is given in Fig. 7. At low frequencies between 10^{-2} and 1 Hz, the loss modulus G'' is higher than the storage modulus G' and, as expected from the slopes of the moduli, it seems they would cross at 1 Hz. It is noteworthy, however, that the moduli do not cross and have at the extrapolated crossing point, independent of the studied systems, a modulus in the range of 1 Pa. This value of 1 Pa must come from the interaction of the bilayers and not from the bulk viscosity of the mixed solvents. The viscosity for an aqueous solvent with 60% glycerol is about 20 times larger than that of the solvent water. However, the loss moduli for a L_α-phase with 5% of the surfactant in the solvent mixtures is about the same as in pure water.

However, at even higher frequencies than 1 Hz the two moduli increase with increasing frequencies and run parallel with each other, with G'' always larger than G'. In this frequency range, the behavior is very different from the behavior of L_1-phases.

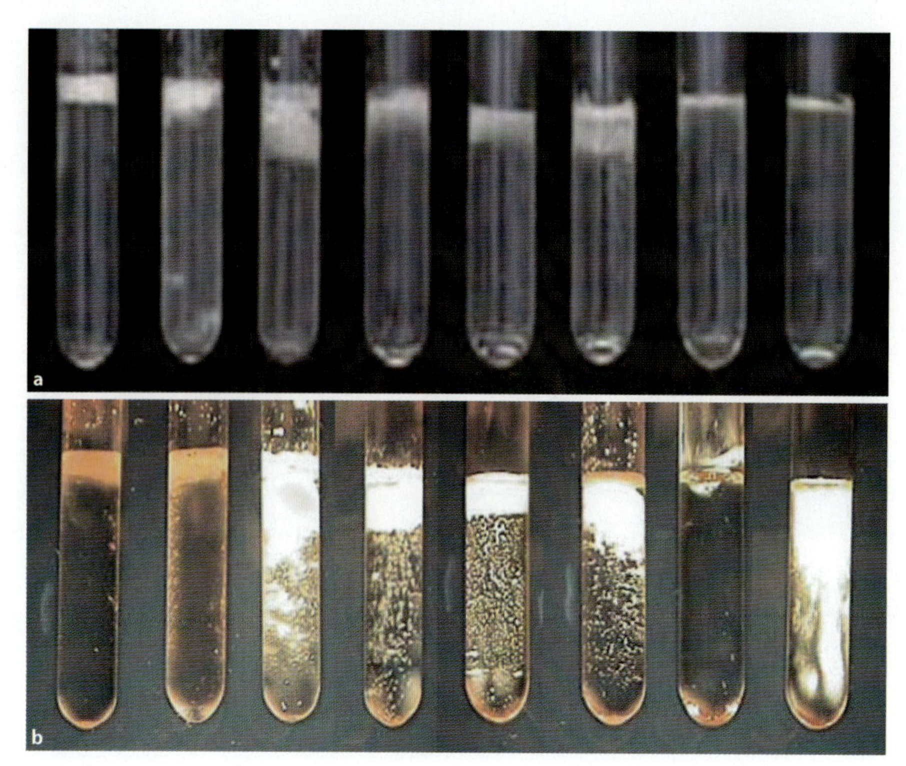

Fig. 6 5% lecithin in glycerin/H_2O mixed solvents with increasing glycerol concentration at room temperature. The concentration of glycerol in mixed solvent is 20, 40, 50, 60, 70, 80, 90, and 100%, respectively. Note that the last sample in 100% glycerol is a one-phase state

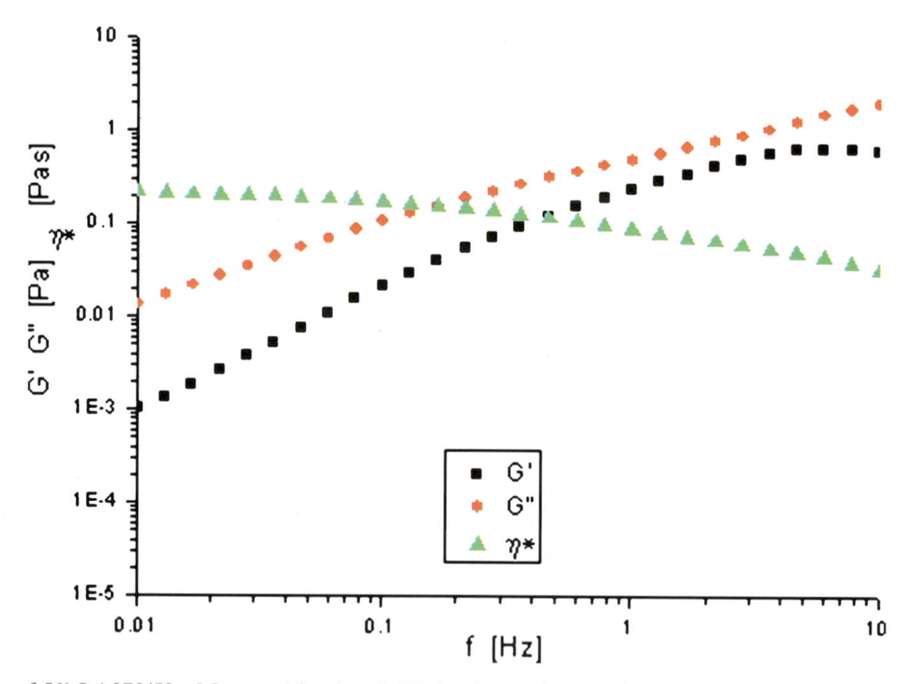

Fig. 7 The rheogram of 5% LA070/50 mM octanol in glycerin/H_2O mixed solvents with the glycerin concentration of 60%

Discussion of the Experimental Results

The effect of glycerol on the phase and aggregation behavior of surfactants and Pluronics has previously been explained as the result of dehydration of amphiphiles [16, 17]. It is well known that the glycerol has a strong ability to form hydrogen bonds with water. Therefore, glycerol competes with water to bind to the surfactant molecules in aqueous solution. In that way, the surfactant molecules, especially nonionic surfactants, such as Pluronics and C_mEO_n, are dehydrated, and their hydrophilicity decreases. This is reflected by the lowering of the cloud point of these systems and the phase separation of surfactants at high glycerol content. In addition, the whole solvent becomes more hydrophobic with the addition of glycerol, resulting in a continuous increase of the critical micelle concentration of the surfactants with the glycerol content. This effect is, however, rather small and can be neglected for the interaction between the bilayers [18, 19].

It could be argued that the swelling of the bilayers is due to the incorporation of glycerol into the bilayers and a change of the composition of the bilayers. The bilayers could become more flexible and the higher flexibility could then result in a larger repulsion. There is, however, no experimental evidence for the incorporation of glycerol into the bilayers. It is known that the addition of glycerol has little effect on the criticial micelle concentration (CMC) of surfactants [18, 19] and liquid crystalline phases are formed even in pure glycerol, as it is shown in this paper. There also is no evidence by SAXS measurements that glycerol is incorporated in the bilayers of L_α-phases of surfactants [11].

The swelling of the L_α-phase in our experiments leads to another explanation for the effect of glycerol on the

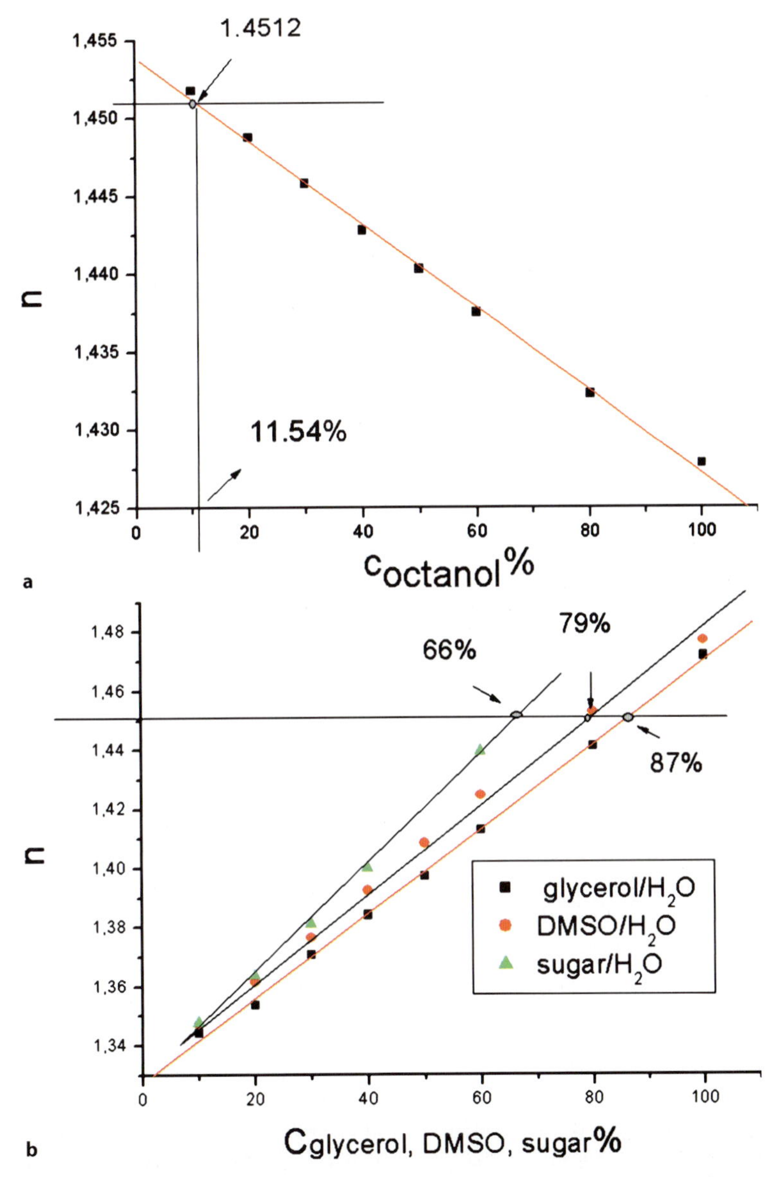

Fig. 8 Refractive indices of mixtures of genapol 070 and octanol in (**a**). Refractive indices of mixtures of water with different cosolvents (**b**)

phase behavior. It suggests that there must be an increase of repulsive or a decrease of attractive forces between the bilayers with increasing glycerol. The bilayers in the L_α-phase have a well-defined spacing which is determined by an equilibrium between the attractive van der Waals forces and repulsive undulation forces. The potential energy, ϕ_A, of attraction per unit area for the situation, when two bilayers of thickness δ approach to a distance d ($d \gg \delta$) is given by:

$$\phi_A = -A/12\pi d^2 \ldots \tag{1}$$

The attractive pressure, P_A, is obtained by differentiating of ϕ_A over distance d:

$$P_A = \partial(\phi_A)/\partial(d) = A/24\pi d^3 , \tag{2}$$

where A is the Hamaker constant, which depends the on material of the bilayers and the medium property;

The Hamaker constant is given by the following approximate expression:

$$A = \frac{3}{4} kT \left[\frac{\varepsilon_d - \varepsilon_m}{\varepsilon_d + \varepsilon_m} \right]^2 + \frac{3}{16\sqrt{2}} \hbar\omega \frac{\left(n_d^2 - n_m^2\right)}{\left(n_d^2 + n_d^2\right)^{3/2}} . \tag{3}$$

Here, k is Boltzmann constant, T the absolute temperature, n_d and n_m are the refractive indices of the bilayers and the medium, respectively. For visible light, ω is the angular frequency of the dominating UV absorption (about $1.7\text{–}2.4 \times 10^{16}$ rad/s) and $h = 2\pi\hbar$ is Planck's constant. The quantities ε_d and ε_m are the dielectric constants of the bilayer and the medium at zero frequency (static field).

The undulation repulsion obeys the Helfrich equation (Eq. 4), which specifies the steric hindrance between adjacent membranes. In a multilayer system, this gives rise to a repulsive pressure, P_R, given by

$$P_R = \frac{3\pi^2}{64} \frac{(kT)^2}{\kappa} \frac{1}{(d-\delta)^3} . \tag{4}$$

The Helfrich equation (Eq. 4) shows that the repulsive pressure between two membranes does not depend on the solvent between the bilayers. The swelling of the L_α-phase must therefore be a result of the lowering of the attraction between the bilayers by the increasing glycerol content. The equation for the Hamaker constant shows that the second term in the equation becomes zero when the refractive indices from the bilayer and the solvent are matched. The reason for the swelling is, therefore, the decrease of the Hamaker constant.

It is known that the light scattering of dispersions disappears when the refractive index of the solvent is matched to the refractive index of the particles. This can be accomplished when the refractive index of the solvent water is changed by the addition of several co-solvents, such as glycerol, DMSO, and the additive sugar. Results of index measurements in water/co-solvents mixtures are given in Fig. 8. In all the studied systems, the refractive index of the solvent mixtures increases linearly from the refractive index of water to the refractive index of the co-solvent, while the refractive index of the surfactant is decreasing with the content of the co-surfactant. It is therefore clear that the two refractive indices match at a particular composition of the solvent.

In order to get a good estimate for the refractive index of the bilayer, we can assume that the bilayers have the same composition as the mixture of the surfactant and co-surfactant in the bulk solution. This is not correct because the bilayers in the L_α-phase and the L_1-phase have a somewhat different composition. The concentration of the two components in the L_α-phase is, however, much higher than in the L_1-phase. Therefore, the assumed approximation is reasonable.

The refractive index of the bilayers should therefore be able to be taken from the plot of the refractive index of the surfactant with increasing co-surfactant concentration. A 5% surfactant concentration with 50 mM octanol corresponds to a mixture of 5 g surfactant and 0.65 g octanol. Such a mixture has a refractive index of 1.455.

This refractive index would corresponds to a 87% glycerol phase, to a 79% DMSO phase, and to a 66% sugar solution. At these compositions, the samples should become completely transparent, and, at the same time, the single L_α-phases should be stable.

However, this is not the case. At these concentrations of the additives the L_α-phases are already unstable. The assumed n-value of 1.455 is probably too high for the bilayer composition, because the bilayer contains water from the solvated ethylenoxide groups. With 30% water, we obtain an average value of 1.425. With this value the stability of the L_α-phases agrees with the composition of the solvent.

Conclusions

Micellar structures from surfactants have a higher refractive index than water. For this reason, micellar solutions scatter light. The differences between the refractive index of micellar structures and the solvent water can be eliminated by adding hydrophilic co-solvents such as glycerol to the aqueous phase. The change of the refractive index of the mixed solvent has also consequences on the interaction between the micelles. With the matching of the refractive index between solvent and micellar structures, the attractive forces between the structures disappear because the Hamaker constant in the DLVO theory decreases towards zero. The consequence of this is that L_α-phases in two-phase L_1/ L_α-systems swell until the L_α-phase takes up the whole volume of the sample.

References

1. Laughlin RG (1994) The Aqueous phase behavior of Surfactants. Academic Press, Harcourt Brace & Company Publishers, London, San Diego, New York, Boston, Sydney, Tokyo, Toronto
2. Meyer HW, Richter W, Brezesinski G (1994) Biochim Biophys Acta 1190:9–19
3. Hoffmann H, Thunig C, Schmiedel P, Munkert U (1994) Langmuir 10:3972–3981
4. Benton WJ, Miller CA (1983) Prog Colloid Polym Sci 68:71–81
5. Miller CA, Ghosh O, Benton WJ (1986) Colloid Surf 19:97–223
6. Helfrich W (1985) J Phys (Paris) 46:1263
7. Jenströmer M, Strey R (1992) J Phys Chem 96:5993
8. Pincus P, Joanny J-F, Andelman D (1990) Europhys Lett 11:763
9. Gräbner D, Matsuo T, Thunig C, Hoffmann H (2001) J Colloid Interf Sci 236:1–13
10. Platz G, Thunig C, Hoffmann H (1990) Colloid Polym Sci 83:167
11. Yun Y, Hoffman H, Richter W, Talmon I, Makarsky E (2007) J Phys Chem B 111:6376–6382
12. Israelachvili JN, Mitchell DJ, Ninham BW (1976) J Chem Soc Faraday Trans II 72:1525
13. Escalante J, Gradzielski M, Mortensen K, Hoffmann H (2000) Langmuir 16(23):8653
14. Ivanova R, Lindman B, Alexandridis P (2000) Langmuir 16:3660–3675
15. Herb CA, Prud'homme RK (eds) (1994) ACS-Symposium Series 578:1–31
16. Lin Y, Alexandridis P (2002) J Phys Chem B 106:12124
17. Iwanaga T, Suzuki M, Kunieda H (1998) Langmuir 14:5775–5781
18. De Errico G, Ciccarelli D, Ortona O (2005) J Colloid Interf Sci 286:747–754
19. Cantú L, Corti M, Degiorgio V, Hoffmann H, Ulbricht W (1987) J Colloid Interf Sci 116:384

Progr Colloid Polym Sci (2008) 134: 120–127
DOI 10.1007/2882_2008_089

Published online: 19 August 2008

NANO-COLLOIDS

Anna Musyanovych
Katharina Landfester

Synthesis of Poly(butylcyanoacrylate) Nanocapsules by Interfacial Polymerization in Miniemulsions for the Delivery of DNA Molecules

Anna Musyanovych (✉) ·
Katharina Landfester
Institute of Organic Chemistry III – Macromolecular Chemistry and Organic Materials, University of Ulm,
Albert-Einstein-Allee 11, 89081 Ulm,
Germany
e-mail: anna.musyanovych@uni-ulm.de

Abstract Monodispersed biodegradable poly(n-butylcyanoacrylate) nanocapsules containing DNA molecules (790 base pairs) within an aqueous core were prepared by anionic polymerization of n-butylcyanoacrylate at the droplets interface in inverse miniemulsion. The aqueous droplets in the size range of 300–700 nm dispersed in the hydrophobic continuous phase were formulated using the miniemulsion technique that allows an easy control of the droplet size and size distribution. After polymerization, the capsules were transferred into an aqueous phase. The effect of several reaction parameters such as the amount of monomer, type of the non-ionic surfactant (i.e. Span®80 and Tween®80) and type of the continuous phase (i.e. Miglyol 812N and cyclohexane) on the shell thickness, capsule size, morphology, polymer molecular weight, and encapsulation efficiency of DNA was investigated. The obtained results indicated that the type of the continuous phase has the largest influence on the capsule average size and polydispersity, whereas the shell thickness and morphology were mainly dependent on the monomer concentration. The encapsulation efficiency of DNA was about 100%, regardless to the surfactant type. At least, about 15% of the total DNA amount was found to be in the form of free chains.

Keywords Biodegradable · DNA encapsulation · Interfacial polymerization · Nanocapsule · Poly(butylcyanoacrylate)

Introduction

Recent advances in the development of drug delivery research area have led to the need for the production of well-defined and characterized nanoparticulates. Especially the particle size and size polydispersity are important factors which influence the distribution of the particles within the body and their interaction and uptake by living cells [1–4]. Colloidal carriers such as liposomes, micelles, dendritic polymers, and nanoparticles are the most promising candidates in regards to the site-specific delivery and controlled drug release. Core-shell particles with a liquid core have gained increased attention in the past years owing to their utilization as sub-μm containers for the encapsulation of biologically active compounds. The main advantages of nanocapsules for drug delivery are the efficient protection of a drug against degradation caused by the influence of the environment, and avoiding desorption of a drug, as it could be observed in the case with physically adsorbed drug molecules onto the particle surface. Moreover, the high entrapment efficiency of the hydrophilic macromolecules could be achieved mainly by encapsulation [5].

Nanocapsules can be formulated from a variety of synthetic or natural monomers or polymers by using different techniques in order to fulfill application requirements. For example, by interfacial deposition of the preformed polymer, poly(D,L-lactide)-based nanocapsules containing an antitumoral agent [6] or poly(methyl methacrylate) capsules with an entrapped antiseptic agent [7] were successfully prepared. The encapsulation of peptides and proteins was achieved by the formulation of microcapsules using the double emulsion technique [8–10], the emulsification/solvent evaporation, the diffusion method [11–13], or the salting-out procedure [14]. Interfacial reactions are one of the well-established methods in the preparation of nanocapsules. The nanometer-sized hollow polymer particles were synthesized employing such interfacial crosslinking reactions as polyaddition/polycondensation [15–18], radical [19, 20] or anionic polymerization [21, 22].

Since the middle of the sixties, polyalkylcyanoacrylates (PACAs) are widely used materials in the biomedical field. First the monomers were used as biodegradable tissue adhesives in surgery, and later as polymeric carriers for biomolecules in a drug (tumor)-targeting system [23]. PACAs have gained an increased interest mainly due to their biodegradability, low toxicity, and enhancement of the drug intracellular penetrations [21, 22]. Several research groups have used the interfacial polymerization approach to synthesize PACA-based capsules containing an oil core [24–26] which are suitable for the entrapment of lipophobic compounds such as insulin [27, 28], indomethacin [29], or antiepileptic drugs [30]. Performing the anionic polymerization of alkylcyanoacrylates at the interface of inverse (water-in-oil) dispersions, capsules with an aqueous core, which is convenient for encapsulation of water soluble compounds, can be successfully prepared [5, 31]. PACA capsules containing aqueous solutions of proteins [32, 33], doxorubicin [34], bovine serum albumin [35], pure oligonucliotides or their complexes with cationic polymers [36–38] have been synthesized.

Generally, the most commonly employed systems for preparing the drug-loaded PACA capsules are either emulsions or microemulsions. The ACA polymerization performed at the emulsion droplets interface usually results in a big size capsules and high polydispersity. Employing the microemulsion system, the size distribution of the capsule can be efficiently improved due to the usage of high amounts of a more hydrophobic surfactant (e.g. sorbitan monooleate) [36], or a more hydrophilic surfactant (polysorbate 80) [35] or a mixture of both types surfactants [31]. However, the size of capsules that can be obtained in microemulsions is limited and usually can not be exceeded more than 200 nm, which might be a problem in order to encapsulate biomolecules of high molecular weights and size. Also, since during the reaction, diffusion processes are effective, an efficient encapsulation of components is often difficult. The high amount of the surfactant is also a disadvantage.

Significant benefits of the miniemulsion technique offer the ability to obtain polymeric nanocapsules in a controlled way with different particle size, morphology and versatility in the polymeric material and surface functionality [39]. Due to the lack of diffusion processes of the monomer throughout the polymerization, an efficient encapsulation can be obtained. The miniemulsion technique generally implies a method that allows one to create stable droplets (50–500 nm) in a continuous phase by applying high shear stress. Moreover, under carefully chosen conditions of miniemulsification, it is possible to entrap dsDNA inside the droplets avoiding damaging of the molecules [40]. The high stability of the system allows to perform the reactions inside the miniemulsion droplets and at their interface [7, 17, 41]. Therefore, this system can be beneficial in the formation of capsules with the desired size.

In these studies, polymeric nanocapsules with encapsulated dsDNA (790 base pairs) were produced via anionic polymerization of *n*-butylcyanoacrylate (BCA) carried out at the interface of homogeneously distributed aqueous droplets in inverse miniemulsion which are in a second step then redispersed in an aqueous continuous phase. The obtained capsules were characterized in terms of size, size distribution, morphology, polymer molecular weight, and encapsulation efficiency of DNA. The effects of surfactant type and concentration, viscosity of the continuous phase, monomer amount, and water-to-oil ratio were investigated and results are discussed in this paper.

Materials and Methods

Materials

The *n*-butylcyanoacrylate monomer was kindly supplied by Henkel, Germany. Miglyol 812N (caprylic/capric triglycerides) was a gift from Sasol, Germany. The surfactants, sorbitan monooleate (Span®80) and polysorbate 80 (Tween®80), were purchased from Aldrich and used as received. The double-stranded deoxyribonucleic acid (dsDNA) template for capsule loading is a product of a polymerase chain reaction and was synthesized from genomic human dsDNA using 5′-CGGCAGCAACAGCAGGT-3′ as forward and 5′-AGCTATAGGCAAA GCCAGAA-3′ as reversible primers. These primers amplify a 790 base pairs (bp) sequence of a housekeeping gene (PGBD; GeneBank AccessionNo. M95623). All other chemicals and solvents used during the work were reagent grade.

Preparation of PBCA Nanocapsules

The PBCA nanocapsules were prepared by interfacial anionic polymerization using the modified procedure previously reported by several groups [36, 42, 43]. Briefly, an organic phase composed of 7.6 g Miglyol 812N (or cyclohexane) and different amounts of a surfactant (Span®80, Tween®80, or a mixture of Span®80 and Tween®80) were added to the 1 or 2 g of 10^{-3} M phosphate saline buffer (pH 7.4). The mixture was vigorously stirred for 10 min

at a room temperature, and then the macroemulsion was subjected to ultrasonication for 60 s at 50% amplitude in a pulse regime (5 s sonication, 10 s pause) using a Branson 450 W sonifier and an inverse cup tip. The temperature was maintained at 4 °C during the sonication step. The obtained miniemulsion was divided between four screw-capped cryo-vials and the monomer *n*-butylcyanoacrylate (70, 100 or 200 μl) was drop-wise added to the miniemulsion under magnetic stirring (600 rpm). After 4 h of polymerization, the PBCA capsules were separated from the oil phase by centrifugation at 14 000 rpm for 20 min. In the case of Miglyol 812N, the capsules were additionally washed with ethanol. After centrifugation, the wet pellet was resuspended in 0.5 wt. % aqueous solution of Lutensol AT50.
For the preparation of DNA loaded PBCA capsules, the oil phase consists of 7.6 g Miglyol 812N and 1 wt. % (corresponding to the continuous phase) of surfactant (Span®80, Tween®80 or a mixture of Span®80 and Tween®80). The DNA was dissolved in 1 g of 10^{-3} M phosphate saline buffer (pH 7.4) at a concentration of 6 nM. The formulation process of PBCA nanocapsules was similar as described above.

Characterization of PBCA Capsules

The average size and size distribution of miniemulsion droplets and polymer capsules were determined by dynamic light scattering using a Zeta Nanosizer (Malvern Instruments, UK), equipped with a detector to measure the intensity of the scattered light at 173° to the incident beam.

Gel permeation chromatography (GPC) was used to determine the average molecular weights and distribution of the PBCA. The measurements were carried out on an apparatus consisting of the Spectra System P2000 pump, an autosampler Agilent 1100, and two detectors, the Shodex differential refractometer RI-71 and the Knauer UV Variable Wavelength Monitor detector. After washing and freeze-drying, the PBCA capsules were dissolved in HPLC-grade THF at a concentration of 5 mg ml^{-1} and separation was carried with two Styragel (Waters) columns with a porosity of 10^3 and 10^4 Å and one from SDV (PSS, Germany) with a porosity of 10^5 Å at a flow rate of 1 ml min^{-1}. The molecular weights were calculated from the elution volume. Polystyrene in the molecular weights range from 685 to 2 470 000 g mol^{-1} (PSS GmbH, Mainz, Germany) were used as standards to generate the calibration curve at concentrations of 2 mg ml^{-1}.

Transmission electron microscopy (TEM) (Philips EM400) was used to study the morphology of the capsules. PBCA capsules redispersed in water were diluted to a solid content of about 0.01% and a drop of the sample was placed on a carbon-coated copper grid. The sample was dried at ambient temperature and observed at an accelerating voltage of 80 kV.

The loading efficiency of dsDNA within the PBCA capsule was estimated using the following procedure. After synthesis, the PBCA capsules were separated from the oil phase and lyophilized. Dry nanocapsules were dissolved in 3 ml of chloroform, and dsDNA was extracted with 15 ml of demineralized water for 5 h. The dsDNA containing water phase was first separated from the polymer/chloroform phase via centrifugation, and then concentrated to the initial volume that was loaded during the capsule formulation. The amount of extracted dsDNA was determined from the resolved agarose gel electrophore-

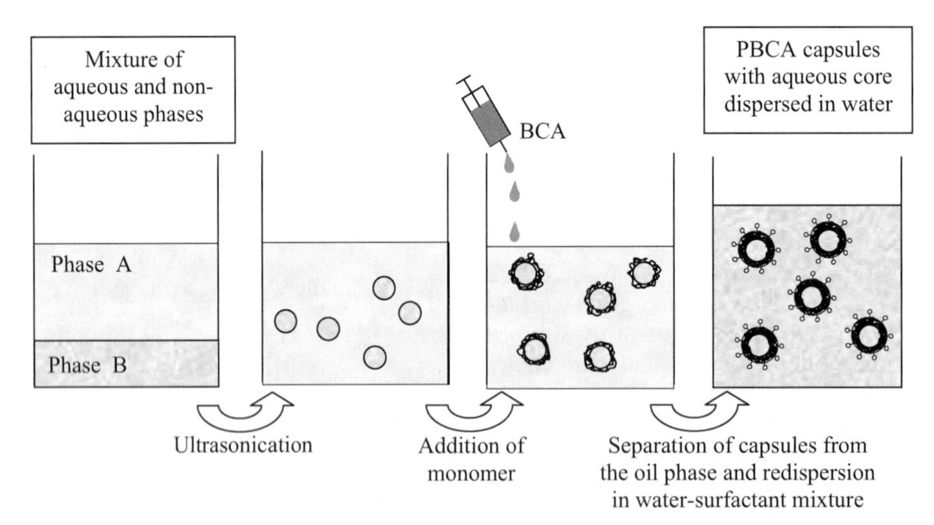

Fig. 1 Schematic representation of the PBCA nanocapsules formulation process in miniemulsion

sis results using the Gene Tools (SynGene) software. The calibration was obtained with different concentrations of dsDNA.

Results and Discussions

Synthesis of PBCA nanocapsules

PBCA nanocapsules were synthesized via anionic polymerization of *n*-butylcyanoacrylate at the water-in-oil droplet interface, which was created by the miniemulsion technique. This technique, in combination with the properly chosen composition of the system, is known to be very efficient in the production of droplets with controlled diameter and monodispersity [44, 45].

The preparative process route is illustrated in Fig. 1.

In the beginning, the aqueous phase (with or without DNA molecules) is miniemulsified by ultrasonication in the non-aqueous phase consisting of the hydrophobic oil and surfactant. Afterwards, the monomer *n*-butylcyanoacrylate is slowly added to the miniemulsion and polymerization takes place at the interface of sterically stabilized aqueous droplets, resulting in the formation of a PBCA shell. Due to the pH, the presence of OH^- ions and additionally also the hydroxyl groups from the surfactant molecule serve as nucleophiles for the initiation of the anionic polymerization. The stability of the droplets and the viscosity of the continuous phase are the key factors influencing the formation of capsules with a low size distribution. Therefore, these effects will be discussed in the following section.

Effect of Surfactant Type, Concentration and Viscosity of the Continuous Phase

Surfactant plays an important role in the stabilization of the droplets before and after polymerization with the capsule formation. Moreover, after polymerization the PBCA nanocapsules are going to be isolated from the oil phase and redispersed in the aqueous phase. Therefore, the choice of surfactant is very crucial. Generally for inverse systems, non-ionic surfactants with a low hydrophilic–lypophilic balance (HLB) values are well-suited. However, after redispersion in water, PBCA capsules stabilized with a water-soluble non-ionic surfactant would better resist the agglomeration process during centrifugation.

Three different stabilizing systems were chosen for the experiments presented in this paper: (i) surfactants for inverse (water-in-oil) systems possessing low HLB values (Span®80, HLB 4.3); (ii) surfactants for oil-in-water systems with high HLB values (Tween®80, HLB 15), and (iii) a mixture of surfactants for both systems (Tween®80 and Span®80 in the ratio of 3 : 2 w/w). Figure 2a shows the average diameter of the miniemulsion droplets in Miglyol 812N, plotted as a function of the surfactant type and concentration.

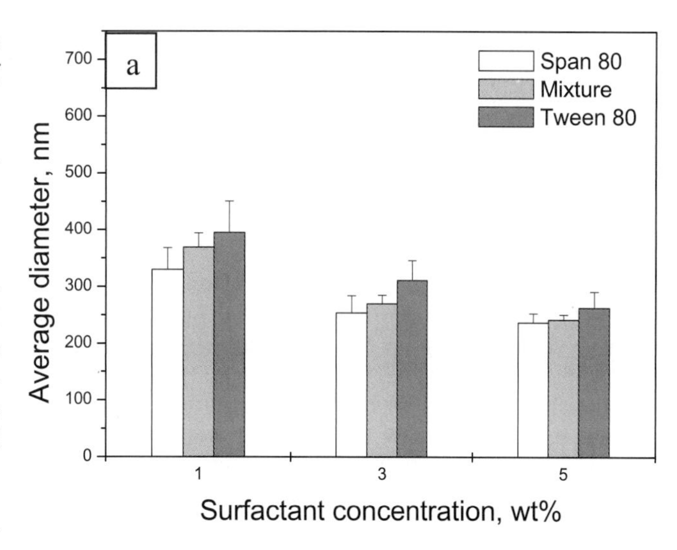

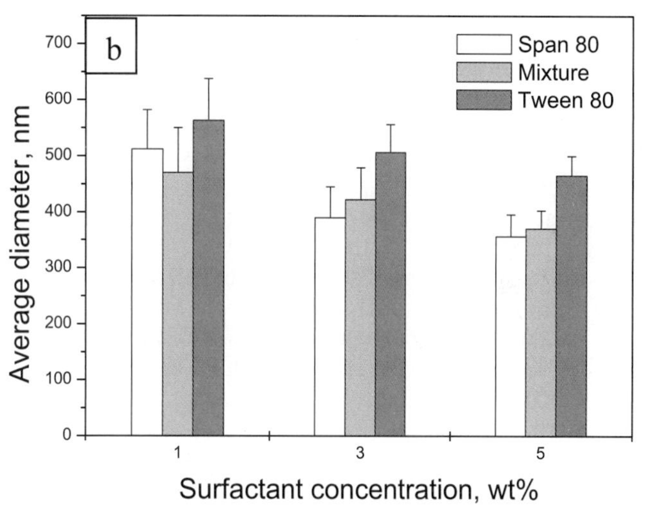

Fig. 2 Average miniemulsion droplet sizes in Miglyol 812N (**a**) or cyclohexane (**b**) as a function of surfactant type and concentration. The concentration of surfactant is given in wt.%, corresponding to the oil phase

From the obtained results it can be seen, that the difference in physico-chemical characteristics of the surfactants, and therefore different emulsifying properties, affect the size of the droplets. In all cases, the decrease in the amount of surfactant leads to larger droplets and a broader size distribution. Generally, the size of droplets was varied between 240 and 400 nm. Droplets with the smallest size were obtained with Span®80 regardless to its concentration indicating that the minimum droplet size for this system is reached. The addition of Tween®80 to Span®80 in a ratio 3 : 2 increases the stability of miniemulsions and improves the monodispersity of the droplets. In this case, both surfactants are involved in the process of the droplet stability. Hydrophobic molecules of Span®80 are oriented at the oil-droplet interface, whereas hydrophilic molecules of Tween®80 preferably stays in an aqueous phase, and bring an additional stability to the droplet from the inner cavity of the aqueous phase. Employing only hydrophilic Tween®80 as a surfactant leads to the formation of large droplets and a high polydispersity (more than 0.3) compared to other stabilizing systems, which can be explained due to a low stabilizing properties of the droplets from the hydrophobic continuous phase.

In order to investigate the influence of the continuous phase physical parameters on the droplets size and size distribution, a series of miniemulsions were formulated using cyclohexane as lipophilic phase. Figure 2b shows that the presence of cyclohexane significantly increases the average size and polydispersity of the initial droplets. In contrast to Miglyol 812N, cyclohexane possesses a lower density and viscosity (Miglyol 812N: density 0.95 g cm^{-3}, viscosity 30 mPa s; cyclohexane: density 0.78 g cm^{-3}, viscosity 1.02 mPa s). Generally, when the amount of surfactant is not enough for preventing the droplets collision, the rate of the droplet coalescence is faster in the system with a low viscous continuous phase. That is why poor droplet stability in cyclohexane is observed.

After the addition of BCA to the miniemulsion, the polymer shell around the nanodroplets is formed resulting in PBCA capsules. Almost in all cases, miniemulsions retain there visually homogeneous state. Small amounts of precipitate on the walls of vials were observed when only Tween®80 was used as a stabilizer and Miglyol 812N as a continuous phase. However, the amount of precipitate was significantly higher in the case of PBCA capsules synthesized in cyclohexane and any kind of surfactant(s). The size of polymer capsules was measured directly after 4 h of the polymerization reaction (in oil phase) and after centrifugation/redispersion (in water phase). The obtained results are shown in Fig. 3.

For both types of continuous phase, the size of nanocapsules is slightly bigger compared to the droplet size before polymerization. In all cases, the capsule size polydispersity decreases, except for using cyclohexane and low amounts of surfactant(s). It should be also pointed out that the capsules obtained in cyclohexane and stabilized by Tween®80, regardless to the concentration, were the largest in the size and contained the highest amount of coagulum. This behaviour may be explained by a possible diffusion of Tween®80 into the continuous phase during the sonication step and the participation in the polymerization of BCA. The Tween®80 molecule has three hydroxyl groups which, as nucleophile, can additionally initiate the anionic polymerization of BCA. As a result, the polymeric chains are formed not only at the interface of the droplets, but also in the continuous phase. The formed polymer can either precipitate onto the droplets surface or form a homo-PBCA particle, if the length of polymeric chain is long enough. The presence of precipitate on the droplets surface or in the continuous phase as well as the presence of newly formed particles leads to a presence of several size populations, and hence the polydispersity increases.

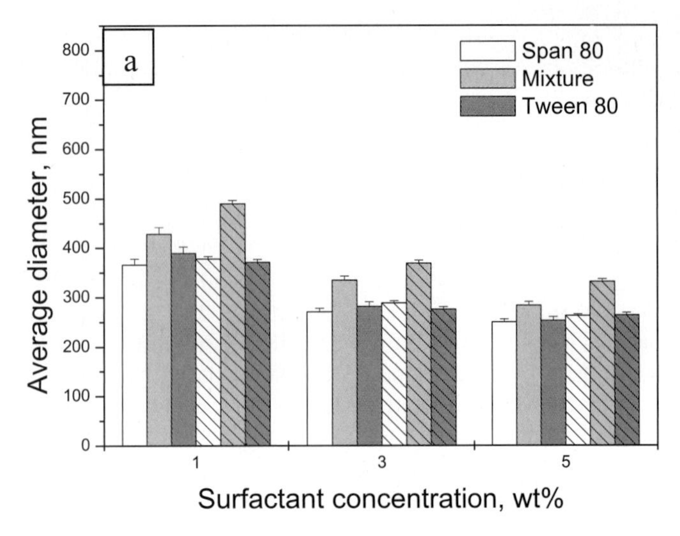

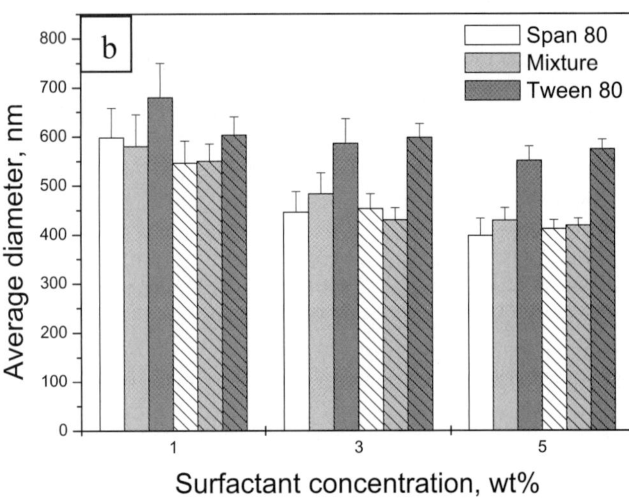

Fig. 3 Average size of PBCA capsules synthesized in Miglyol 812N (**a**) or in cyclohexane (**b**) as a function of surfactant type and concentration. Concentration of surfactant is given in wt.%, corresponding to the oil phase. Columns without pattern correspond to the capsule size measured directly after polymerization, and columns with pattern correspond to the capsule size after redispersion in water

Effect of Monomer Concentration

The effect of monomer concentration on the size and morphology of PBCA capsules was studied on miniemulsions stabilized with 5 wt. % of Span®80 (or Tween®80 or their mixture) in Miglyol 812N. The size of the capsules was measured after purification and redispersion them in (aqueous) non-ionic 0.5 wt. % Lutensol AT50 solution. No significant influence on the size and size distribution of PBCA nanocapsules was observed by varying the amount of monomer that was added (Fig. 4). Similar results were also obtained during the preparation of poly(isohexylcyanoacrylate) nanocapsules with an oil core [26]. The amount of introduced monomer is directly related to the

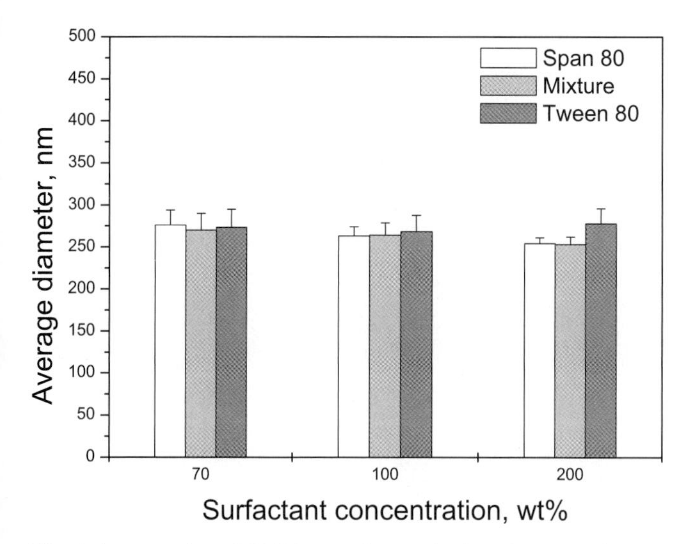

Fig. 4 Average size of PBCA capsules synthesized in Miglyol 812N as a function of surfactant type and amount of introduced *n*-butylcyanoacrylate. Concentration of surfactant(s) was 5 wt. %, corresponding to the oil phase

amount of polymer per capsule. Other authors, however, found a significant increase in the capsule size with the increase of the added monomer amount [31]. The difference in the obtained results could be due to the difference in the capsule formulation systems. In the first case, the capsules were obtained from direct (oil-in-water) emulsions, and in the second case, poly(ethylcyanoacrylate) nanocapsules were formed from inverse (water-in-oil) microemulsions. The high concentration of the surfactants, that is usually required for preparation of microemulsion, might be a key point that has an influence on the monomer polymerization process, affecting the final capsule size.

The increase in the monomer amount results in nanocapsules with different morphology and shell thickness. In Fig. 5, the TEM images of PBCA capsules obtained with 5 wt. % of Span®80 and different amounts of butylcyanoacrylate are presented. From the TEM images one can see that the polymeric shell formed with a lowest amount of butylcyanoacrylate (70 μl) is thin and has a porous structure. A more dense and regular PBCA shell can be observed for capsules obtained with higher amounts of monomer. The shell thickness depends on the monomer concentration and was in the range of 5–15 nm and 20–40 nm for 100 and 200 μl of butylcyanoacrylate, respectively.

Gel Permeation Chromatography (GPC)

In principle, the interfacial polymerization of butylcyanoacrylate can be initiated by any nucleophilic group. Keeping in mind that either Span®80 or Tween®80 contains hydroxyl groups in their molecule, one could expect that the molecular weight of poly(butylcyanoacrylate) would depend on the surfactant concentration as well as on the amount of the added monomer. The influence of these parameters in the presence of a surfactant mixture was studied by GPC and the obtained results are summarized in Table 1.

From the obtained results it can be concluded that (as expected) a higher concentration of the used surfactant results in the formation of significantly shorter polymeric chains showing clearly the involvement of the surfactant in the initiation process. With the increase in the amount of added monomer, a polymer with lower molecular weight is formed, although the difference is not very significant. Higher molecular weight polymer was also obtained by decreasing the oil to water ratio.

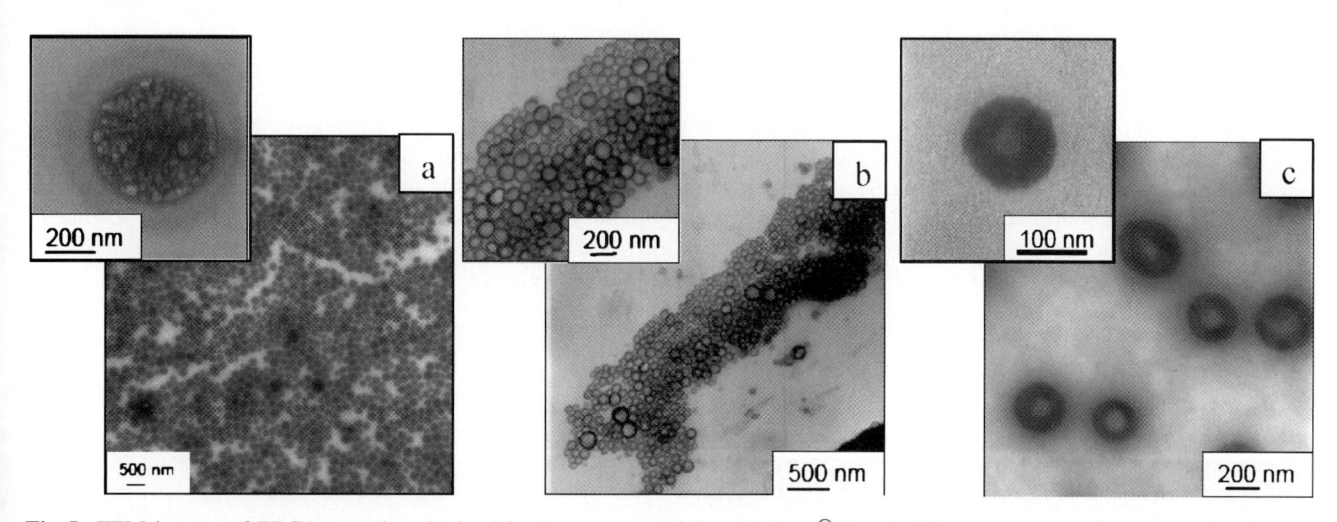

Fig. 5 TEM images of PBCA capsules obtained in the presence of 5 wt. % Span®80 and different amounts of monomer butylcyanoacrylate: **a** 70 μl, **b** 100 μl, **c** 200 μl

Table 1 Molecular weights of PBCA capsules prepared with a mixture of surfactants (Span®80 and Tween®80) under different conditions. The amount of the continuous phase (cyclohexane) was kept constant at 7.6 g

Sample name	Amount of surfactant, (wt. %)	Amount of BCA, (μl)	Amount of aqueous phase, (g)	M_n, (g mol^{-1})	M_w, (g mol^{-1})
AM-M3	5	100	1	39450	76100
AM-M10	12	100	1	15100	41600
AM-M11	19	100	1	7600	11300
AM-M4	5	70	1	45250	79530
AM-M8	5	100	2	54900	93800
AM-M7	5	200	1	43670	68770
AM-M9	5	200	2	71380	102200

Encapsulation of dsDNA in the PBCA Capsules

The PBCA nanocapsules have a great potential in the drug delivery research. Especially the delivery of the DNA molecules is of high interest. Lambert et al. [36] encapsulate 20-mer oligonucleotides into the poly(*iso*butylcyanoacrylate). In the present experiments for encapsulation a well defined 790 bp long dsDNA template was used. The template has a length of about 280 nm and is still resistant to the applied sonication procedure in contrast to the long-size genomic DNA [40]. The encapsulation of dsDNA molecules was achieved using miniemulsions stabilized with 5 wt. % of Span®80 (or Tween®80 or their mixture) in Miglyol 812N. The amount of encapsulated dsDNA was analyzed by agarose gel electrophoresis. No dsDNA was found in the continuous phase after the nanocapsule formation. In contrast, a characteristic band around 800 bp was detected after analysis of the capsule interior and confirmed the presence of dsDNA products (Fig. 6). The last lane (lane 4) corresponds to the dsDNA amount of unpolymerized miniemulsion and serves as a proof that nearly 95% of the introduced dsDNA molecules could be recovered from the system. As it can be seen from the same figure, the amount of encapsulated dsDNA into PBCA capsules was approximately the same, regardless to the surfactant type. The concentration of free (and easily extractable) dsDNA chains inside the capsule was found to be $15 \pm 5\%$, thus revealing that about 85% of dsDNA molecules could not be detected since they were probably associated with the nanocapsules wall.

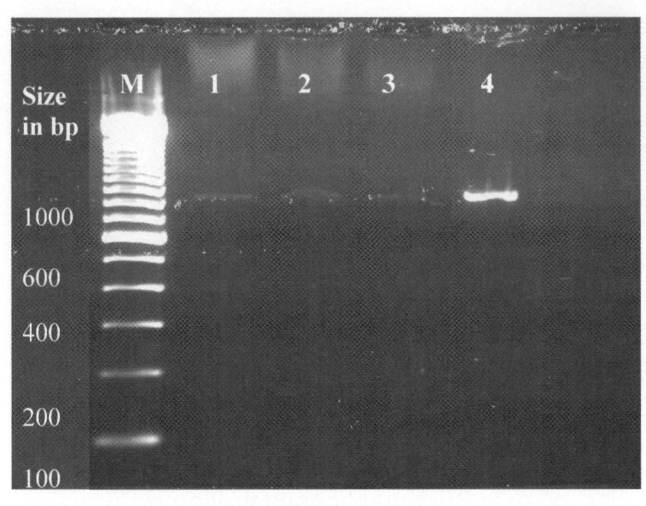

Fig. 6 Agarose gel electrophoresis of dsDNA recovered from the PBCA capsules obtained with Miglyol 812N and different surfactants. Lane M: 100-bp ladder; lane 1: 5 wt. % of Span®80; lane 2: 5 wt. % of Tween®80; lane 3: 5 wt. % of Span®80 and Tween®80; lane 4: unpolymerized miniemulsion prepared with 5 wt. % of Span®80 and Tween®80

Conclusion

In this work, we have shown that PBCA nanocapsules with an aqueous core containing dsDNA can be successfully prepared by miniemulsion technique. After encapsulation, at least about 15% of the initially loaded dsDNA molecules, that does not change their structural integrity, could be recovered from the capsules. The average capsule size and polydispersity are mainly determined by the type and concentration of the surfactant(s) as well as by the viscosity of the continuous phase. Thus the mixture of both hydrophilic and hydrophobic surfactants in the continuous phase with higher viscosity leads to monodisperse capsules. The capsule morphology can be controlled by varying the monomer amount. The polymeric shell thickness from 5 up to 40 nm can be obtained. In order to form the polymeric shell with high molecular weight, the use of low surfactant concentrations and monomer amounts are of the best choice.

References

1. Jung T, Kamm W, Breitenbach A, Kaiserling E, Xiao JX, Kissel T (2000) Eur J Pharm Biopharm 50:147
2. Zauner W, Farrow NA, Haines AMR (2001) J Control Release 71:39
3. Ogawara K, Yoshida M, Higaki K, Kimura T, Shiraishi K, Nishikawa M, Takakura Y, Hashida M (1999) J Control Release 59:15
4. Lamprecht A, Schäfer U, Lehr CM (2001) Pharm Res 18:788
5. Pitaksuteepong T, Davies NM, Tucker IG, Rades T (2002) Eur J Pharm Biopharm 53:335
6. De Faria TJ, De Campos AM, Senna EL (2005) Macromol Symp 229:228

7. Paiphansiri U, Tangboriboonrat P, Landfester K (2006) Macromol Biosci 6:33
8. Hildebrand GE, Tack JW (2000) Int J Pharm 196:173
9. Bilati U, Allémann E, Doelker E (2005) AAPS Pharm Sci Technol 6
10. Conway BR, Oya Alpar H (1996) Eur J Pharm Biopharm 42:42
11. Quintanar-Guerrero D, Allémann E, Doelker E, Fessi H (1998) Pharm Res 15:1056
12. Song CX, Labhasetwar V, Murphy H, Qu X, Humphrey WR, Shebuski RJ, Levy RJ (1997) J Control Release 43:197
13. Hariharan S, Bhardwaj V, Bala I, Sitterberg J, Bakowsky U, Ravi Kumar MNV (2006) Pharm Res 23:184
14. Allémann E, Leroux JC, Gurny R, Doelker E (1993) Pharm Res 10:1732
15. Arshady R (1989) J Microencapsulation 6:13
16. Danicher L, Frere Y, Le Calve A (2000) Macromol Symp 151:387
17. Crespy D, Stark M, Hoffmann-Richter C, Ziener U, Landfester K (2007) Macromolecules 40:3122
18. Torini L, Argillier JF, Zydowicz N (2005) Macromolecules 38:3225
19. Scott C, Wu D, Ho CC, Co CC (2005) J Am Chem Soc 127:4160
20. Sarkar D, El-Khoury J, Lopina ST, Hu J (2005) Macromolecules 38:8603
21. Allémann E, Leroux J-C, Gurny R (1998) Adv Drug Deliv Rev 34:171
22. Vauthier C, Dubernet C, Fattal E, Pinto-Alphandary H, Couvreur P (2003) Adv Drug Deliv Rev 55:519
23. Kante B, Couvreur P, Lenaerts V (1980) Int J Pharm 7:45
24. Al Khouri Fallouh N, Roblot-Treupel L, Fessi H, Devissaguet JP, Puisieux F (1986) Int J Pharm 28:125
25. Wohlgemuth M, Mächtle W, Mayer C (2000) J Microencapsulation 17:437
26. Chouinard F, Kan FWK, Leroux J-C, Foucher C, Lenaerts V (1991) Int J Pharm 72:211
27. Aboubakar M, Puisieux F, Couvreur P, Vauthier C (1999) Int J Pharm 183:63
28. Damge C, Michel C, Aprahamian M, Couvreur P (1988) Diabetes 37:246
29. Miyazaki S, Takahashi A, Kubo W, Bachynsky J, Löbenberg R (2003) J Pharm Pharm Sci 6:240
30. Fresta M, Cavallaro G, Giammona G, Wehrli E, Puglisi G (1996) Biomaterials 17:751
31. Watnasirichaikul S, Rades T, Tucker IG, Davies NM (2002) Int J Pharm 235:237
32. Florence AT, Haq ME, Johnson JR (1976) J Pharm Pharmacol 28:539
33. Wood DA, Whateley TL, Florence AT (1981) Int J Pharm 8:35
34. El-Samaligy MS, Rohdewald P, Mahmoud HA (1986) J Pharm Pharmacol 38:216
35. Li S, He Y, Li C, Liu X (2005) Colloid Polym Sci 283:480
36. Lambert G, Fattal E, Pinto-Alphandary H, Gulik A, Couvreur P (2000) Pharm Res 17:707
37. Lambert G, Fattal E, Pinto-Alphandary H, Gulik A, Couvreur P (2001) Int J Pharm 214:13
38. Hillaireau H, Le Doan T, Chacun H, Janin J, Couvreur P (2007) Int J Pharm 331:148
39. Landfester K (2001) Macromol Rapid Commun 22:896
40. Musyanovych A, Mailänder V, Landfester K (2005) Biomacromolecules 6:1824
41. Tiarks F, Landfester K, Antonietti M (2001) Langmuir 17:908
42. Watnasirichaikul S, Davies NM, Rades T, Tucker IG (2000) Pharm Res 17:684
43. Chouinard F, Kan FWK, Leroux J-C, Foucher C, Lenaerts V (1991) Int J Pharm 72:211
44. Landfester K (2000) Macromol Symp 150:171
45. Bechthold N, Tiarks F, Willert M, Landfester K, Antonietti M (2000) Macromol Symp 151:549

Progr Colloid Polym Sci (2008) 134: 128–133
DOI 10.1007/2882_2008_086

Published online: 14 August 2008

Christina Diehl
Sabine Fluegel
Karl Fischer
Michael Maskos

Oligo-DNA Functionalized Polyorganosiloxane Nanoparticles

Christina Diehl · Sabine Fluegel · Karl Fischer · Michael Maskos (✉)
Institute of Physical Chemistry, University Mainz, Jakob-Welder-Weg 11,
55128 Mainz, Germany
e-mail: maskos@uni-mainz.de

Abstract Different oligo-DNA surface functionalized polyorganosiloxane nanoparticles have been synthesized and characterized. The core–shell polyorganosiloxane nanoparticles possess a hydrodynamic radius of 14.1 (sample code CS_1) and 12.1 nm (sample code CS_2), respectively. Coupling of different short chain poly(ethylene oxide)s resulted in water soluble nanospheres. For CS_2-PEO_{110}, a number of approximately 500 PEO per siloxane nanosphere has been determined. Two different (partially fluorescently labelled) oligo-DNA sequences (12 and 20 bases, respectively) have been coupled to the PEO grafted on the nanoparticles with high coupling efficiency. MDCKII cells were incubated using a solution of these bio-hybrid nanoparticles and the successful intracellular uptake was visualized by fluorescence microscopy.

Keywords Cellular uptake · Nanoparticles · Oligo-DNA · PEO · Polyorganosiloxane

Introduction

Polyorganosiloxane nanoparticles have recently shown to possess the capability to be utilized as nano-containers and nano-reactors [1–10]. The colloidal particles are synthesized via a modified sol–gel process employing alkoxysilanes as monomers and a surfactant for the stabilization of the growing particles. Sequential addition of monomers and monomer mixtures allows the formation of core–shell particles. The particles are redispersable after subsequent saturation of the reactive surface. The incorporation of additional functional groups on the surface of the nanoparticles will open the possibility to further modify the surface properties of the particles, e.g. hydrido-silane functionalities can be used for subsequent coupling of heterotelechelical end-functionalized poly(ethylene oxide) PEO via a hydrosilation reaction. This will lead to water soluble nanospheres, which additionally can be further chemically modified by subsequent reaction of carboxy groups as second PEO end-groups with amino-terminated oligo-DNA via EDC coupling. The resulting nanoparticles thus contain the polyorganosiloxane nanospheres as core and the corona is built of PEO and attached oligo-DNA. The schematic picture of the multi-functional nanoparticles is given in Fig. 1.

Experimental

Materials and Methods

Toluene, tetrahydrofurane, chloroform, 2-(4-hydroxyphenylazo)benzoic acid (HABA), potassium trifluoroacetate, allyl alcohol, succinic anhydride, naphthaline, potassium, and (1-ethyl-3-(3-dimethylaminopropyl)carbodiimid hydrochloride (EDC) were purchased from Sigma-Aldrich, Germany, in highest quality. Diethoxydimethylsilane (D), methyltrimethoxysilane (T), triethoxysilane (T-H) and tetramethyldisiloxane (M-H) were obtained from Wacker Chemie, Germany. Dodecylbenzenesulfonic acid (DBS), *p*-chloromethylphenyltrimethoxysilane (ClBz-T) and the Karstedt-Catalyst were purchased from ABCR, Germany.

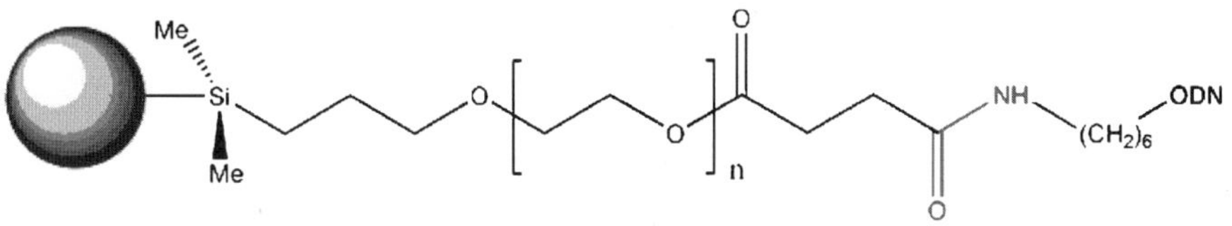

Fig. 1 Schematic representation of the ODN-modified siloxane-based nanoparticles. *Blue*: polyorganosiloxane nanoparticle, with attached PEO (only one out of many shown) and subsequent coupling of oligo-DNA (ODN)

Ethylene oxide, morpholinethansulfonic acid was obtained from Fluka, Germany. The dialysis tubing (regenerated cellulose) was obtained from SpectrumLab. The oligo-DNA (ODN) was obtained from Biomers, Germany. The fluorescent dyes used were 6-carboxyfluoresceine (FAM) and 6-carboxytetramethylrhodamine (TAMRA).

Water was deionized by a MilliQ system, Millipore.

IR measurements were performed with a Bruker Vector 33 using NaCl plates. UV/Vis spectroscopy was done with a Varian Cary 100 Bio. For the GPC measurements, columns from MZ Analysentechnik, Germany, were used (pore sizes 10^3, 10^4, 10^5, 10^6 Å), in combination with a Waters 2487 UV-detector (254 nm) and a Waters 2410 refractive index detector. MALDI-TOF mass spectrometry was done with a Micromass ToFSpecE. The matrix used was HABA with the addition of potassium trifluoroacetate. Osmometry was performed with an Osmomat090 from Gonotec, Germany (membrane cut of 20 000 g/mol). The density measurements were performed by a DMA 602 HP from Anton Paar, Austria. Light scattering has been done with an argon ion laser (SpectraPhysics, Beam-Lok 2080, 514 nm, 500 mW), an ALV SP125/5N-39 goniometer with single photon detector SO-SIPD and an ALV-5000 multiple-tau correlator at 20 °C and between 30 and 150° in steps of 10°. Data analysis was done with the programme Simplexe, V3.2. The diffusion coefficients were obtained by extrapolation to zero scattering angle and application of the Stokes–Einstein relation results in the corresponding hydrodynamic radius. In addition, the radius of gyration and the weight average of the molecular weight were determined according to Zimm by extrapolation to zero scatting angle and zero concentration [11]. The refractive index increment dn/dc was determined by an in-house built instrument (based on interferometry) [12]. A Philips CM-12, operating at 120 kV, was used to get the TEM pictures of the sample deposited on carbon coated copper grids. A multimode scanning probe microscope nanoscope IIIa from Vecco Metrology was used to obtain AFM pictures on mica. A fluorescence microscope from Leica Microsystems was used, equipped with an oil immersion objective (40-fold magnification).

The cell culture experiments were done using MDCKII cells. The cells were incubated over night with 1 ml of the nanospheres ($c = 700$ mg/ml) at 37 °C, and subsequently washed with PBS buffer to remove non-internalized nanospheres.

Synthesis

The polyorganosiloxane nanospheres were synthesized in aqueous dispersion as described before, using the monomer mixtures given in Table 1. Subsequent surface saturation with M-H led to nanospheres that could be re-dispersed in organic solvents such as toluene. The Si–H functionalities are readily detected by IR spectroscopy (2133 and 2195 cm^{-1}).

The heterotelechelic PEO was synthesized via anionic polymerisation as described before. The hydrosilation reaction was performed at 40 °C in toluene using Karstedt's catalyst. To remove excess of PEO, the reaction solution was dialysed against THF and subsequently against water (MWCO 10 000 g/mol). The coupling of the PEO modified nanospheres and the ODN was performed at 20 °C in 0.1 M MES buffer using a 300-fold excess of EDC. After 2 h, the reaction mixture was dialyzed (PEO_{52}: MWCO 10 000 g/mol, PEO_{110}: MWCO: 50 000 g/mol).

Results and Discussion

The synthesis of the polyorganosiloxane nanoparticles has been performed in aqueous dispersion as described before [1–10]. The nanoparticles possess a core-shell architecture and the employed monomer mixtures forming the core and shell are given in Table 1.

The obtained polyorganosiloxane nanoparticles are redispersable in organic solvents such as toluene and the results of the characterization in solution are given in Table 2.

Table 1 Monomers employed for the synthesis of the polyorganosiloxane core–shell nanoparticles

	Amount used in the mixture (g)	
Core monomers	CS_1	CS_2
T	9	7
D	5	2
ClBz-T	1	1
Shell monomers		
T	6	8
D	3	6
T-H	1	1

Table 2 Characterization of the polyorganosiloxane nanoparticles in solution

Sample	M_n (10^6 g/mol)	M_w (10^6 g/mol)	R_g (nm)	R_h (nm)	ρ-ratio	ρ (g/cm^3)
CS_1	1.1	1.7	12.3	14.1	0.87	1.22
CS_2	1.2	1.3	< 10	12.1	< 0.83	1.21

Table 3 α-Allyl-PEO-ω-carboxy polymers as determined by MALDI-TOF MS

Sample	M_n (g/mol)	M_w (g/mol)	M_w/M_n
PEO_{52}	2450	2500	1.02
PEO_{110}	4800	5000	1.04

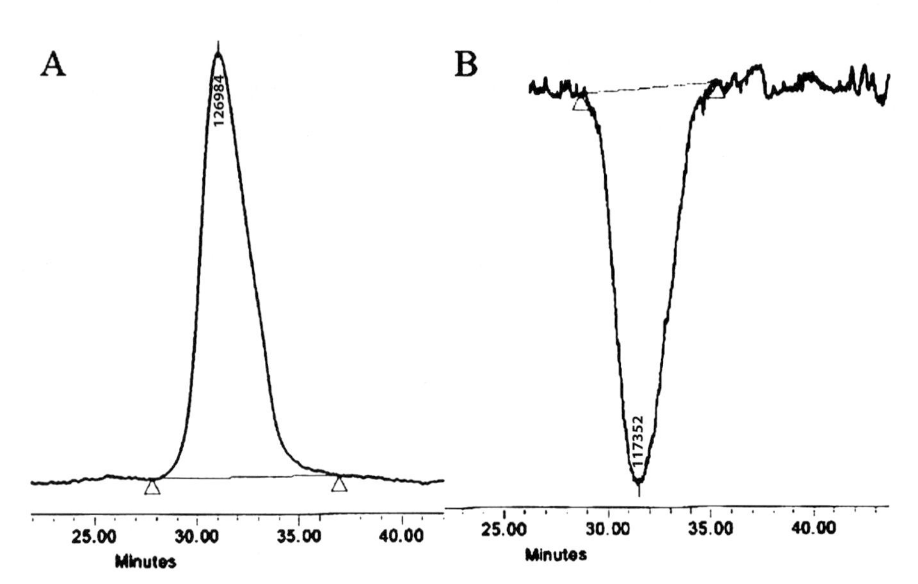

Fig. 2 GPC trace of CS_1 in toluene ($c = 10$ g/l), **A** UV detector (254 nm), **B** refractive index detector

As example, the GPC trace of CS_1 in toluene is presented in Fig. 2, which shows a monomodal, narrow distribution of the particle sizes.

The nanoparticles possess hydrido-functionalities on the surface and in the outer shell, which can be employed in a hydrosilation reaction in order to modify the surface of the particles. For this purpose, two different heterotelechelic poly(ethylene oxides) have been synthesized. Both have an allylic and a carboxylic end-group, respectively. The data describing the characteristic parameters of the polymers are given in Table 3.

From MALDI-TOF MS it is also seen that the end functionalization is 100% within experimental errors.

Subsequently, the PEO was coupled to the polyorganosiloxane nanospheres via hydrosilation. After the reaction, IR spectroscopy shows nearly no vibrations corresponding to Si–H. A TEM and an AFM picture of the sample CS_1-PEO_{110} is shown in Fig. 3.

The average size of the CS_1-PEO_{110} nanoparticles is $R = 15$ nm, which nicely corresponds to the hydrodynamic radius of the CS_1 spheres of $R_h = 14.1$ nm, indicating a slight increase in the size of the spheres upon surface modification with PEO. The increase is a result of the commonly observed decrease of the size of the polyorganosiloxane nanoparticles upon drying and analysis by TEM or AFM, and the increase of the size due to the attachment of the PEO. The degree of surface functionalization has been determined for CS_2-PEO_{110} by the measurement of the refractive index increment dn/dc. The dn/dc of PEO_{110} in water has been determined to $(dn/dc)_{PEO_{110},H_2O} = 0.135$ ml/g, corresponding nicely to the reported value of 0.132 ml/g [13]. The dn/dc for CS_2 in water is not directly accessible due to the insolubility of the nanospheres in water. Therefore, the dn/dc has been determined in organic solvents with different refractive indices and a linear extrapolation has been performed to determine the dn/dc in water. The determined values are given in Table 4.

This results in a

$$\left(\frac{dn}{dc}\right)_{CS_2,H_2O} = 0.107\ \text{ml/g}$$

for water ($n = 1.333$). Finally, the measurement of dn/dc of the coupled spheres in water gives

$$\left(\frac{dn}{dc}\right)_{CS_2\text{-}PEO_{110},H_2O} = 0.125\ \text{ml/g}\,.$$

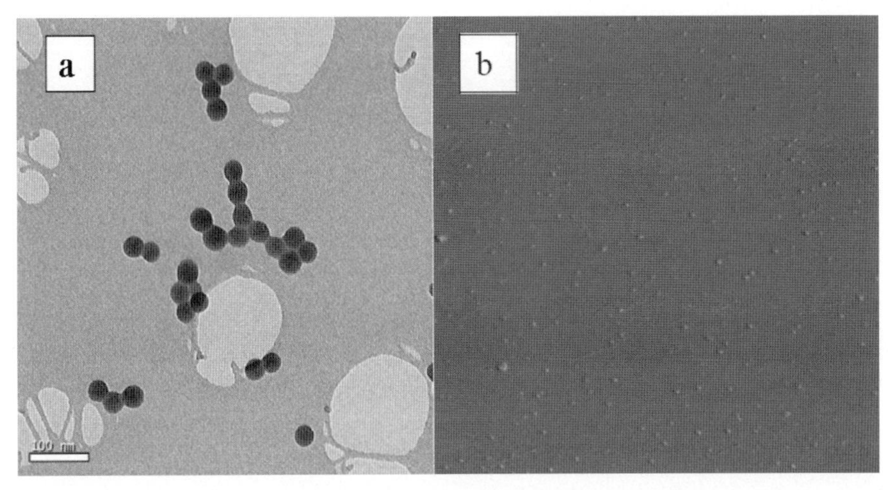

Fig. 3 Pictures of sample CS_1-PEO_{110}, from aqueous solution; **a** TEM, **b** AFM (phase, picture size: $1.5 \times 1.5\,\mu m^2$)

Table 4 dn/dc values of CS_2 determined in different solvents

Solvent	Refractive index n	dn/dc (ml/g)
Tetrahydrofurane	1.406	0.031
Chloroform	1.445	−0.010
Toluene	1.494	−0.061

According to

$$\left(\frac{dn}{dc}\right)_{CS_2\text{-}PEO_{110},H_2O} = \left(1 - w_{PEO_{110}}\right)\left(\frac{dn}{dc}\right)_{CS_2,H_2O} + w_{PEO_{110}}\left(\frac{dn}{dc}\right)_{PEO_{110},H_2O},$$

the mass fraction $w_{PEO_{110}}$ of PEO_{110} is determined to $w_{PEO_{110}} = 0.64$, which corresponds to an average number of approximately 500 PEO_{110} per CS_2 nanosphere.

Table 5 Number of PEO per particle, estimated from UV/Vis absorption (260 nm)

Sample	Number of PEO per particle [a]
CS_1-PEO_{110}-DNA_{20}	720
CS_1-PEO_{110}-DNA_{20}-FAM	810
CS_2-PEO_{110}-DNA_{12}	450
CS_2-PEO_{110}-DNA_{12}-TAMRA	630
CS_2-PEO_{52}-DNA_{20}-FAM	720

[a] Estimated assuming 100% DNA coupling.

The subsequent EDC mediated coupling of oligo-DNA to the carboxy groups of the PEO leads to the formation of functionalized bio-hybrid-nanoparticles. Two different oligo-DNA sequences have been used. A 20mer (5′-TCC ATG ACG TTC CTG ACG TT-3′) and a 12mer (5′-

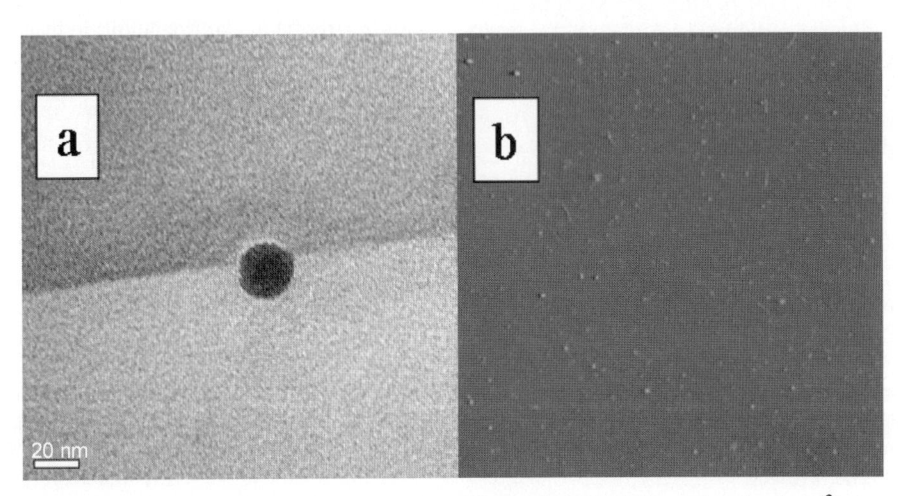

Fig. 4 Pictures of CS_2-PEO_{110}-DNA_{12}-TAMRA; **a** TEM (*scale bar*: 20 nm), **b** AFM (phase, size: $1 \times 1\,\mu m^2$)

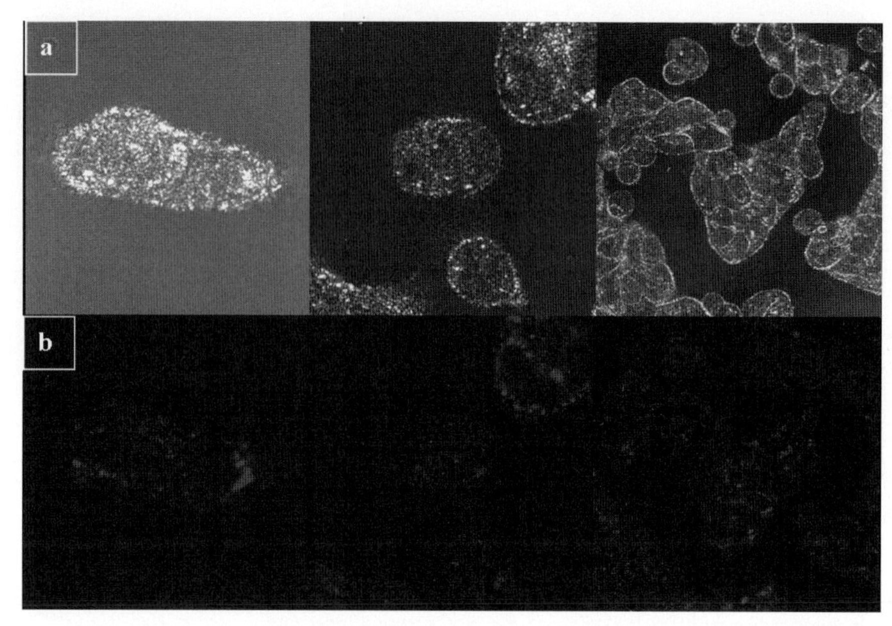

Fig. 5 Light microscopy (**a**) and fluorescence microscopy (**b**) of MDCKII cells incubated with CS_2-PEO_{110}-DNA_{12}-TAMRA, 40-fold magnification

TCC ATG ACG TTC-3′), having the same but shorter sequence. An amino-linker (C6-Amino) has been added in both cases at the 5′ end, which is used for the coupling reaction. In addition, two different fluorescent dyes have also been coupled to the 3′ end of either the 20mer (FAM), or the 12mer (TAMRA), respectively. The degree of coupling (ratio of oligo-DNA per PEO) has been determined for CS_2-PEO_{110}-DNA_{12} (0.90) and CS_2-PEO_{110}-DNA_{12}TAMRA (1.26) by UV/Vis absorption at 260 nm (corresponding to the DNA, neglecting possible, but if present, very tiny contributions of the chlorobenzylic-functionalities in the siloxane networks). Assuming (within experimental error and for a crude approximation) 100% conversion, the number of PEO per particle can be estimated. The results are presented in Table 5 and appear to be reasonable [14].

A TEM and AFM picture of CS_2-PEO_{110}-DNA_{12}-TAMRA is shown in Fig. 4. The radius in TEM is $R = 13$ nm, and the average radius in AFM is $R = 15$ nm, both in good agreement with the parental size of the corresponding CS_2 of $R_h = 12.1$ nm.

The interaction of the bio-hybrid-nanospheres CS_2-PEO_{110}-DNA_{12}-TAMRA with living MDCKII cells has been studied. A rather high concentration of CS_2-PEO_{110}-DNA_{12}-TAMRA ($c = 700$ mg/ml) was applied onto the cells and incubated over night. The result was observed by a combination of light and fluorescence microscopy and is shown in Fig. 5.

The pictures show still living cells and from the fluorescence microscopy it can be deduced that the nanoparticles entered the cells, most likely due to endocytosis, as indicated by the many point-like structures, especially close to the cell membrane, visible within the cells. In addition, it follows that the nanoparticles show no toxicity at the applied concentration of approximately $c = 0.135$ mM.

Acknowledgement We would like to thank Manfred Schmidt for the discussions, Marco Tarantola for the help with the cells, Simon Faiß for the fluorescence microscopy, Waltraut Müller for the TEM measurements, Korinna Krohne for the AFM pictures, Eva Wächtersbach for the help with the GPC and dn/dc measurements, and Hans-Michael Orfgen for the MALDI-TOF MS and the DFG SPP1313 for financial support.

References

1. Jungmann N, Schmidt M, Maskos M, Weis J, Ebenhoch J (2002) Macromolecules 35:6851
2. Jungmann N, Schmidt M, Maskos M (2001) Macromolecules 34:8347
3. Jungmann N, Schmidt M, Ebenhoch J, Weis J, Maskos M (2003) Angew Chem Int Ed 42:1714
4. Baumann F, Schmidt M, Deubzer B, Geck M, Dauth J (1994) Macromolecules 27:6102
5. Baumann F, Deubzer B, Geck M, Dauth J, Schmidt M (1997) Macromolecules 30:7568
6. Roos C, Schmidt M, Ebenhoch J, Baumann F, Deubzer B, Weis J (1999) Adv Mater 11:761
7. Emmerich O, Hugenberg N, Schmidt M, Sheiko S, Baumann F, Deubzer B, Weis J, Ebenhoch J (1999) Adv Mater 11:1299

8. Jungmann N, Schmidt M, Maskos M (2003) Macromolecules 36:3974
9. Graf C, Schärtl W, Maskos M, Schmidt M (2000) J Chem Phys 112:3031
10. Kramer T, Scholz S, Maskos M, Huber K (2004) J Colloid Interface Sci 279:447
11. Zimm B (1948) J Chem Phys 16:1093
12. Becker A, Köhler W, Müller B (1995) Ber Bunsenges Phys Chem 99:600
13. Lide DR (2003) CRC Handbook of Chemistry and Physics, 84th ed. CRC Press Inc., Boca Raton
14. Walsh MK, Wang X, Weimer BC (2001) J Biochem Biophys Methods 47:221

Progr Colloid Polym Sci (2008) 134: 134–140
DOI 10.1007/2882_2008_079

Published online: 26 June 2008

Renate Messing
Annette M. Schmidt

Heat Transfer from Nanoparticles to the Continuum Matrix

Renate Messing · Annette M. Schmidt (✉)
Institut für Organische und Makromolekulare Chemie,
Heinrich-Heine-Universität Düsseldorf,
Universitätsstr. 1, 40225 Düsseldorf,
Germany
e-mail:
schmidt.annette@uni-duesseldorf.de

Abstract Magnetocalorimetry is a new method to determine transition temperatures and the corresponding enthalpies in enthalpic phase transitions by an intrinsic heating process. We present first results on the magnetically induced melting process of a system of magnetically activated iron oxide nanoparticles embedded in an ice matrix in order to investigate the mechanisms involved in heat generation and heat transfer at the nanometer scale. By high frequency (HF) irradiation, magnetic field energy is transferred to heat locally in the particles, and the endothermic melting process of the ice matrix can be followed by recording temperature development throughout the matrix. Significant differences can be found when the results are compared to extrinsic melting processes in the absences of a field both at ambient temperature and in differential scanning calorimetry (DSC) experiments. By recording the temperature development throughout the ice matrix, first insights on the heat transfer process in magnetic heating are obtained.

Keywords Heat transport · Magnetic heating · Magnetic nanoparticles · Magnetocalorimetry · Melt transition

Introduction

The property to transfer electromagnetic energy into heat is an interesting feature in nanoparticles with permanent or inducible dipoles, such as metal, semiconductor or magnetic colloids [1–7]. Upon irradiation with electromagnetic waves of appropriate frequency, heat power is produced locally and specifically, being of potential for biomedical applications in hyperthermia [8, 9] and drug delivery [10, 11], as well as for technical use in adhesives. An issue of great significance is the transfer of the heat power from the particle to the surrounding medium. Macroscopic models are found to give an incomplete image of the heat transfer on the nanoscale, due to a high surface-to-volume ratio and the impact of a hydrodynamic interphase.

While much of the research in the field has emphasized the gradual heating effects by many nanoparticles dispersed homogeneously throughout a macroscopic volume [7, 12, 13], recent work has suggested the ability of a single nanoparticle to locally and selectively raise the temperature of molecules either attached to the particle or nearby [14]. However, theoretical studies based on the diffusive heat flow equation indicate that the localization of heating of soft materials by electromagnetic excited nanoparticles is limited by heat diffusion [2, 3, 6]. The resulting local temperature rise is shown to be negligible, or at least limited to some K.

The aim of the study is to obtain more detailed information on the mechanisms involved in the magnetic heating process under electromagnetic irradiation. For this purpose, stable magnetic colloids are prepared in aqueous medium, and the melt transition of ice is investigated in detail. The experimental setup allows the detection of energy consumption connected with the endothermic phase transition, giving a convenient method to evaluate the transition temperature and enthalpy. This way, the characteristics of the melting transition in AC magnetic fields by magnetic heating of the incorporated particles (*intrinsic* heating) can be compared with respective behaviour by conventional (*extrinsic*) heating. A main hypothesis assumes that the heat transport results in a temperature gradient on the interface between the particle surface and the medium. By using particles of different heating characteristics in media with distinct thermal diffusion behaviour, experimental results on the heat transfer at the nanoscale are obtained and compared to theoretical considerations that are of impact for fundamental research and the application of nanoparticle heating.

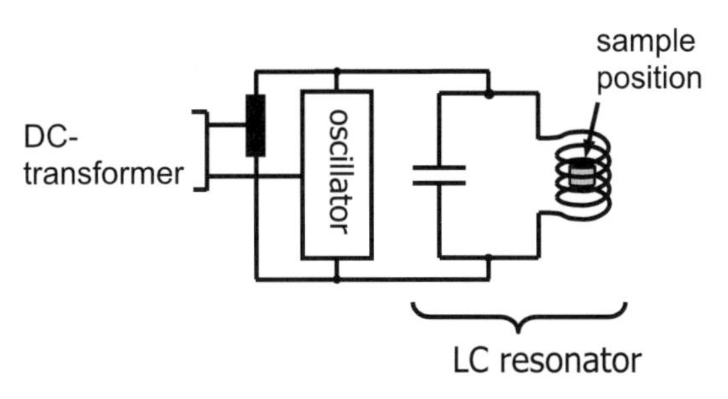

Fig. 1 Schematic diagram of the LC resonant circuit-based HF generator used for magnetic heating experiments

Experimental Part

Materials and Synthesis Procedures

$FeCl_3 \cdot 6H_2O$ (Baker) and $FeCl_2 \cdot 4H_2O$ (Baker) are used as received. 25% NH_3 aq. (Fluka) and 37% HCl aq. (Riedel–De Haën) are of p.a. quality. Tetramethyl ammonium hydroxide (NEt_4OH, 25%, Acros), 65% nitric acid (Fluka) and citric acid (Normapur) are used without purification. The water used for synthesis and dilution is of Millipore quality.

All synthetic procedures are performed under nitrogen atmosphere.

Magnetite (Fe_3O_4) nanoparticles are prepared by alkaline precipitation of ferrous and ferric chloride (molar ratio 1 : 2) aqueous solution using Cabuil and Massart's method [15]. The resulting suspension is washed several times with water to remove excess salts. Afterwards the precipitate is treated with nitric acid (2 M), and stabilized in aqueous dispersion according to the method of Philipse et al. [16]. The slow addition of 0.01 M citric acid leads to particle flocculation followed by the redispersion through the addition of tetramethyl ammonium hydroxide in order to increase the pH of the dispersion to 7.

Instrumentation

To investigate the colloidal properties of the prepared water-based ferrofluid, dynamic light scattering (DLS) experiments are implemented using a Malvern HPPS-ET. The samples are diluted with water ($V_{water} : V_{ferrofluid} = 19 : 1$), and measurements are carried out at 25 °C. Surface composition is investigated by ATR-IR spectroscopy on a Nicolet ATR-FTIR 5SXB equipped with a diamond. The magnetic properties of the ferrofluid are analyzed by using a Princeton vibrating sample magnetometer (VSM) MicroMag. Transmission electron microscopy images are obtained on a LEO 922 OMEGA electron microscope at 200 kV. The melting process of the water-based dispersions by extrinsic heating is investigated on a Mettler Toledo TC15 DSC with the ability to adjust the heating rate.

The magnetic heating experiments are performed on a Hüttinger HF generator TIG 5.0/300, equipped with a copper coil. The schematic diagram of the used experimental setup is shown in Fig. 1. The system is operated at 300 kHz and a magnetic field strength H between $4\,kA\,m^{-1}$ and $42.6\,kA\,m^{-1}$. The samples are placed in 5 ml glass vessels isolated with polyurethane foam and placed inside the copper induction coil. The sample temperature is detected with an Opsens fiber optic temperature sensor OTG-A-62 placed inside the experimental container. The same sample container and temperature sensing is used to record the thawing process in the absence of AC irradiation at ambient temperature.

Results

Characterization of Magnetite Colloids

In this study, the magnetite (Fe_3O_4) nanoparticles used for magnetocalorimetry investigations are obtained by alkaline precipitation after Massarts's method [15] and stabilized electrostatically with citric acid [16] to result in a water-based ferrofluid. To obtain a good comparability, the particles in all experiments are from one single synthetic batch and characterized carefully with respect to their surface characteristics, the dispersion state and magnetic properties.

The TEM image of electrostatically stabilized magnetite dispersion (Fig. 2b) shows irregular, nearly spherical particles with diameters between approximately 5 nm to 15 nm and an average core diameter d_c of 12 nm. The observed particle aggregation is possibly formed during

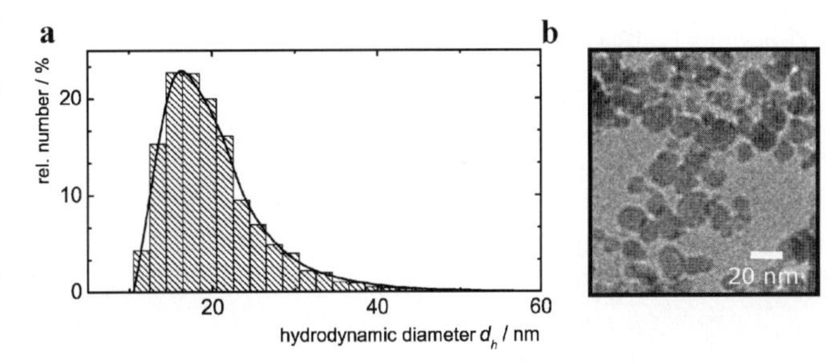

Fig. 2 **a** DLS results and **b** TEM image of the stock ferrofluid

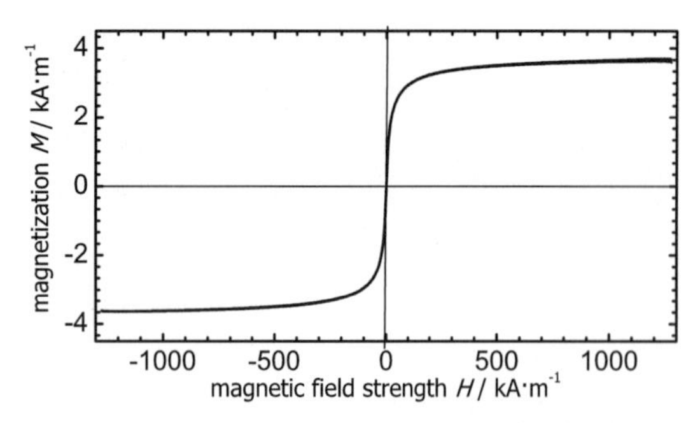

Fig. 3 VSM magnetization curve of the superparamagnetic stock ferrofluid

sampling, but separated and single shaped particles are visible. The particle diameters are well below the limit for the appearance of single domains in magnetite and therefore superparamagnetic behaviour of magnetite ferrofluids in magnetic fields are expected.

The adsorption of citric acid on the surface of the magnetite particles is proved in ATR-IR-spectroscopy experiments. Therefore the sample dispersion is precipitated several times in acetone to eliminate non-surface bound salts and dried in vacuum before measurement. The vibration absorption of protonated carboxylic acid groups of citric acid are found at 1743 cm^{-1}, whereas the deprotonated surface bound carboxylate groups show a peak at 1561 cm^{-1}. Both values correspond to the literature [17].

In vibrating sample magnetometry (VSM) of the undiluted stock dispersion (Fig. 3), the magnetization curve shows no hysteresis confirming the superparamagnetic character of the particles, and gives information on the magnetite content and the volume average magnetite core size. The calculated magnetite content $\mu_{Fe_3O_4}$ amounts of 4.65 mass % and the average core diameter d_c is calculated to 11.7 nm.

The DLS results of the stock dispersion (Fig. 2a) display a hydrodynamic diameter d_h of 18 nm and a small particle size distribution. A predominantly single-disperse particle state can be assumed from the comparison with the average core diameter d_c of 12 nm from TEM respectively 11.7 nm from VSM, taking into account that the hydrodynamic diameter d_h includes the stabilizing layer, while only the core diameter d_c is of impact in TEM and VSM experiments.

To confirm the stability of ferrofluid against dilution, freeze-thaw cycles and magnetic heating, DLS measurements on samples of diverse magnetite content were executed prior to and after the respective procedures. The results are shown in Table 1 and confirm in all cases the excellent dispersion stability of the investigated samples, with hydrodynamic diameters d_h between 17 nm and 21 nm in all cases. This finding is an important sign for the reversibility of our experiments.

Table 1 Number-average hydrodynamic diameter d_h of ferrofluids with different magnetite content $\mu_{Fe_3O_4}$, measured before and after shock freeze and melting in MC experiments

Dilution (V_{water}/ $V_{ferrofluid}$)	$\mu_{Fe_3O_4}$ (mass %)	d_h before MC (nm)	d_h after MC (nm)
0 : 1	4.65	18.8 ± 0.6	18.8 ± 1.1
1 : 5	3.72	18.7 ± 0.1	18.9 ± 0.3
2 : 3	2.79	17.6 ± 0.6	18.8 ± 0.4
3 : 2	1.86	19.4 ± 0.6	18.9 ± 0.4
5 : 1	0.93	19.1 ± 0.3	18.9 ± 0.3
9 : 1	0.47	20.5 ± 0.3	20.6 ± 0.5
49 : 1	0.09	17.2 ± 0.6	17.8 ± 0.3

Magnetic Heatability of Magnetite Nanoparticles in Water

The basic heating characteristics of the magnetic Fe_3O_4 nanoparticles under electromagnetic radiation of an AC magnetic field of 300 kHz are investigated in aqueous dispersions of different states of dilution and under variation of the magnetic field strength H. In these experiments, the samples are subjected to the field at room temperature, and the heating process is monitored by a glass fiber optic sensor system. At the maximum magnetic field strength H of 42.6 kA m^{-1}, the dispersions reach temperatures of up to 90 °C (when the experiment was stopped to avoid irreversible effects) in short time depending on the particle concentration.

The magnetocalorimetry (MC) heating curves show a decreasing heating effect with decreasing magnetite content respectively magnetic field strength H. At low magnetite content and at low magnetic field strength the dispersions do not reach the high temperature of the more concentrated samples in higher magnetic fields strength.

To analyze the heating efficiency of magnetite nanoparticles in AC magnetic fields, the specific heat power (SHP) is calculated from the initial slope of the heating curve by taking into account the heat capacity c_p and the magnetite content $\mu_{Fe_3O_4}$ of the system (Fig. 5). In magnetite dependency experiments a linear relationship between heat power $(c_p \cdot (dT/dt))$ and magnetite content $\mu_{Fe_3O_4}$ is observed. From the slope, the average SHP is calculated to (104.2 ± 2.9) W g^{-1}. The calculated heat power $(c_p \cdot (dT/dt))$ in magnetic field strength dependency experiments shows square law behaviour at low fields that levels in when the magnetic field strength H is increased.

In summary it can be said that the investigated Fe_3O_4 nanoparticles exhibit a combination of properties that is useful and desirable in magnetocalorimetric experiments. They possess a high heat power with scalable characteristics together with the required stability against dilution with water and heating in electromagnetic fields.

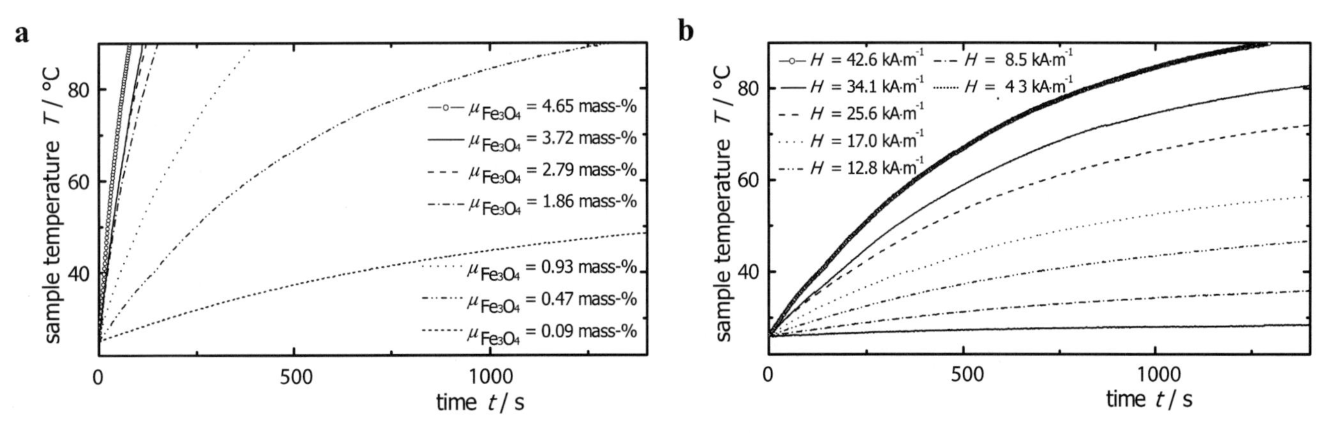

Fig. 4 MC heating curves of water-based magnetite colloids **a** with a varying magnetite content $\mu_{Fe_3O_4}$ measured at the maximum magnetic field strength $H = 42.6\,\mathrm{kA\,m^{-1}}$; **b** with a constant magnetite content $\mu_{Fe_3O_4} = 0.47$ mass % measured at varying magnetic field strength H

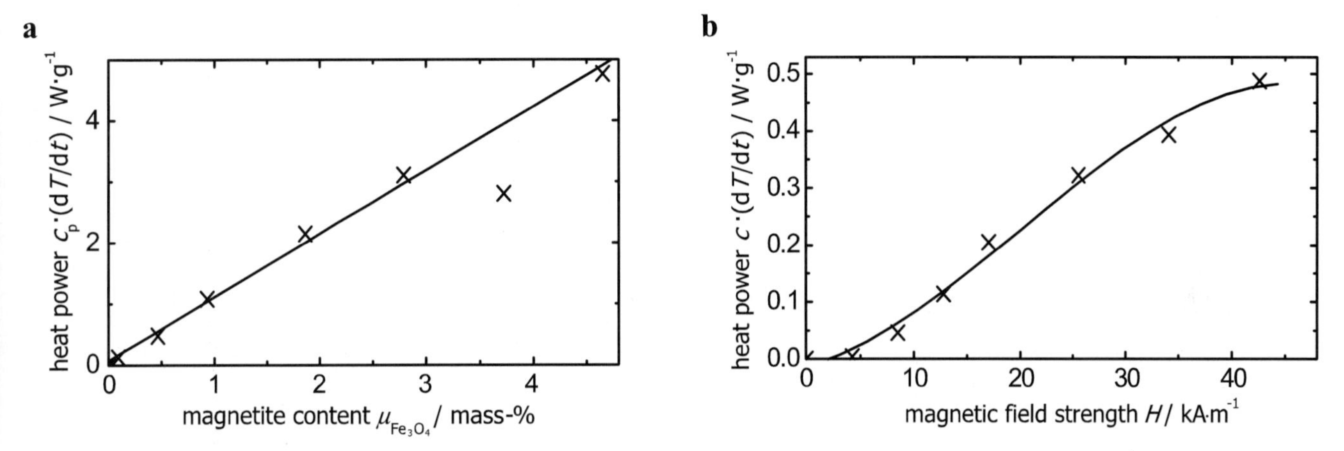

Fig. 5 Calculated heat power **a** versus magnetite content $\mu_{Fe_3O_4}$ at maximum magnetic field strength $H = 42.6\,\mathrm{kA\,m^{-1}}$; **b** versus magnetic field strength H by constant magnetite content $\mu_{Fe_3O_4} = 0.47$ mass %

Magnetic Field Generated Phase Transition of Ice

In order to utilize the generation of heat that can be initialized locally and selectively in the direct environment of the particles by magnetic heating, for magnetocalorimetric investigations, heating experiments in the absence and the presence of an AC magnetic field are performed and compared on the phase transition of ice.

Melting of ice is an endothermic process that consumes thermal energy by transformation in molecular motion. When a steady flow of thermal energy is released by magnetic heating to the particles' environment, a rise in temperature in the ice matrix can be observed, with two steps in the temperature growth rate. The first step is connected to the crystal modification transition of ice at −25 °C, while the bigger one occurs to the melting of ice at 0 °C (Fig. 6). We investigate the ice-to-water melting transition in more detail with respect to the dependence on magnetic particle concentration and magnetic field strength, and compare the results to the respective melting processes in the absence of the field.

For this purpose, a method is developed to extract information on the melting temperature T_m and the melting enthalpy ΔH_m from the MC heating curve. A schematic representation can be found in Fig. 7a.

The melting enthalpy ΔH_m is obtained as follows:

$$\Delta H_m = c_{p,ice} \left(\frac{dT}{dt} \right)_{ice} \Delta t \tag{1}$$

and

$$\Delta t = t_2 - t_1 \,. \tag{2}$$

The detection of the melting temperature T_m occurs by differentiation of the heating curve whereas T_m is found in the minimum of the derivatives (Fig. 7b).

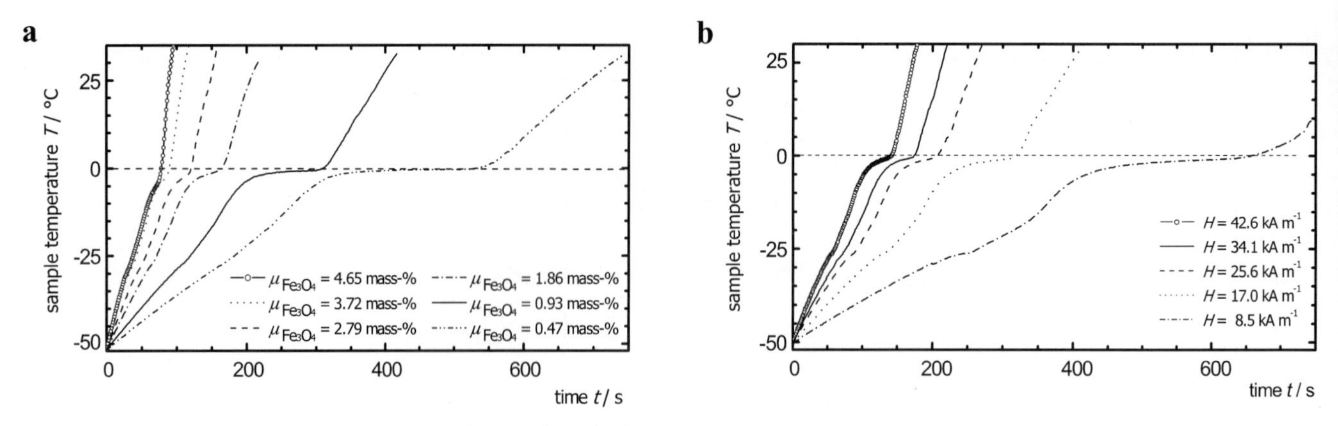

Fig. 6 MC heating curves of frozen water-based magnetite ferrofluids **a** with varying magnetite content $\mu_{Fe_3O_4}$ at maximum magnetic field strength $H = 42.6\,\text{kA}\,\text{m}^{-1}$; **b** of a sample with a constant magnetite content $\mu_{Fe_3O_4} = 0.09$ mass % at varying magnetic field strength H

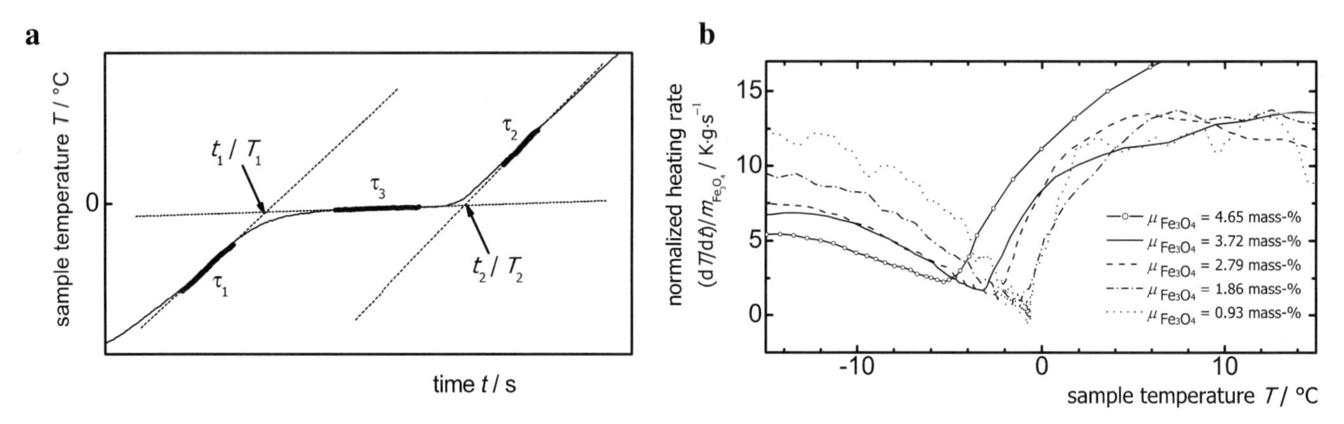

Fig. 7 **a** Conception of analysis of magnetocalorimetry results from AC field-induced melting of ice; **b** MC heating curve derivatives for dispersions with varying magnetite content $\mu_{Fe_3O_4}$ at maximum magnetic field strength $H = 42.6\,\text{kA}\,\text{m}^{-1}$

The endothermic melting process of ice to water is also analyzed by differential scanning calorimetry (DSC).

$$\Delta H_m = \int_{T_1}^{T_2} \left(\frac{dQ}{dt}\right) \Delta t \,. \tag{3}$$

Both methods allow the determination of the melting enthalpy ΔH_m and the melting temperature T_m. From comparison of the results, information on differences in the melting process by extrinsic (DSC) and intrinsic (MC) heating are expected.

The melting temperature T_m in dependence of the magnetite content is analyzed by MC, DSC at constant heating rate (dT/dt) and by thawing at ambient temperature (Fig. 8a). To analyze the magnetic field strength dependency of the melting process two samples of different magnetite content are measured in MC. We performed the corresponding DSC experiments with the analogue heating rates (dT/dt) of previously executed MC experiments. The resulting melting temperatures T_m are compared in Fig. 8b.

A clear melting point depression and a decrease of the melting enthalpy ΔH_m in dependence of increasing magnetite content is found, caused by the presence of Fe_3O_4 nanoparticles and the accompanied salts resulting from electrostatical stabilized stock ferrofluid. Apart from the well-known melting point depression by solutes [18], an influence of the nanoparticle/matrix interface on the crystallization behaviour of the matrix can be expected due to the Gibbs–Thomson effect, similar to the observation on ice crystallization in micro- and mesoporous solids [19–21].

This phenomenon is clearly observed as well in MC as in DSC experiments. Differences in the actual values can be due to the differences in the methods and calibration. However, the measured T_m in MC experiments show the lowest values whereas the decrease of T_m proceeds faster with increasing magnetite content $\mu_{Fe_3O_4}$ than in thawing at ambient. This is in accordance with the hypothesis of heat transfer by temperature gradient at the interphase between the particle surface and the surrounding.

A significant difference between extrinsic and intrinsic heating behaviour becomes evident when the melting temperature T_m of a single sample that is subjected to a row of reversible freezing-magnetic heating cycles is compared. It is observed that the samples measured

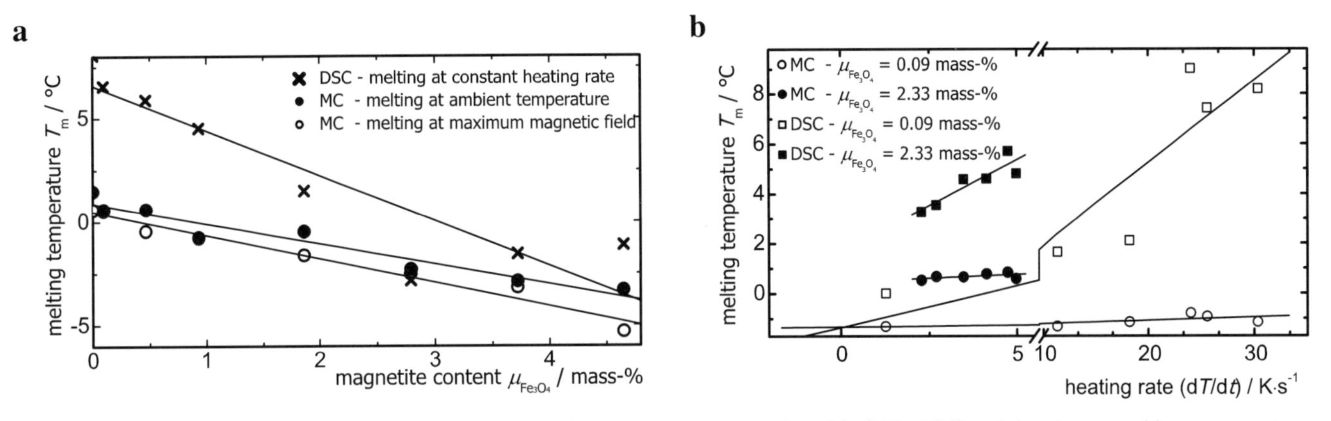

Fig. 8 Resulting melting temperature T_m **a** versus magnetite content $\mu_{Fe_3O_4}$ analyzed in MC, DSC and thawing at ambient temperature **b** versus heating rate (dT/dt) of two samples with magnetite contents $\mu_{Fe_3O_4}$ of 2.33 mass % and 0.09 mass %, analyzed in MC and DSC

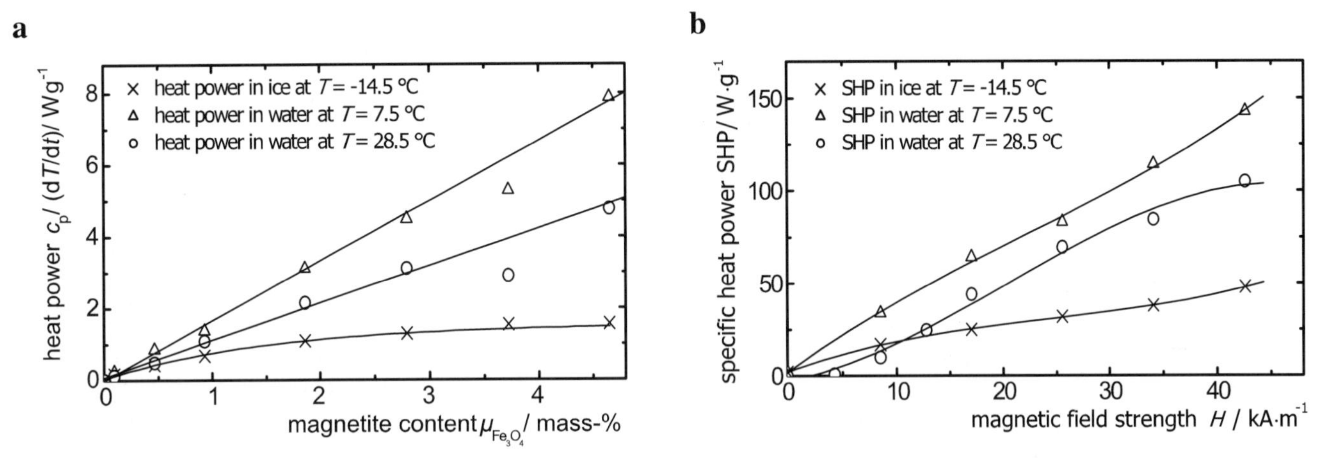

Fig. 9 Analysis of magnetic heating efficiency of magnetic dispersions in MC experiments at three different states: in ice at −14.5 °C, in cold water at 7.5 °C and in warm water at 28.5 °C **a** by means of heat power $(c_p \cdot (dT/dt))$ versus magnetite content $\mu_{Fe_3O_4}$; **b** by means of SHP versus magnetic field strength H in samples with magnetite content of $\mu_{Fe_3O_4} = 2.33$ mass % (ice, cold water) and 0.47 mass % (warm water)

under electromagnetic radiation (MC) show nearly constant values in contrast to obtained DSC results, which depend strongly on applied heating rate. The results suggest that in contrast to DSC experiments, the heat diffusion process is not the rate determining process for melting in MC experiments.

By determination of the heat power of different samples at different temperatures, a linear dependence of the heat power on magnetite content is found for the liquid samples at 7.5 °C and 28.5 °C, similar to the findings described above, and indicating a concentration-independent specific heat power. However, for samples at −14.5 °C, thus in the frozen state, a significantly lower heat power is found that is levelling in at high magnetite concentrations. In ice, the particles are immobilized, and thus the remagnetization by Brownian rotation (one of the prominent mechanisms of magnetic heating especially for bigger, blocked particles) is inhibited. At the same time, thermal convection and diffusive heat conduction are suppressed, which eventually leading to an enhanced superposition of the particles' thermal fields and the observed behaviour.

In order to obtain more information, more detailed experiments are planned, that cover the investigation of phase transitions in organic media, i.e. polymeric gels or organic solvents. To achieve a higher accuracy in the experimental setup we work on an enhanced isolation and sample geometry to minimize the thermal transfer to the environment.

Conclusion

In this study we introduced magnetocalorimetry as a method to investigate the heat generation from magnetic nanoparticles. Using ice as a matrix and observing the melting effect, we can determine the apparent melting temperature T_m and the melting enthalpy ΔH_m of the ice matrix and compare the results to melting processes in the absence of the field driven by heat exchange with the environment. We found fully reversible freezing-heating-thawing cycles in our experiments and detected significant differences in the course of melting temperature on salt content and on heating rate, as compared to conventional heating processes. By determination of the specific heat power of dispersed nanoparticles in dependence of the applied magnetic field strength H, different trends are found for liquid and immobilized samples. The observations give rise to the assumption that heat transfer may play a crucial rule in local heat loss processes as of impact in magnetic fluid hyperthermia and bond-on-demand applications.

Acknowledgement We thank Prof. Dr. H. Ritter for his steady interest in this work. For help with TEM analysis, we acknowledge Dr. Y. Lu and Prof. Dr. M. Ballauff, Universität Bayreuth, and for VSM experiments, we thank Prof. Dr. W. Gawalek, Dr. R. Müller and S. Dutz, IPHT Jena. We gratefully acknowledge the DFG (SPP1259, and Emmy Noether Program) and FCI (Liebig grant for A.M. Schmidt) for financial support.

References

1. Richardson HH, Hickman ZN, Govorov AO, Thomas ACV, Zhang W, Kordesch ME (2006) Nano Lett 6:783
2. Govonov AO, Zhang W, Skeini T, Richardson H, Lee J, Kotov NA (2006) Nanoscale Res Lett 1:84
3. Nyborg W (1988) Phys Med Biol 33:785
4. Andrä W, d'Ambly CG, Hergt R, Hilger I, Kaiser WA (1999) J Magnetism Magn Matter 194:197
5. Keblinski P, Cahill DG, Bodapati A, Sullivan CR, Taton TA (2006) J Appl Phys 100:054305
6. Rabin Y (2002) J Hyperthermia 18:194
7. Rosensweig RE (2002) J Magnetism Magn Mater 252:370–374
8. Glöckl G, Hergt R, Zeisberger M, Dutz S, Nagel S, Weitschies W (2006) J Phys Condens Matter 18:S2935
9. Andrä W (1998) Magnetic Hyperthermia. In: Andrä W, Nowak H (eds) Magnetism in Medicine. Wiley-VCH, Weinheim, p 455
10. Müller-Schulte D, Schmitz-Rode T (2006) J Magnetism Magn Mater 302:267
11. Schmidt AM (2005) J Magnetism Magn Mater 289C:5
12. Lowe LB et al. (2003) J Am Chem Soc 125:14258
13. Jones CD, Lyon LA (2003) J Am Chem Soc 125:460
14. Hamad-Schifferli K, Schwartz JJ, Santos AT, Zhang S, Jacobson JM (2002) Nature 415:152
15. Massart R, Cabuil V (1987) J Chem Phys 84:967
16. van Ewijk GA, Vroege GJ, Philipse AP (1999) J Magnetism Magn Mater 201:31
17. Rochiccioli-Deltcheff C, Franck R, Cabuil V, Massart R (1987) J Chem Res 18:1209
18. Wedler G (2004) Lehrbuch der physikalischen Chemie. Wiley-VHC, Weinheim
19. Mori A, Maruyama M, Furukawa Y (1996) J Phys Soc Japan 65:2742
20. Rennie GK, Clifford J (1977) J Chem Soc Faraday Trans 73:680
21. Hirama Y, Takahashi T, Hino M, Sato T (1996) J Colloid Interf Sci 184:349

Progr Colloid Polym Sci (2008) 134: 141–148
DOI 10.1007/2882_2008_093

Published online: 2 August 2008

NANO-COLLOIDS

Susann Schachschal
Andrij Pich
Hans-Juergen Adler

Growth of Hydroxyapatite Nanocrystals in Aqueous Microgels

Susann Schachschal · Andrij Pich (✉) ·
Hans-Juergen Adler
Department of Macromolecular Chemistry and Textile Chemistry, Technische Universität Dresden, 01062 Dresden, Germany
e-mail: andrij.pich@chemie.tu-dresden.de

Abstract In present paper we demonstrate that aqueous microgels can be used as containers for in-situ synthesis of hydroxyapatite. The hydroxyapatite nanocrystals (HAp NCs) become integrated into microgels forming hybrid colloids. The HAp NCs loading in the microgel can be varied in a broad range. The HAp NCs are homogeneously distributed within the microgel corona. The deposition of the inorganic nanocrystals decreases colloidal stability of the microgels and leads to particle aggregation at high HAp NCs loading. Due to the strong interactions between HAp NCs and polymer chains the swelling degree of microgel decreases and temperature-sensitive properties disappear at high loading of the inorganic component. The hybrid colloids can be used as building blocks for the preparation of nano-structured films on solid substrates.

Keywords Hydroxyapatite · Microgel · Nanoparticles · Thermosensitive particle

Introduction

The biomimetic preparation of composite materials in synthetic systems received recently a strong interest and stimulated intensive research in this field. The hydroxyapatite (HAp, $Ca_{10}(PO_4)_6(OH)_2$) belongs to the most important biominerals which can be found in natural hard tissues [1, 2]. However, hydroxyapatite as well as other calcium phosphates exhibit poor mechanical properties (low elasticity and high brittleness). The combination of HAp with polymeric matrix leads to the hybrid materials exhibiting high flexibility, mechanical strength, biocompatibility and good processing/shaping [3]. The interaction of the HAp with polymeric matrix is realized by covalent bonds, hydrogen bonds [4], dipole–dipole interactions or complexation of Ca^{2+}-ions by functional groups such as amine, acetylamine or hydroxyl [5]. In this way polymeric matrix plays an important role in the nucleation, growth of HAp crystals and determines their size, morphology and orientation in the composite material. The most important application fields for hybrid materials containing HAp are tissue engineering [6], implants [3], drug delivery systems [7, 8], catalysis [9, 10], adsorbents [11] and protein chromatography [12, 13].

A large variety of polymeric materials have been used as templates for the synthesis of HAp such as protein collagen [14], poly(L-lactic acid) [15], poly(aspargic acid) [16], alginates [17], gelatine [18], chitosan [4, 5, 19], chitin [20], dendrimers [21, 22], hydrogels [23–26], etc. The use of polymeric particles as templates for the growth of HAp NCs has not been intensively studied. Polymeric particles as templates provide several advantages such as uniform size, extremely large surface area and enormous possibilities for the surface functionalization. In this way, well defined hybrid particles can be prepared for use in chromatography columns or as catalyst in different technical systems. An applicability of polymeric particles coated by Pd^0 for the growth of HAp NCs has been demonstrated by Tamai et al. [27]. The homogeneous growth of the hydroxyapatite layer on the particle surface led to the formation of polystyrene core-HAp shell hybrids. Recently, Sukhorukov and co-workers [28] demonstrated the utilization of the pH-sensitive polyelectrolyte capsules for the growth of hydroxyapatite nanocrystals and fabrication of hybrid hollow spheres.

The aim of present study was to demonstrate that aqueous microgel particles can be used for the growth of HAp NCs. Due to the porous microgel structure and presence of functional groups the HAp NCs can be effectively stabilized in the microgel interior. Obtained hybrid colloids can be used for the injectable tissue replacement or as building blocks for the preparation of scaffolds for tissue engineering in form of nanostructured films or hydrogels.

Experimental Part

Materials

N-vinylcaprolactam (VCL) (Aldrich) was purified by conventional methods and then distilled under vacuum. Acetoacetoxyethyl methacrylate (AAEM), vinylimidazole (VIm), initiator 2,2′-azobis(2-methylpropyonamidine) dihydrochloride (AMPA) and crosslinker *N*,*N*′-methylenebisacrylamide (BIS) (Aldrich) have been used as received. Calcium nitrate tetrahydrate ($Ca(NO_3)_2 \cdot 4H_2O$) (Grüssing, 99%), diammonium hydrogenphosphate ($(NH_4)_2HPO_4$) (Riedel-de Haën, 99%), and ammonia solution (Fisher Chemicals, 25%) have been used as received.

Microgel Synthesis [29]

The polymerization procedure can be described as follows. Appropriate amounts of AAEM (0.321 g), VCL (1.783 g), VIm (0.071 g) and BIS cross-linker (0.06 g) were dissolved in 145 g of deionized water. Double-wall glass reactor equipped with stirrer and reflux condenser was purged with nitrogen. The monomers and crosslinker were dissolved in water by stirring in the reactor for 1 h at 70 °C under continuous purging with nitrogen. After that the 5 ml aqueous solution of APMA initiator (0.05 g) was added under continuous stirring. Reaction was carried out for 8 h. Microgel dispersion was purified by dialysis with Millipore Dialysis System (cellulose membrane, MWCO 100 000). The VIm content (4.91%) in the microgel has been determined by potentiomentrical titration. Final solid content of the microgel dispersion was 1.59%.

Preparation of Hydroxyapatite Nanocrystals

Appropriate amount of $Ca(NO_3)_2 \cdot 4H_2O$ was dissolved in microgel dispersion and the pH value has been adjusted to 10 by addition of 25% NH_4OH. In a separate flask appropriate $(NH_4)_2HPO_4$ amount was dissolved in water and pH value was adjusted to 10 by addition of 25% NH_4OH (detailed amounts of ingredients are listed in Table 1). The $(NH_4)_2HPO_4$ solution has been added slowly under continuous stirring to the polymeric dispersion containing $Ca(NO_3)_2$ and mixture was stirred for 24 h. Samples were purified by dialysis (SpectraPor 3 membrane MWCO 3500, Millipore) to remove non-reacted reagents and by-products.

Characterization Methods

Dynamic light scattering DLS measurements were performed using commercial laser light scattering spectrometer (ALV/DLS/SLS-5000) equipped with an ALV-5000/EPP multiple digital time correlator and laser goniometer system ALV/CGS-8F S/N 025 was used with a He-Ne laser (Uniphase 1145P, output power of 22 mW and wavelength of 632.8 nm) as the light source. The DLS experiments were carried out in the range of scattering angles $\theta = 30–90°$. All solutions were filtrated using a 5.0 μm membrane filter before measurements.

Stability measurements of microgel dispersions were performed with separation analyser LUMiFuge 114 (LUM GmbH, Germany). Measurements were made in glass tubes at acceleration velocities from 500 to 3000 rpm. The slope of sedimentation curves was used to calculate the sedimentation velocity and to get information about stability of the samples.

Table 1 Reagents used for the synthesis of hydroxyapatite within microgels and some characteristics of obtained hybrid particles

N	Microgel [g]	$Ca(NO_3)_2 \cdot 4H_2O$ [g]	$(NH_4)_2HPO_4$ [g]	HAp^T [wt.%]	HAp^M [wt.%]	Ca/P	R_h [nm]
0	–	–	–	–	–	–	475
1	0.2725	0.0342	0.0114	5.1	4.7	1.54	366
2	0.2725	0.0683	0.0229	9.7	7.9	1.52	388
3	0.2725	0.2049	0.0686	24.2	23.6	1.90	331
4	0.2180	0.2733	0.0915	34.7	34.2	1.65	agg
5	0.1635	0.3080	0.1031	44.5	44.4	1.86	agg
6	0.1635	0.4099	0.1372	51.6	52.2	1.58	agg

T – theoretical value; M – measured value; agg – microgel aggregation

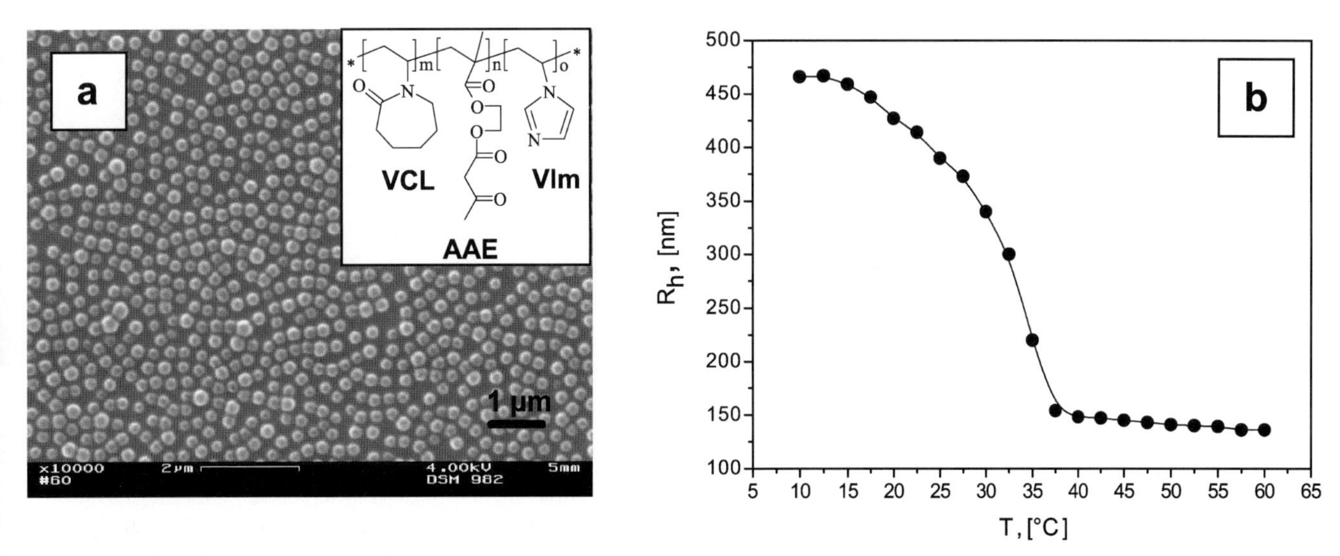

Fig. 1 SEM image of VCL/AAEM/VIm microgel particles (*inset* shows a copolymer structure) (**a**); hydrodynamic radius (R_h) as a function of the temperature for VCL/AAEM/VIm microgels (**b**)

SEM images were taken with Gemini microscope (Zeiss, Germany). Samples were prepared in the following manner. Microgel dispersions were diluted with deionized water, dropped onto cleaned glass support and dried at room temperature. Samples were coated with thin Au/Pd layer to increase the contrast and quality of the images. Pictures have been obtained at voltage of 4 kV.

TEM images have been made with a Hitachi HD 2000 instrument operating at 200 kV. Diluted microgel dispersions were placed onto carbon-coated copper grids and dried at room temperature.

To determine the hydroxyapatite content in composite particles the TGA 7 Perkin Elmer instrument (Pyris-Software Version 3.51) was used. Samples were analyzed in temperature range 25–800 °C (heating rate 10 K/min in nitrogen atmosphere).

The XRD measurements have been performed with Siemens P5005 powder X-ray diffractometer (source – Cu K_α; $\lambda = 1.54$ Å).

IR spectra were recorded with Mattson Instruments Research Series 1 FTIR spectrometer. Dried polymer samples were mixed with KBr and pressed to form a tablet.

Results and Discussion

The microgel particles used in present study are based on copolymer of vinylcaprolactam (VCL) and acetoacetoxyethyl methacrylate (AAEM). As described in our previous studies the microgel particles possess heterogeneous structure and consist of AAEM-rich core and VCL-rich shell due to some peculiarities of the polymerization process [30]. The use of small fraction of vinylimidazole (VIm) during VCL/AAEM microgel synthesis allows selective incorporation of VIm units in the swollen VCL corona [29]. The VCL/AAEM/VIm microgels are characterized by narrow particle size distribution (see Fig. 1a) and superior colloidal stability. The temperature-sensitivity of the microgels is provided by the phase transition temperature (T_{tr}) of PVCL chains which occurs at approx. 33 °C (the lower critical solution temperature (LCST) value for the linear PVCL is around 32 °C). This phenomenon occurs due to the different solvatation of PVCL chains by water molecules at the temperatures below and above the phase transition temperature (below T_{tr} polymer is swollen and above T_{tr} polymer is collapsed). It is believed that the volume phase transition in the microgel occurs as a result of reduced hydrogen bonding between water molecules and the polymer and hydrophobic aggregation of the polymer that leads to the microgel shrinkage (see Fig. 1b). This process is reversible and if the temperature decreases below critical value the microgel swells.

The porous microgel structure has been used in present study to deposit hydroxyapatite nanocrystals (HAp NCs). The preparation of hybrid microgels containing HAp NCs is shown in Scheme 1. The first step is addition of Ca^{2+}-ions to the microgel dispersion and the formation of HAp in aqueous phase can be performed by precipitation reaction of calcium and hydrophosphate ions in basic aqueous medium.

As shown in Scheme 1 the ionic strength in the reaction mixture increases considerably after addition of calcium salt. The presence of the ions in aqueous phase influences strongly the solubility of the polymer chains in water. This effect induces change of the microgel size as well as the volume phase transition temperature and colloidal stability. The complexity of such transformations is enhanced by different influence originating from the nature and charge of the ions. As shown in Fig. 2a the addition of Ca-salt to

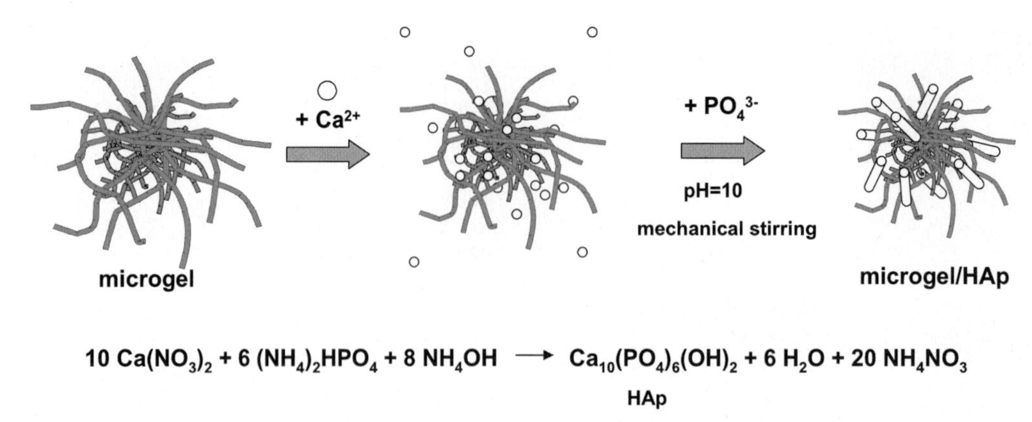

Scheme 1 Synthesis of HAp NCs in microgel particles

the microgel solution led to some shrinkage of microgels (the calcium nitrate concentrations used in this experiment correlate with that used for the synthesis of HAp NCs). Figure 2a indicates that the decrease of the microgel hydrodynamic radius occurs after addition of salt; however no particle aggregation occurs in the investigated salt concentration range.

The VCL/AAEM/VIm microgels are sterically stabilized and exhibit no charge at pH = 6. Therefore, the reduction of the microgel size is probably not a consequence of the charge compensation but can be explained by the partial reduction of the hydrogen bonds between polymer chains and water molecules. This leads to some "dehydratation" of polymer segments and more coiled conformation that induces shrinkage of the microgel shell. The additional reason for the microgel shrinkage could be complexation of Ca^{2+}-ions by β-diketone groups of AAEM units. In this case Ca^{2+}-ions can form intermolecular complexes that also provoke the partial shrinkage of the microgels. However, it should be noted that the VCL/AAEM/VIm microgels do not shrink completely after salt addition (in the collapsed state as shown in Fig. 1b microgels possess R_h of 150 nm).

After addition of diammonium hydrophosphate the formation of HAp takes place in the reaction mixture (see Scheme 1). The formation of HAp is followed by the strong reduction of the colloidal stability of the microgels. As shown in Fig. 2b the sedimentation velocity of the composite particles increases with the increase of HAp content. The samples with high HAp loading show clear phase separation after the reaction has been completed (Fig. 2b). These results indicate that the presence of HAp increases microgel affinity to form aggregates and precipitate from the solution.

The morphology of the composite particles is demonstrated in Fig. 3. The SEM images indicate that the deposition of HAp NCs occurs presumably in the microgels, however secondary HAp NCs can be also found. The increase of the HAp content induces some morphology change of the microgels. They appear as flattened

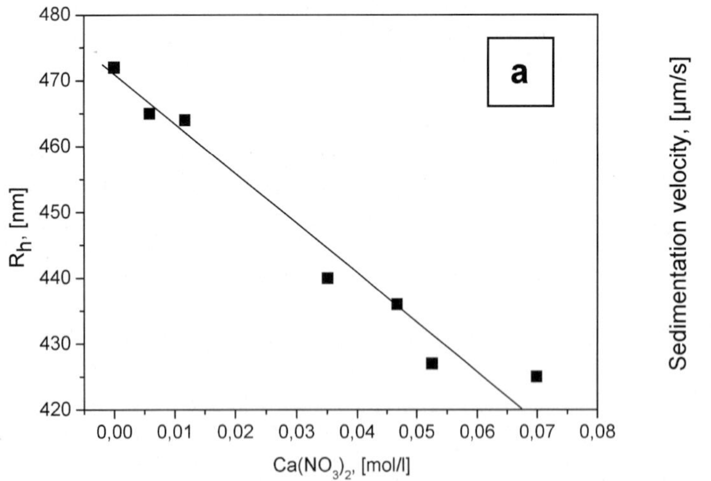

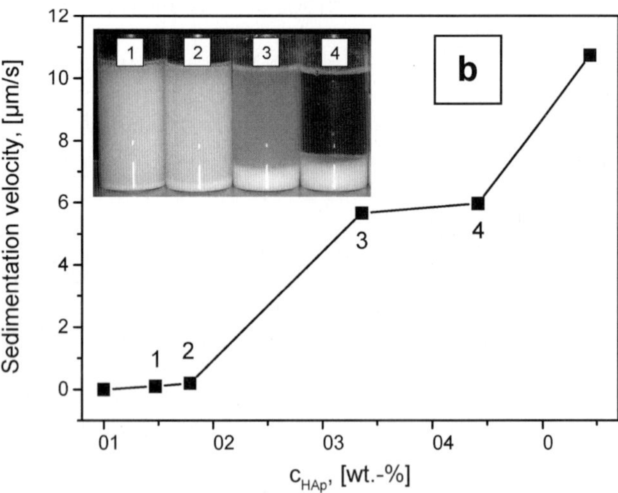

Fig. 2 The sedimentation velocity of hybrid microgels with different HAp contents

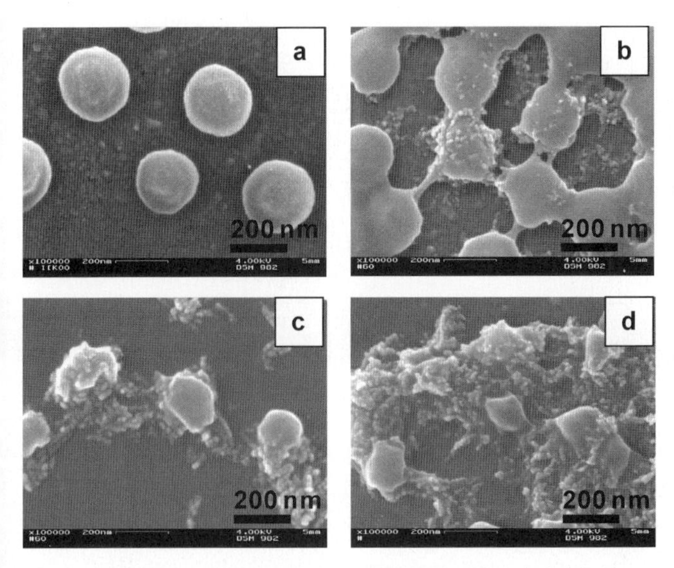

Fig. 3 SEM images of PVCL/PAAEM/PVIm-HAp hybrid microgels prepared with different HAp contents: **a** microgel; **b** 7.9% HAp; **c** 34.2% HAp; **d** 52.2% HAp

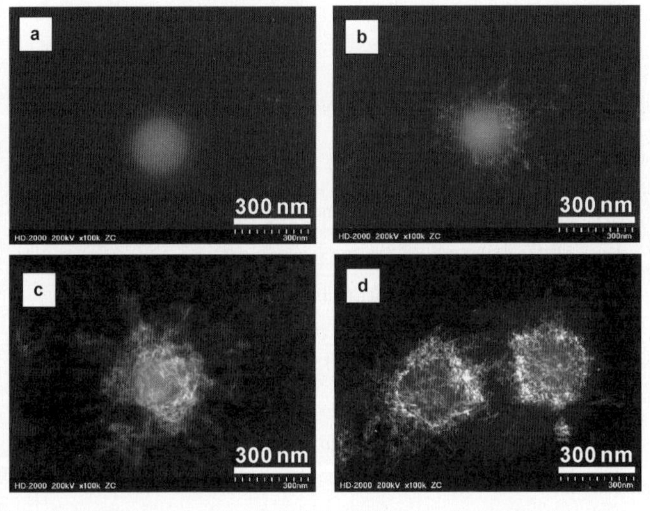

Fig. 4 TEM images of PVCL/PAAEM/PVIm-HAp hybrid microgels containing different amount of HAp: **a** 0%; **b** 7.9%; **c** 34.2%; **d** 52.2%

discs with coronas filled by HAp NCs. The composite particles shrink after HAp loading. This has been also confirmed by DLS measurements for colloidally stable samples (see Table 1). Some reduction of the microgel size can be expected since HAp NCs deposited in the microgel shell reduce the hydratation and mobility of the polymer chains due to the adsorption of the polymer segments on inorganic particle surface. Due to this reason the thermosensitivity of the microgels decreases considerably and disappears at high HAp loadings (data not shown here). The increase of the HAp loading reduces the swelling degree of the microgels and they behave as compact spheres.

The dark-field TEM images presented in Fig. 4 give more detailed insight into the structure of composite microgels. Note that in these images only compact microgel core is clearly visible. The HAp NCs possess needle-like morphology and are located in the outer microgel layer and randomly distributed within microgels. The increase of the HAp loading does not change considerably the morphology and size of NCs.

The results of EDX line scan performed on single microgel particle with deposited HAp NCs presented in Fig. 5 indicate presence of Ca and P in the microgel outer layer that confirms the effective deposition of HAp NCs into the microgels.

It should be noted that the controlled deposition of HAp NCs takes place only in VCL/AAEM microgels functionalized by VIm units. The use of VCL/AAEM microgels with other functional groups such as primary or secondary amines led to the accumulation of the HAp on the microgel surface and non-homogeneous distribution of inorganic NCs in the microgel network. Additionally the formation of HAp aggregates in aqueous phase was detected. The VCL/AAEM microgel without additional functionalization showed similar results. Based on the experimental results discussed above we assume that the controlled growth and stabilization of the HAp NCs in the microgel shell occurs due to the effective interactions between polymer segments and HAp NCs surface. In basic medium (reaction conditions) the hydroxyl groups on the HAp surface participate in the formation of hydrogen bonds with carbonyls of VCL units. The VIm units can interact with Ca^{2+}-ions on the HAp surface by formation of coordination bonds by donating the electron pair available on N atom. Contrary, in weakly acidic range we can assume the electrostatic attraction between phosphonate groups of HAp and imidazolium ions located in the polymer chains. This indicates that highly swollen functionalized shell of the VCL/AAEM/VIm microgels control to some extent the growth of the HAp NCs and provides their stabilization in the polymer network.

The elemental analysis gives a possibility to determine Ca/P ratio which is quite important characteristic of HAp. The experimental results summarized in Table 1 suggest that in all samples Ca/P ratio oscillate between 1.5 and 1.9 (theoretical value for HAp is 1.67). The Ca/P ratio higher then theoretical value indicates that some substitutions of other ions are probably present in the crystal structure instead of phosphate (we assume this to be carbonate initially present in aqueous phase). The XRD data for selected samples presented in Fig. 6a suggest that hybrid materials exhibit characteristic signals which correlate with the reference spectra for HAp. The signal intensity is strongly dependent on the HAp amount in the sample and this is the reason for bad signal/noise ratio in the spectra of composite microgel. The most important absorption bands of the phosphate group in the HAp structure appear at 565, 603, 1034 und 1094 cm^{-1} in the IR-spectrum. Figure 6b shows that increase of the HAp content in the microgels

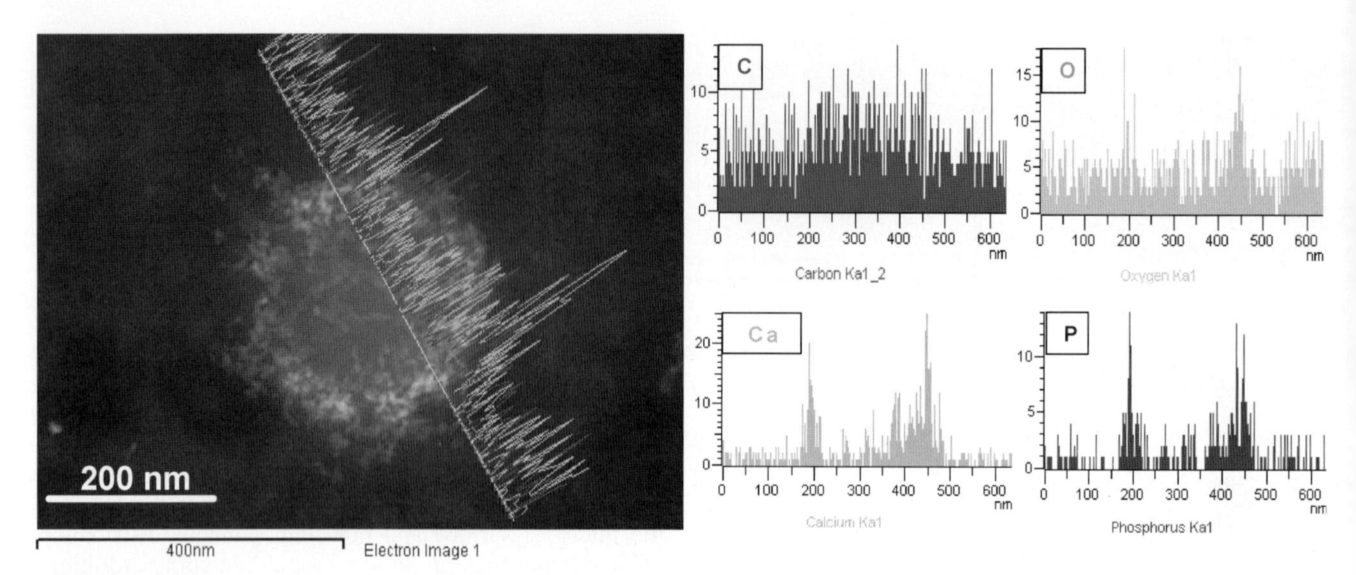

Fig. 5 TEM image with EDX line scan of single microgel containing 52.2% HAp. The element distribution curves showed as separate windows (colour of the lines in the image corresponds to the colour in element distribution spectra)

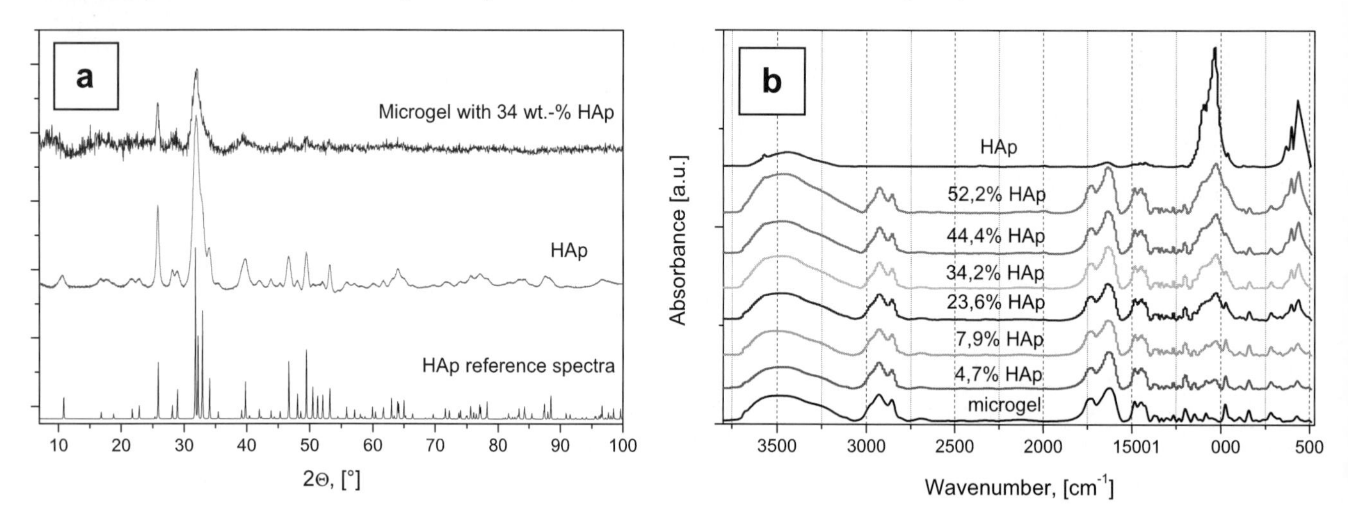

Fig. 6 XRD (**a**) and IR spectra (**b**) of VCL/AAEM/VIm-HAp hybrid microgels

induces the stepwise increase of the band intensity for the inorganic component.

Our preliminary results demonstrate that hybrid microgels can be used for the preparation of the well-defined composite films (see Fig. 7).

The white line in Fig. 7b shows the area used for the EDX mapping to determine the element distribution. Figure 7c indicates that the strongest Ca signal is detected around microgel core, so the composite structure of the microgel is preserved after film-formation. The results presented in Fig. 7 demonstrate that hybrid microgels can be used for the preparation of the well-defined composite films. In this case the amount of inorganic material as well as its localization in the bulk material is determined by the loading and morphology of the microgel particles respectively.

We assume that obtained hybrid microgels containing HAp NCs can be effectively used as injectable tissue replacement materials. The simple microgel functionalization allows additional modification of the particle surface and conjugation with bio-active molecules. HAp-containing microgels can be used as building blocks for the preparation of scaffolds for tissue engineering in form of nanostructured thin films or hydrogels.

Conclusions

In summary, we demonstrate that aqueous microgels can be used for growth of HAp nanocrystals (HAp NCs). By using conventional precipitation method the HAp NCs become incorporated into the microgel particles. The

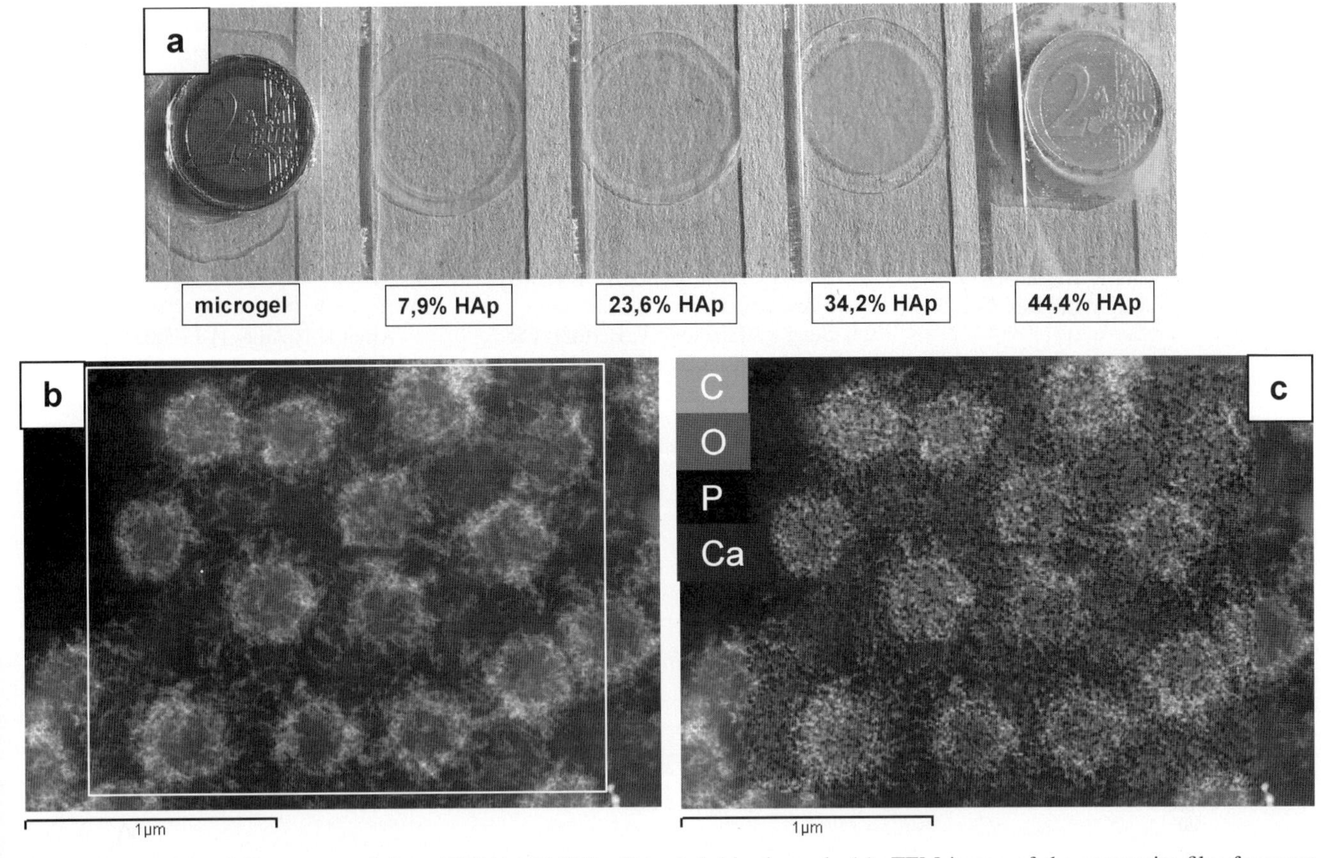

Fig. 7 Photograph of films prepared from VCL/AAEM/VIm-HAp hybrid microgels (**a**); TEM image of the composite film fragment obtained by drying VCL/AAEM/VIm-HAp (52.2%) sample (**b**); EDX element mapping showing distribution of different elements (**c**)

HAp NCs are homogeneously distributed in the microgel shell and the loading of inorganic material in the composite particles can be varied in a broad range. The incorporation of HAp NCs reduces the colloidal stability of the microgels and decreases their ability of changing the size as response to the temperature variation. The chemical structure of HAp NCs incorporated into microgels has been confirmed with XRD and IR-spectroscopy. The incorporation of HAp NCs into polymeric particles allows preparation of nano-structured composite films on solid supports by simple casting procedure.

Acknowledgement Authors thank Deutsche Forschungsgemeinschaft (DFG) for financial support.

References

1. Mann S, Webb J, Williams RJP (1989) Biomineralization: Chemical and Biological Perspectives. VCH Publishers, New York
2. Dorozhkin SV, Epple M (2002) Angew Chem 114:3260
3. Neumann M, Epple M (2006) Eur J Trauma 2:125
4. Zhang L, Li YB, Yang AP, Peng XL, Wang XJ, Zhang X (2005) J Mater Sci Mater Med 16:213
5. Rusu VM, Ng CH, Wilke M, Tiersch B, Fratzl P, Peter MG (2005) Biomaterials 26:5414
6. Rodríguez-Lorenzo LM, Ferreira JMF (2004) Mater Res Bull 39:83
7. Kim H-W, Knowles JC, Kim H-E (2005) J Mater Sci Mater Med 16:189
8. Barroug A, Glimcher MJ (2002) J Ortopaedic Res 20:274
9. Zahouily M, Bahlaouan W, Bahlaouan B, Rayadh A, Sebti S (2005) Arkivoc 13:150
10. Wakamura M (2005) Fujitsu Sci Tech J 41:181
11. Kawai T, Ohtsuki C, Kamitakahara M, Tanihara M, Miyazaki T, Sakaguchi Y, Konagaya S (2006) Environ Sci Technol 40:4281
12. Felício-Fernandes G, Laranjeira M (2000) Quím Nova 23:441
13. Jungbauer A, Hahn R, Deinhofer K, Luo P (2004) Biotechnol Bioeng 87:364
14. Hsu F, Tsai S, Lan C, Wang Y (2005) J Mater Sci Mater Med 16:341
15. Zhang R, Ma PX (1999) J Biomed Mater Res 45:285
16. Bigi A, Boanini E, Gazzano M, Rubini K (2005) Cryst Res Technol 40:1094

17. Malkaj P, Pierri E, Dalas E (2005) J Mater Sci Mater Med 16:733
18. Montemegro RVD (2003) PhD thesis. Crystallization, Biomimetics and Semiconducting Polymers in Confined Systems. Golm, Germany
19. Kong L, Gao Y, Cao W, Gong Y, Zhao N, Zhang X (2005) J Biomed Mater Res A 75:275
20. Wan ACA, Khor E, Hastings GW (1997) J Biomed Mater Res 38:235
21. Boduch-Lee KA, Chapman T, Petricca SE, Marra KG, Kumta P (2004) Macromolecules 37:8959
22. Chen H, Holl MB, Orr BG, Majoros I, Clarkson BH (2003) J Dent Res 82:443
23. Kaneko T, Ogomi D, Mitsugi R, Serizawa T, Akashi M (2004) Chem Mater 16:5596
24. Song J, Saiz E, Bertozzi CR (2003) J Am Chem Soc 125:1236
25. Song J, Malathong V, Bertozzi CR (2005) J Am Chem Soc 127:3366
26. Ho E, Lowman A, Marcolongo M (2007) J Biomed Mater Res A 83:249
27. Tamai H, Yasuda H (1999) J Colloid Interf Sci 212:585
28. Schukin DG, Sukhorukov GB, Möhwald H (2003) Angew Chem Int Ed 42:4472
29. Pich A, Tessier A, Boyko V, Lu Y, Adler H-J (2006) Macromolecules 39:7701
30. Boyko V, Pich A, Lu Y, Richter S, Arndt K-F, Adler H-J (2003) Polymer 44:7821

Progr Colloid Polym Sci (2008) 134: 149–155
DOI 10.1007/2882_2008_082

Published online: 14 August 2008

S. Lutter
J. Koetz
B. Tiersch
S. Kosmella

Formation of Cadmium Sulfide Nanoparticles in Poly(ethylene Glycol)-Modified Microemulsions

S. Lutter · J. Koetz (✉) · B. Tiersch · S. Kosmella
Institut für Chemie, Universität Potsdam, Karl-Liebknecht-Straße 24–25, Haus 25, 14476 Potsdam (Golm), Germany
e-mail: koetz@rz.uni-potsdam.de

Abstract This paper is focused on the formation of cadmium sulfide (CdS) nanoparticles in the poly(ethylene glycol) (PEG)-modified microemulsion consisting of sodium dodecylsulfate (SDS), xylene, pentanol, and water. Due to the presence of the polymer a bicontinuous, sponge-like microemulsion is developed, which is used as a template phase for the particle formation beside the classical w/o microemulsion. The stability and size of the particles is strongly influenced by polymer concentration and molecular weight of PEG and aggregation seems to be favoured in the sponge-like template phase. However, the process of particle aggregation in the bicontinuous phase can be hindered by increasing the concentration and molecular weight of the polymer. After solvent evaporation CdS nanoparticles with diameters of 10 nm can be redispersed, which tend to build up larger aggregate clusters.

Keywords Cadmium sulfide · Microemulsion · Nanoparticles · Poly(ethylene glycol)

Introduction

The formation of nanometer-sized semiconductor materials has received a lot of attention during the last years because of numerous, highly promising applications, e.g. in electronic devices like solar cells or LEDs [1–4], medical diagnostics [5–7] and photo catalysis [8–10]. Nearly all applications benefit from the strongly size dependent optical, optoelectronical and magnetic properties of such materials. The reason therefore is the well investigated size quantization effect [11–14], which can be observed for nanoscalic semiconductors like cadmium sulfide or cadmium selenid. With decreasing particle size the band gap increases and a red-shift of the fluorescence bands and the onset of absorption of light occur. Particles that show this effect are frequently called Q-particles or quantum dots. However, the quite special properties can be only observed, when monodisperse particles with particle dimensions below 10 nm are produced. Therefore, a lot of activities were focused on the formation of monodisperse nanocrystals. For example sol–gel methods [15, 16], solvothermal processes [17, 18] and irradiation [19, 20] have been used for the preparation of such materials. Furthermore, the preparation has been successfully realized in block copolymer micelles and polymer microgels [21–24]. Another interesting approach is the use of water-in-oil microemulsions, whose nanometer sized water droplets can act as nanoreactors [25–29]. The inverse microemulsion droplets are suitable reaction media, which favour the formation of nanoparticles with a narrow size distribution. The size and the shape of the particles formed are affected by the droplets size, and the presence of additives, e.g. polymers [30–33]. Our own investigations have shown that polyelectrolyte-modified water-in-oil microemulsions can be used as a template phase for the formation of ZnS and CdS nanoparticles [34, 35].

The aim of the present study was to investigate the formation of cadmium sulfide nanoparticles in a water-in-oil microemulsion in comparison to a bicontinuous, sponge-like microemulsion template phase. Therefore, the pseudo-ternary microemulsion system sodium dodecylsulfate (SDS)/xylene-pentanol (1 : 1)/water was modified with the non-ionic polymer poly(ethylene glycol) (PEG). Previously, we have reported that a bicontinuous microemulsion can be observed at a polymer concentration of $\geq$ 10 wt. % and nearly equal amounts of oil and water [36]. We found that the structure of the template phase did not only influence the size and shape, but also the aggregation of the primilary nanoparticles. Therefore, the role of the polymer concentration and molecular weight was investigated in more detail by means of UV-Vis spectroscopy, dynamic light scattering, and transmission electron microscopy.

Experimental

Material

Poly(ethylene glycol) (PEG), PEG I and PEG II are purchased from Fluka and used without further purification. The average molecular weights of the two PEG samples are determined by gel permeation chromatography (GPC) (compare Table 1). The commercially available pentanol (> 99%, Fluka), xylene (> 99%, Roth) and sodium dodecylsulfate (SDS, > 99%, Fluka) are used as obtained. $CdCl_2$ and $(NH_4)_2S$ are purchased by Merck-VWR, and water is purified with the Modulab PureOne water purification system (Continental).

Phase Diagram

The partial phase diagram at 25 °C is determined optically by titration of the oil-alcohol/surfactant mixtures with the aqueous poly(ethylene glycol) solution. The samples are stirred until the system became optically clear and thermostated at 25 °C in a water bath. The region of the isotropic phase is determined by dropwise addition of the polymer solution to the system. More than 20 data points are received for each phase diagram.

Synthesis of CdS Nanoparticles

CdS nanoparticles are prepared by mixing two adequate microemulsions, which contain the precursor $CdCl_2$ (5 mM) and $(NH_4)_2S$ (5 mM), respectively. The samples are stored for 3 days in a vacuum oven at 35 °C, to evaporate the solvent (water, xylene, pentanol). After the evaporation process the remaining solid fraction includes an excess of surfactant and polymer, while the particle mass accounts of about 1%. Unfortunately the precursor concentration is constrained by the stability of the microemulsion template phase and can only be increased to 5 mM. The amounts of SDS and PEG depend on the composition of the microemulsion template phases used, especially on the percentage of the aqueous phase and on the polymer concentration. For example at point A and a polymer concentration of 10 wt. % the dried residue contains about 90% of SDS and 9% of PEG. At point B the percentages changed to about 62 and 37% respectively. Afterwards, the crystalline samples are redispersed in water using an ultrasonic finger for further characterization.

Table 1 Average molecular weights of poly(ethylene glycol) determined by GPC

Polymer	M_n (g/mol)	M_w (g/mol)
PEG I	2800	3000
PEG II	20 300	28 300

Characterization of the Nanoparticles

UV-Vis absorption measurements are realized by means of a Cary 5000 UV-Vis NIR spectrometer (Varian) in a wave length range between 200 and 700 nm. Therefore, the samples were placed in a quartz cuvette with a path length of 1 cm. Dynamic light scattering measurements are used to obtain the size and size distribution of the CdS particles in the microemulsion, as well as the redispersed particles in water. The measurements are carried out at a fixed angle of 173° using the Zetasizer Nano ZS (Malvern), equipped with a He-Ne laser and a digital autocorrelator. The particle size distributions were obtained by using the "multiple narrow modes" result calculation, which is provided by the DTS (nano) software (Malvern). This model allows especially the characterization of multi modal distributions and uses the non-negative least squares (NNLS) algorithm. Note that especially large particle dimensions on the micrometer scale have to be regarded very critically.

All measurements are performed using the refractive index of CdS. In the case of cadmium sulfide nanoparticles produced in the w/o microemulsion the viscosity η and the refractive index n_D of the continuous oil phase, namely the xylene-pentanol (1 : 1) mixture ($\eta = 1.454$ cP, $n_D = 1.165$) are used. Consequently η and n_D for water are used when the CdS nanoparticles are redispersed in the aqueous phase. Morphology and size of the redispersed CdS particles are also determined by transmission electron microscopy. Therefore, a small amount of the aqueous solutions is dropped on copper grids, dried and examined in the EM 902 transmission electron microscope (Zeiss) (acceleration voltage 90 kV). The high amount of surfactant brings also difficulties for the preparation of the samples for TEM measurements and consequently samples have to be washed with water to reduce the amount of surfactant.

Results

Phase Behaviour

Figure 1 shows the partial phase diagram of the pseudo-ternary system SDS/xylene-pentanol (1 : 1)/water. In absence of any further additive the system shows the seperated w/o (L2 phase) and o/w (L1 phase) microemulsions. As already described before, due to the presence of the nonionic polymer poly(ethylene glycol) an enlargement of the isotropic phase is induced and a phase channel, connecting the L1 and L2 phase is formed at 10 wt. % of PEG [36]. Our own investigations have shown that the dimensions of the phase channel are influenced by the polymer concentration, molecular weight, and also by the temperature [37].

The cryo scanning electron micrograph in Fig. 1 shows that the isotropic phase channel is characterised by a sponge-like structure, which can be typically found for bicontinuous microemulsions. If interactions between polymer molecules and the surfactant film occur, the bending elasticity and the spontaneous curvature H_0 of the surfactant film of the microemulsion can be influenced. In special cases H_0 can be changed to zero, which leads to the formation of a bicontinuous structure. It is well known that SDS and poly(ethylene glycol) can develop strong interactions [38–40]. These interactions can induce the structural changes observed for the microemulsion system discussed here. We have already shown that the bicontinuous structure is changed by varying the polymer concentration and the molecular weight of PEG [41]. The dimensions of the inverse microemulsion droplets can also be checked by means of cryo SEM and diameters of about 60 nm can be determined as shown in [41], too.

Taken into account these former results, the nanoparticle formation was realized at point A and B marked in the phase diagram, which represent the w/o and the bicontinuous microemulsion, respectively. The microemulsions were modified by using the poly(ethylene glycol) samples PEG I and PEG II at polymer concentrations of 10, and 20 wt. %.

Nanoparticles Formation

CdS Nanoparticles in the Microemulsion Template Phase. It has already been described in detail that CdS nanoparticles show a UV-Vis spectrum, which strongly depends on the size of the particles [42–44]. Figure 2 shows the UV-Vis spectra of CdS particles in the microemulsions at point A and B in presence of 10 wt. % PEG I at certain times after mixing the both precursor microemulsions. Independent of the template phase used, one can observe a typical absorption shoulder of the CdS particles between 420 and 450 nm already after a reaction time of 1 min. Hence, the particle formation process in the microemulsion template phase is a very fast process, which is nearly finished after mixing of the two adequate microemulsions.

The absorption behaviour of the CdS particles formed in the w/o microemulsion (Fig. 2a) is not dramatically changed during the first hour of reaction time. Just after more than 60 min a clear decrease of the absorption intensity between 425 and 450 nm occurs. At the same time the onset of the absorption band is significantly increased. In contrast, point B shows a more continuous decrease of the absorption intensity over the time range of 2 h. Furthermore, the absorption shoulder is red-shifted to 470 nm, which indicates larger particle diameters. This assumption can be also confirmed by the more pronounced

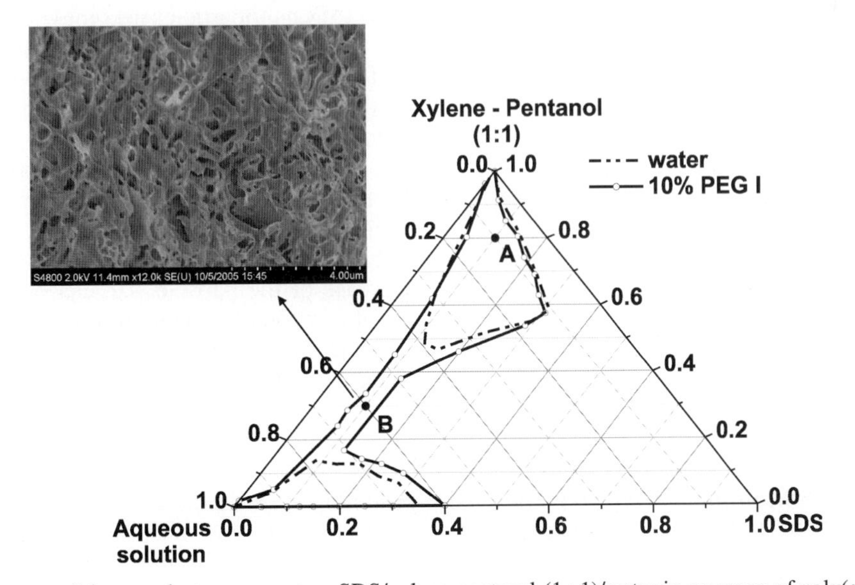

Fig. 1 Partial phase diagram of the pseudo-ternary system SDS/xylene-pentanol (1 : 1)/water in presence of poly(ethylene glycol)

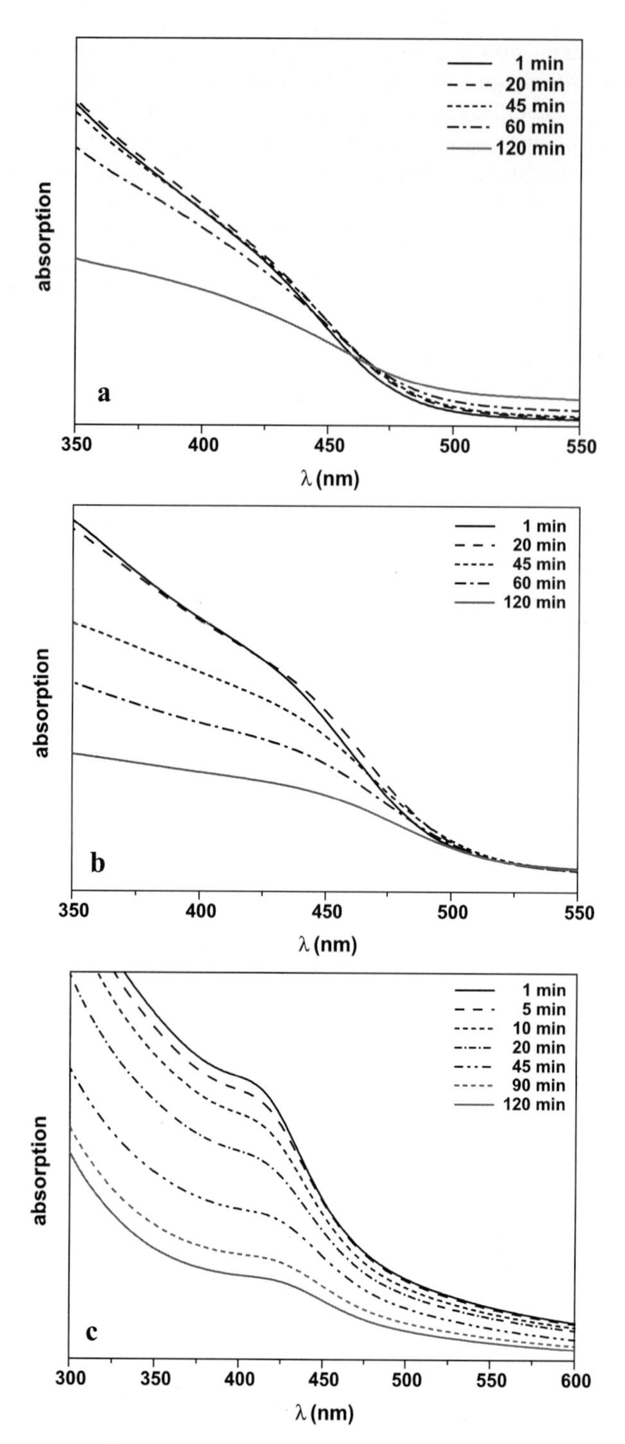

Fig. 2 UV-Vis absorption spectra of CdS nanoparticles in the PEG-modified microemulsions at point A in presence of 10 wt. % PEG I (**a**) and at point B in presence of 10 wt. % PEG I (**b**) and PEG II (**c**)

yellowish colour of the solution. The results reveal a time-dependent aggregation of the primilary formed particles in the microemulsion. It seems to be plausible, that CdS particles formed in the inverse microemulsion droplets show a higher stability against these aging effects, than particles formed in the water-continuous domains of a sponge-like phase.

To check the influence of the molar mass of the polymer the same experiments were realized with PEG II. The absorption curve in Fig. 2c shows remarkable differences in comparison to Fig. 2b. The absorption shoulder is much more pronounced and clearly shifted to 420 nm. The absorption intensity is also continuously decreased after starting the reaction by mixing both microemulsions. However, during this aging process the absorption maximum is only slightly shifted to 430 nm. Furthermore, the particle size can be directly obtained from the UV-Vis absorption spectra. Therefore, the wavelength of the absorption threshold (λ_s), which is a function of the nanoparticle size, has to be determined. In detail, λ_s can be obtained by plotting $(A/\lambda)^2$ vs. $1/\lambda$ according to [27]. The UV-Vis spectra of point B in presence of PEG I reveal λ_s values between 475 and 505 nm, which increase with proceeding reaction time. With regard to references [27,45] these values can be related to CdS particle diameters between 7 and 10 nm. In presence of PEG II λ_s is shifted from 465 to 480 nm representing particle sizes of 6 to 7 nm.

In addition, the particle diameters are calculated from dynamic light scattering measurements and summarized in Table 2. It has to be mentioned, that the contrast of the polymer-modified microemulsion in absence of the nanoparticles is not sufficient enough for dynamic light scattering and therefore a characterisation of the droplet size by this method is not practicable. To avoid the previously discussed aging of the samples, the measurements are performed immediately after mixing the both microemulsions. The particle size distributions obtained by peak analysis by intensity show a main fraction on the micrometer scale and a second fraction of significant smaller diameters of about 10 nm. However, this fraction of 10 nm-sized CdS nanoparticles becomes dominant in the number-weighed size distribution. One can assume that a few dust particles or aggregates affect the intensity distribution on the micrometer scale, while the majority of the CdS nanoparticles are characterized by much smaller diameter on the nanometer scale. Unfortunately a filtration of the microemulsion to remove dust particles or aggregates is not possible. In general, similar diameters between 7 and 11 nm can be determined for both points. A small decrease of the particle size can be observed by increasing the molecular weight in the L2 phase, and by increasing the polymer concentration in the bicontinuous phase.

CdS Nanoparticle Redispersed in Aqueous Medium. In a second step, the solvent compounds (water/xylene/pentanol) are evaporated in a vacuum oven and the remaining crystalline material is redispersed in water. The redispersed CdS particles are also characterized by dynamic light scattering. First of all, a significant increase of

Table 2 Particle size distribution of CdS nanoparticles in the PEG-modified microemulsion determined by dynamic light scattering

	Microemulsion Point A						Microemulsion Point B					
	Intensity [a]			Number [b]			Intensity [a]			Number [b]		
	*d** [nm]	DW** [nm]	%	*d** [nm]	DW** [nm]	%	*d** [nm]	DW** [nm]	%	*d** [nm]	DW** [nm]	%
10 wt. % PEG I	1700	100	57	11	1	> 99	1600	320	70	11	1	98
	11	1	43				13	1	25			
20 wt. % PEG I	2400	345	78	10	1	95	8	1	100	8	1	> 99
	10	1	22									
10 wt. % PEG II	1300	353	90	7	2	> 99	1140	230	86	11	2	95
	12	6	7				13	2	10			

[a] Intensity size distribution;
[b] Number size distribution;
* *d* – diameter;
** DW – distribution width.

Table 3 Particle size distribution of the redispersed CdS nanoparticles after solvent evaporation obtained from DLS measurements

	Point A						Point B					
	Intensity [a]			Number [b]			Intensity [a]			Number [b]		
	*d** [nm]	DW** [nm]	%	*d** [nm]	DW** [nm]	%	*d** [nm]	DW** [nm]	%	*d** [nm]	DW** [nm]	%
10 wt. % PEG I	112	8	93	95	12	> 99	124	28	84	31	3	98
							32	4	6			
20 wt. % PEG I	134	16	70	126	15	> 99	109	11	92	34	2	98
	> 1000		30				35	3	8			
10 wt. % PEG II	124	20	65	125	21	> 99	116	22	75	33	4	98
	329	44	29				37	5	17			

[a] Intensity size distribution;
[b] Number size distribution;
* *d* – diameter;
** DW – distribution width.

the particle diameters can be observed for all the samples as it is illustrated in Table 3.

The diameter of the CdS particles prepared in the microemulsion point A (w/o microemulsion) increases to about 110 to 130 nm and also a second fraction of significant larger particles of several hundred nanometres can be observed. It has to be pointed out that in contrast to the CdS nanoparticles in the microemulsion at point A no significant differences appear between the intensity or number size distribution of the redispersed nanoparticles. However, when the polymer concentration or the molar mass is increased in the L2 phase larger particle aggregates were detected. In the bicontinuous phase a quite opposite effect becomes reasonable. This means the average particle size is decreasing and the small particle fraction (ca. 35 nm in size) becomes more relevant. Again the fraction of particles with smaller particle diameter becomes dominant in the number size distribution.

In addition, the size and the shape of the redispersed cadmium sulfide particles are characterized by transmission electron microscopy. Figure 3 shows the electron micrographs of the particles at point A. One can see, that spherical particle with diameters between 10 and 20 nm are formed, which are partly arranged to larger aggregates. In some cases larger geometrical structures like triangles can be observed. We have already shown that the nanoparticle formation in liquid crystalline template phases leads to well-organized geometrical structures like rods or triangles [46, 47]. It seems to be plausible, that the microemulsion also passes liquid crystalline states during the evaporation of the solvent compounds, due to the increase of the relative amount of surfactant. For this reason the CdS nanoparticles primarily formed in the microemulsion can also aggregate to triangular structures during the evaporation process as to be seen in Fig. 3b.

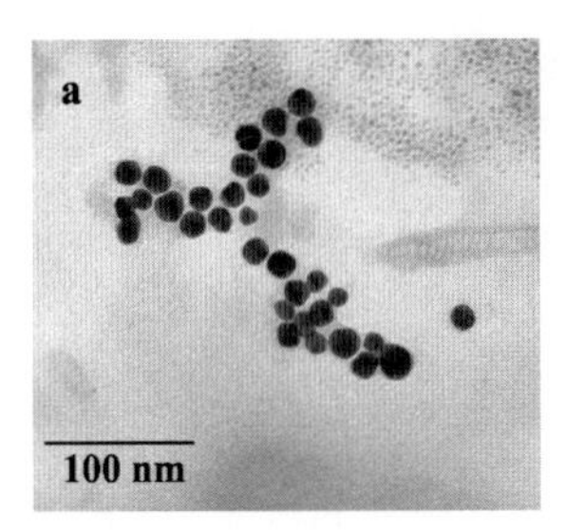

Fig. 3 Electron micrographs of the redispersed CdS particles synthesized in the microemulsion at point A in presence of 10 wt. % PEG I (**a**) , 10 wt. % PEG II (**b**)

In agreement to the DLS measurements, at point B the electron micrographs show that compact aggregates of 20 to 30 nm sized, cubic shaped CdS particles are formed in presence of 10 wt. % PEG I. The size of these aggregates is several hundred nanometres. An increase of the polymer concentration to 20 wt. % results in the formation of spherical CdS particles with diameters between 15 and 26 nm, which partly aggregate to more or less compact structures of about 100 to 200 nm as to be seen in Fig. 4a. However, the tendency to aggregate seems to be less pronounced due to the higher polymer concentration.

When the molecular weight of the polymer is increased we have already discussed the shift of the UV-Vis absorption band to smaller wavelength indicating smaller particles diameters. In fact, the micrograph in Fig. 4b shows spherical CdS particles with diameter of about 5 to 15 nm. In addition, the particles are well separated and aggregation is not recognizable.

Conclusions

Our investigations show that the poly(ethylene glycol)-modified microemulsion system SDS/xylene-pentanol (1 : 1)/water exhibits different microemulsion structures, e.g. the classical w/o microemulsion and a bicontinuous microemulsion. The inverse microemulsion droplets, as well as the sponge-like bicontinuous microemulsion can be used as a template phase for the formation of CdS nanoparticles. Independent of the template phase used 10 nm sized particles can be determined by dynamic light scattering. However, by means of time-dependent UV-Vis measurements one can observe an aging effect of the samples, which leads to a shift of the absorption band to higher wavelengths and indicates the proceeding growth of the particles or aggregation phenomena. This effect is much more pronounced in the bicontinuous template phase than in the w/o microemulsion. The particles formed in the water domains of the bicontinuous phase seem to underlie stronger particle–particle interactions, because there is no restriction like in the inverse microemulsion droplets.

When the CdS particles are redispersed in water after solvent evaporation an increase of the particle diameters can be observed. Furthermore, larger aggregates of single CdS particles can be observed especially in presence of PEG of lower molar mass. One can conclude that the aggregation already starts in the microemulsion, but is much more intensified during the evaporation process. In good agreement to the UV-Vis measurements, the strongest aggregation can be observed for the sponge-like template phase. However, an increase of the molecular weight seems to increase the stability of the particles and avoids aggregation. Surprisingly, the diameters of the CdS particles formed in the bicontinuous microemulsion in presence of higher molar mass of PEG are decreased. This can be explained by the simultaneous decrease of the diameter of the water channels, which has already been shown for the system in former investigations [41]. Accordingly, the shoulder in the UV-Vis absorption spectra and λ_S are blue-shifted. Earlier investigations show the formation of supramolecular nanoparticle–polymer composites, when $BaSO_4$ particles are formed in the PEG-modified system [36, 41]. However, in the case of cadmium sulfide such composites can not be observed. The interactions between the polymer chains and the particle surface, which are the key parameter in this con-

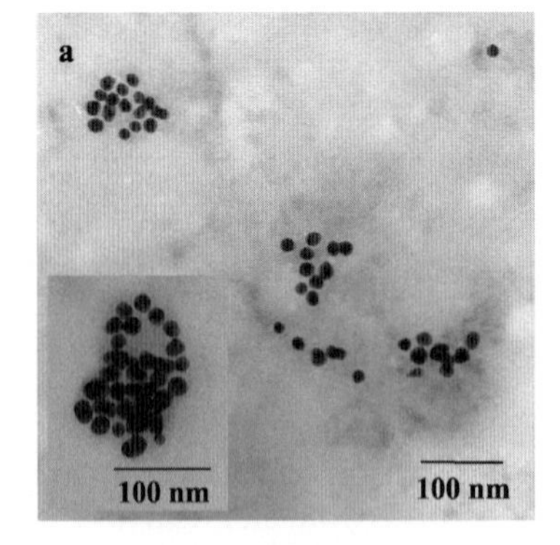

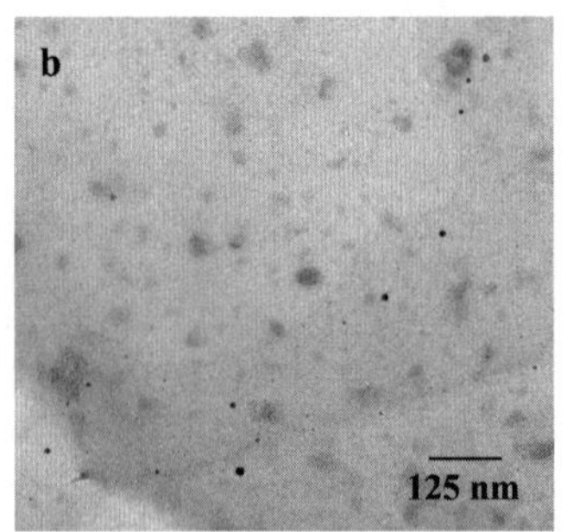

Fig. 4 Electron micrographs of the redispersed CdS particles synthesized in the microemulsion at point B in presence of 20 wt. % PEG I (**a**) and 10 wt. % PEG II (**b**)

text seems to be quite different depending on the type of material.

One can conclude, that not only the often used w/o microemulsion is a useful template phase for the formation of CdS nanoparticles, but also the polymer-induced bicontinuous phase provides semiconductor nanoparticles of about 10 nm. However, higher polymer concentrations and molecular weights of PEG are necessary to increase the stability of the particles and avoid aggregation phenomena.

Acknowledgement The authors thank the Graduate School "Confined Interactions and Reaction in Soft Matter" program of the University of Potsdam for financial support. A. Laschewsky, University of Potsdam & Fraunhofer Institut für Angewandte Polymerforschung Golm, is gratefully acknowledged for providing access to the UV-Vis spectroscopy instrument.

References

1. Han LL, Qin DH, Jiang X, Liu YS, Wang L, Chen JW, Cao Y (2006) Nanotechnology 17(18):4736
2. Huynh WU, Dittmer JJ, Alivisatos AP (2002) Science 295:2425
3. Wei C, Grouquist D, Roark J (2002) J Nanosci Nanotechnol 2:47
4. Matsui I (2005) J Chem Eng Japan 38(8):535
5. Nie SM, Xing Y, Kim GJ, Simons JW (2007) Annu Rev Biomed Eng 9:257
6. Rhyner MN, Smith AM, Gao XH, Mao H, Yang LL, Nie SM (2006) Nanomedicine 1(2):209
7. Santra S, Xu JS, Wang KM, Tan WH (2004) J Nanosci Nanotechnol 4(6):590
8. Serpone N, Khairutdinov RF (1997) Semiconductor Nanoclusters – Physical Chemical Catalytical Aspects Studies. Surf Sci Catal 103:417
9. Zou JJ, Chen C, Liu CJ, Zhang YP, Han Y, Cui L (2005) Mater Lett 59(27):3437
10. Zhang WU, Zhong Y, Fan J, Sun SQ, Tang N, Tan MY, Wu LM (2003) Sci China B 46(2):196
11. Brus LE (1983) J Chem Phys 79(11):5566
12. Brus LE (1984) J Chem Phys 80(9):4403
13. Henglein A (1989) Chem Rev 89:1861
14. Weller H (1993) Angew Chem Int Ed Engl 32:41
15. Lifshitz E, Dag I, Litvin I, Hodes G, Gorer S, Reisfeld R, Zelner M, Minti H (1998) Chem Phys Lett 288(2–4):188
16. Bhattacharjee B, Ganguli D, Chaudhuri S, Pal AK (2003) Mater Chem Phys 78(2):372
17. Li Y, Huang FZ, Zhang QM, Gu ZN (2000) J Mater Sci 35(23):5933
18. Lu QY, Gao F, Zhao DY (2002) Nanotechnology 13(6):741
19. Yin YD, Xu XL, Ge XW, Lu Y, Zhang ZC (1999) Radiat Phys Chem 55(3):353
20. Shao MW, Li Q, Xie B, Wu J, Qian YT (2003) Mater Chem Phys 78(1):288
21. Hamley IW (2003) Nanotechnology 14(10):39
22. Moffit M, Vali H, Eisenberg A (1998) Chem Mater 10(4):1021
23. Zhang JG, Xu SQ, Kumacheva E (2004) J Am Chem Soc 126(25):7908
24. Mandal D, Chatterjee U (2007) J Chem Phys 126:134507
25. Towey TF, Khan-Lodhi A, Robinson BH (1990) J Chem Soc Faraday Trans 86(22):3757
26. Agostiano A, Catalano M, Curri ML, Della Monica M, Manna L, Vasanelli L (2000) Micron 31:253
27. Caponetti E, Pedone L, Chillura Martino D, Panto V, Turco Liveri V (2003) Mater Sci Eng C 23:531
28. Khiew PS, Radiman S, Huang NM, Soot Ahmad M (2003) J Cryst Growth 254:235
29. Khiew PS, Huang NM, Radiman S, Soot Ahmad M (2004) Mater Lett 58:516
30. Capek I (2004) Adv Colloid Interface Sci 110(1/2):49
31. Eastoe J, Hollamby MJ, Hudson L (2006) Adv Colloid Interface Sci 128:5
32. Curri ML, Agostiano A, Manna L, Della Monica M, Catalano M, Chiavarone L, Spagnolo V, Lugara M (2000) J Phys Chem B 104(35):8391
33. Koetz J, Bahnemann J, Lucas G, Tiersch B, Kosmella S (2004) Colloids Surf A 250(1–3):423
34. Koetz J, Jagielski N, Kosmella S, Friedrich A, Kleinpeter E (2006) Colloids Surf A 288(1–3):36
35. Koetz J, Baier J, Kosmella S (2007) Colloid Polym Sci 285(15):1719
36. Koetz J, Andres S, Kosmella S, Tiersch B (2006) Compos Interfaces 13(4/6):461
37. Lutter S, Koetz J, Koeth A (2007) Proceedings PARTEC 2007, International Congress for Particle Technology
38. Bloor DM, Wyn-Jones E (1982) J Chem Soc Faraday Trans 78:657
39. Nikas YJ, Blankschtein D (1994) Langmuir 10:3512
40. Gjerde MI, Nerdal W, Hoiland H (1998) J Colloid Interface Sci 197:191
41. Lutter S, Koetz J, Tiersch B, Kosmella S (2009) J Dispers Sci Technol 30(6), accepted
42. Spanhel L, Haase M, Weller H, Henglein A (1987) J Am Chem Soc 109:5649
43. Vossmeyer T, Katsikas L, Giersig M, Popovic IG, Diesner K, Chemdessine A, Eychmüller A, Weller H (1994) J Phys Chem 98:7665
44. Weller H, Schmidt HM, Koch U, Fojtik A, Baral S, Henglein A (1986) Chem Phys Lett 124(6):557
45. Rosetti E, Ellison JL, Gibson JM, Brus LE (1984) J Chem Phys 80(9):4464
46. Robertson D, Tiersch B, Kosmella S, Koetz J (2007) J Colloid Interface Sci 305(2):345
47. Tong Q, Kosmella S, Koetz J (2006) Progr Colloid Polym Sci 133:152

AUTHOR/TITLE INDEX

Progr Colloid Polym Sci (2008) 134: 157–158

KEY WORD INDEX

Printing: Krips bv, Meppel, The Netherlands
Binding: Stürtz, Würzburg, Germany